Web Content Caching and Distribution

Web Content Caching and Distribution

Proceedings of the 8th International Workshop

Edited by

Fred Douglis

IBM T.J. Watson Research Center,
Hawthorne, NY, U.S.A.

and

Brian D. Davison

Department of Computer Science and Engineering,
Lehigh University, Bethlehem, PA, U.S.A.

8TH INTERNATIONAL WORKSHOP
WCW 2003
ON WEB CONTENT
CACHING AND DISTRIBUTION

SPRINGER-SCIENCE+BUSINESS MEDIA, B.V.

Library of Congress Cataloging-in-Publication Data

ISBN 978-90-481-6626-8 ISBN 978-1-4020-2258-6 (eBook)
DOI 10.1007/978-1-4020-2258-6

Printed on acid-free paper

Contents

A Message from the Workshop Chairs ix

Credits xi

Contributing Authors xiii

Part 1 – **Mobility**

Mobility-aware server selection for mobile streaming multimedia content distribution networks 1
Muhammad Mukarram Bin Tariq, Ravi Jain, Toshiro Kawahara

Performance of PEPs in cellular wireless networks 19
Pablo Rodriguez, Vitali Fridman

Part 2 – **Applications**

Edge caching for directory based Web applications: Algorithms and performance 39
Apurva Kumar, Rajeev Gupta

Computing on the edge: A platform for replicating Internet applications 57
Michael Rabinovich, Zhen Xiao, Amit Aggarwal

Scalable consistency maintenance for edge query caches: Exploiting templates in Web applications 79
Khalil Amiri, Sara Sprenkle, Renu Tewari, Sriram Padmanabhan

Part 3 – **Architectures**

Proxy+: Simple proxy augmentation for dynamic content processing 91
Chun Yuan, Zhigang Hua, Zheng Zhang

Synopsis: Multicast cloud with integrated multicast and unicast content distribution routing 109
Dan Li, Arun Desai, Zheng Yang, Kenneth Mueller, Stephen Morris, Dmitry Stavisky

Synopsis: A large enterprise content distribution network: Design, implementation and operation 119
Jacobus Van der Merwe, Paul Gausman, Chuck Cranor, Rustam Akhmarov

Synopsis: Architectural choices for video-on-demand systems 129
Anwar Al Hamra, Ernst W. Biersack, Guillaume Urvoy-Keller

Part 4 – **Multimedia**

Dynamic cache reconfiguration strategies for a cluster-based streaming proxy 139
Yang Guo, Zihui Ge, Bhuvan Urgaonkar, Prashant Shenoy, Don Towsley

Stream engine: A new kernel interface for high-performance Internet streaming servers 159
Jonathan Lemon, Zhe Wang, Zheng Yang, Pei Cao

Streaming flow analyses for prefetching in segment-based proxy caching to improve delivery quality 171
Songqing Chen, Bo Shen, Susie Wee, Xiaodong Zhang

Part 5 – **Customization**

Subscription-enhanced content delivery 187
Mao Chen, Jaswinder Pal Singh, Andrea LaPaugh

Cooperative architectures and algorithms for discovery and transcoding of multi-version content 205
Claudia Canali, Valeria Cardellini, Michele Colajanni, Riccardo Lancellotti, Philip S. Yu

Synopsis: User specific request redirection in a content delivery network 223
Sampath Rangarajan, Sarit Mukherjee, Pablo Rodriguez

Part 6 – **Peer-to-Peer**

Friendships that last: Peer lifespan and its role in P2P protocols 233
Fabian E. Bustamante, Yi Qiao

Synopsis: A fine-grained peer sharing technique for delivering large media files over the Internet 247
Mengkun Yang, Zongming Fei

Part 7 – **Performance and Measurement**

Proxy-cache aware object bundling for Web access acceleration 257
Chi Hung Chi, HongGuang Wang, William Ku

Synopsis: A case for dynamic selection of replication and caching strategies 275
Swaminathan Sivasubramanian, Guillaume Pierre, Maarten van Steen

Synopsis: Link prefetching in Mozilla: A server-driven approach 283
Darin Fisher, Gagan Saksena

Synopsis: A generalized model for characterizing content modification dynamics of Web objects 293
Chi Hung Chi, HongGuang Wang

Part 8 – **Delta Encoding**

Server-friendly delta compression for efficient Web access 303
Anubhav Savant, Torsten Suel

Evaluation of ESI and class-based delta encoding 323
Mor Naaman, Hector Garcia-Molina, Andreas Paepcke

Author Index 345

A Message from the Workshop Chairs

Dear Participant:

Welcome to the 8th International Web Caching and Content Delivery Workshop. Since our first meeting in 1996, this workshop has served as the premiere forum for researchers and industry technologists to exchange research results and perspectives on future directions in Internet content caching and content delivery. This year we received 46 submissions, of which 15 have been selected as full-length papers and 8 as synopses. We extend our thanks to the authors of the selected papers, all of which are included in these proceedings. In addition to technical presentations, we are pleased to have Bill Weihl of Akamai to present the keynote address, and a panel discussion on uncachable content organized by Zhen Xiao of AT&T Labs – Research.

While originally scheduled to be held in Beijing, China, the workshop moved to the US this year as a result of the concerns over the SARS virus. We are indebted to our industrial sponsor, IBM, for providing the facilities in which to hold the workshop. The T.J. Watson Research Center that serves as our venue spans three sites across two states, and is the headquarters for the eight IBM research labs worldwide. We are also grateful to the members of the program committee for helping to select a strong program, and to the members of the steering committee who continue to provide advice and guidance, even as plans are made for next year's workshop.

In past years, we have found great topics and fruitful discussion as people from industry and academia interact. We are confident that you will experience the same at this year's workshop.

Brian D. Davison	Fred Douglis
General Chair	*Program Chair*

Credits

General Chair
Brian D. Davison, *Lehigh University*

Program Chair
Fred Douglis, *IBM T.J. Watson Research Center*

Program Committee
Martin Arlitt, *University of Calgary*
Remzi Arpaci-Dusseau, *University of Wisconsin*
Chi-Hung Chi, *National University of Singapore*
Mike Dahlin, *University of Texas at Austin*
Fred Douglis, *IBM T.J. Watson Research Center*
Zongming Fei, *University of Kentucky*
Leana Golubchik, *University of Southern California*
Jaeyeon Jung, *MIT LCS*
Dan Li, *Cisco Systems, Inc.*
Guillaume Pierre, *Vrije Universiteit, Amsterdam*
Weisong Shi, *Wayne State University*
Oliver Spatscheck, *AT&T Labs – Research*
Renu Tewari, *IBM Almaden Research Center*
Amin Vahdat, *Duke University*
Geoff Voelker, *University of California, San Diego*
Zhen Xiao, *AT&T Labs – Research*

Steering Committee
Azer Bestavros, *Boston University*
Pei Cao, *Cisco*
Jeff Chase, *Duke University*
Valentino Cavalli, *Terena*
Peter Danzig, *University of Southern California*
John Martin, *Network Appliance*
Michael Rabinovich, *AT&T Labs – Research*
Wojtek Sylwestrzak, *Warsaw University*
Duane Wessels, *The Measurement Factory*

Keynote Speaker
William Weihl, *Akamai Technologies, Inc.*

Panel Moderator
Zhen Xiao, *AT&T Labs – Research*

Panelists
Indranil Gupta, *University of Illinois, Urbana-Champaign*
Arun Iyengar, *IBM Research*
Michael Rabinovich, *AT&T Labs – Research*
Torsten Suel, *Polytechnic University*
William Weihl, *Akamai Technologies, Inc.*

Session Chairs
Chi-Hung Chi, *National University of Singapore*
Brian D. Davison, *Lehigh University*
Fred Douglis, *IBM T.J. Watson Research Center*
Zongming Fei, *University of Kentucky*
Michael Rabinovich, *AT&T Labs – Research*
Pablo Rodriguez, *Microsoft Research, Cambridge*
Oliver Spatscheck, *AT&T Labs – Research*
Torsten Suel, *Polytechnic University*

External Reviewers
Benjamin Atkin
Yan Chen
Subhabrata Sen
Andrew Tridgell
Limin Wang
Craig Wills
Kun-Lung Wu

Contributing Authors

Amit Aggarwal	*Microsoft*
Anwar Al Hamra	*Institut Eurecom*
Rustam Akhmarov	*AT&T Labs – Research*
Khalil Amiri	*Imperial College London*
Ernst W. Biersack	*Institut Eurecom*
Fabian E. Bustamante	*Department of Computer Science, Northwestern University*
Pei Cao	*Cisco Systems, Inc.*
Claudia Canali	*University of Parma*
Valeria Cardellini	*University of Roma "Tor Vergata"*
Mao Chen	*Department of Computer Science, Princeton University*
Songqing Chen	*College of William and Mary*
Chi Hung Chi	*National University of Singapore*
Michele Colajanni	*University of Modena and Reggio*
Chuck Cranor	*AT&T Labs – Research*
Arun Desai	*Cisco Systems, Inc.*
Zongming Fei	*Department of Computer Science, University of Kentucky*

Darin Fisher *IBM*

Vitali Fridman *Microsoft Research, Cambridge*

Hector Garcia-Molina *Department of Computer Science, Stanford University*

Paul Gausman *AT&T Labs – Research*

Zihui Ge *Department of Computer Science, University of Massachusetts at Amherst*

Yang Guo *Department of Computer Science, University of Massachusetts at Amherst*

Rajeev Gupta *IBM India Research Lab*

Zhigang Hua *Institute of Automation, Chinese Academy of Sciences*

Ravi Jain *DoCoMo Communications Laboratories USA*

Toshiro Kawahara *DoCoMo Communications Laboratories USA*

William Ku *National University of Singapore*

Apurva Kumar *IBM India Research Lab*

Riccardo Lancellotti *University of Roma "Tor Vergata"*

Andrea LaPaugh *Department of Computer Science, Princeton University*

Jonathan Lemon *Cisco Systems, Inc.*

Dan Li *Cisco Systems, Inc.*

Stephen Morris *Cisco Systems, Inc.*

Kenneth Mueller *Cisco Systems, Inc.*

Sarit Mukherjee *Microsoft Research, Cambridge*

Mor Naaman *Department of Computer Science, Stanford University*

Sriram Padmanabhan *IBM Santa Teresa Lab*

Andreas Paepcke *Department of Computer Science, Stanford University*

Guillaume Pierre *Department of Mathematics and Computer Science, Vrije Universiteit, Amsterdam*

Yi Qiao *Department of Computer Science, Northwestern University*

Michael Rabinovich *AT&T Labs – Research*

Sampath Rangarajan *Lucent Technologies Bell Laboratories*

Pablo Rodriguez *Microsoft Research, Cambridge*

Gagan Saksena *AOL*

Anubhav Savant *CIS Department, Polytechnic University*

Bo Shen *Hewlett-Packard Laboratories*

Prashant Shenoy *Department of Computer Science, University of Massachusetts at Amherst*

Jaswinder Pal Singh *Department of Computer Science, Princeton University*

Swaminathan Sivasubramanian *Department of Mathematics and Computer Science, Vrije Universiteit, Amsterdam*

Sara Sprenkle *Duke University*

Dmitry Stavisky *Cisco Systems, Inc.*

Torsten Suel *CIS Department, Polytechnic University*

Muhammad Mukarran Bin Tariq *DoCoMo Communications Laboratories USA*

Renu Tewari *IBM Almaden Research Center*

Don Towsley *Department of Computer Science, University of Massachusetts at Amherst*

Bhuvan Urgaonkar *Department of Computer Science, University of Massachusetts at Amherst*

Guillaume Urvoy-Keller *Institut Eurecom*

Jacobus Van der Merwe *AT&T Labs – Research*

Maarten van Steen *Department of Mathematics and Computer Science, Vrije Universiteit, Amsterdam*

HongGuang Wang *National University of Singapore*

Zhe Wang *Cisco Systems, Inc.*

Susie Wee *Hewlett-Packard Laboratories*

Zhen Xiao *AT&T Labs – Research*

Zheng Yang *Cisco Systems, Inc.*

Mengkun Yang *Department of Computer Science, University of Kentucky*

Philip S. Yu *IBM T.J. Watson Research Center*

Chun Yuan *Microsoft Research Asia*

Xiaodong Zhang *College of William and Mary*

Zheng Zhang *Microsoft Research Asia*

MOBILITY AWARE SERVER SELECTION FOR MOBILE STREAMING MULTIMEDIA CONTENT DISTRIBUTION NETWORKS

Muhammad Mukarram Bin Tariq, Ravi Jain, and Toshiro Kawahara
DoCoMo Communications Laboratories USA, Inc.

Abstract We propose a Content Delivery Network (CDN) with servers arranged hierarchically in multiple tiers. Lower-tier servers are topologically closer to the clients, and hence can deliver better QoS in terms of end-to-end delay and jitter. On the other hand, higher-tier servers have a larger coverage area and hence their clients incur fewer server handoffs. We present a server selection scheme that reduces the number of server handoffs while meeting differentiated QoS requirement for each client. The scheme dynamically estimates the client's residence time and uses a simple algorithm to assign clients to the appropriate tier. The scheme also caters for traditional server selection criteria, such as the expected QoS from the servers, bandwidth consumption, and the server load. We show through simulations that the scheme can achieve up to 15% reduction in handoffs, at the cost of minimal increases in delay and jitter while ensuring that clients of different QoS classes experience different delays.

1. Introduction and Overview

We expect that multimedia and data traffic will surpass the traditional voice traffic in mobile networks by the year 2005 [18]. High quality streaming multimedia content is likely to form a significant portion of this traffic. It is therefore important that large mobile networks find ways to manage client traffic and data efficiently.

Content distribution networks (CDN) have proven effective for managing content-based traffic for large numbers of clients in the Internet. CDN consist of surrogate servers that replicate the content of the origin servers and serve it to the clients. CDN employ server selection and request redirection methods for selecting an appropriate (surrogate) server and redirecting the client's request to that server. CDN reduce load on both the network and origin server by localizing the traffic and providing many alternate sources of content. Iyengar et al., [8] provide an overview of existing CDN technologies.

Although there has been much work on the Internet-wide CDN, CDN for mobile

F. Douglis and B.D. Davison (eds.), Web Content Caching and Distribution, 1-18.

networks have received little attention so far. Mobile CDN proposals, such as [4], target static WAP or HTML content for mobile clients, while only [16, 17] consider streaming multimedia content distribution for mobile networks. This lack of attention has largely been because of the limited Internet access capabilities of most mobile terminals of recent past. However, this is changing quickly with deployment and acceptability of 3G services [11] that allow mobile terminals to access Internet and other data services at speeds comparable to traditional wired access. In this paper, we consider server selection algorithms for streaming multimedia content distribution in networks with mobile users.

The large size of most multimedia content, long-lived nature of typical multimedia sessions, client mobility, and the capricious nature of wireless communication medium, put together, present an interesting challenge. A CDN for mobile networks must address all of these issues. In [16], Tariq et al. show that QoS, in terms of delay and jitter, can be significantly improved using server handoff; a process of changing the server as the client moves so that the client continues to receive content from a nearby server.

Server handoff itself, however, can have adverse effects. It can disrupt the streaming from the server, causing glitches at the client. Moreover, changing the server can be an expensive process for the network. Before a session can be handed over to a new server, sufficient content must be pre-fetched (or placed) at the server to ensure smooth delivery [15]. Random and frequent client movements can cause significant signaling and content placement overhead.

One approach to the problem of stream disruption is to mitigate it by sufficient buffering at client equipment at the cost of increased playback delay. Another approach is using make-before-break handoffs. However, reducing the actual number of handoffs is not trivial. In [16], authors propose to delay the server handoff process to reduce the number of handoffs and resulting overhead. In this paper we present a more sophisticated server selection scheme.

Our scheme reduces the number of server handoffs for mobile clients by using client mobility information and selecting a server with which the client can remain associated for an extended period, thus reducing the need for server handoffs. We also cater for traditional server selection criteria such as expected QoS (in terms of delay and jitter) from the server, traffic localization, and server load. The basic trade-off that we explore is how to maintain QoS service differentiation for clients while reducing the network cost due to server handoffs.

The rest of this paper is organized as follows. Section 2 describes our mobility-based server-selection scheme. Sections 3 and 4 are dedicated to simulation and results. In section 5, we discuss related work, and conclude the paper in section 6.

2. Mobility Based Server Selection

Our approach is to use the client's mobility information as an additional metric in the server selection process, along with traditional metrics such as client-server proximity, expected QoS, content availability, and server load. We try to select a server that will remain suitable for content delivery to a client for an extended period of time and thus eliminating the need for frequent server handoff.

2.1 Layout of servers in content distribution network

We consider a content distribution network with a relatively large number of servers arranged in a logically hierarchical topology, where the servers closest to the access network belong to the lowest tier (see Figure 1). This topology allows us to maximize the traffic localization and obtain desired trade-off between the number of server handoffs and the QoS to the clients. These advantages become further apparent in proceeding sections.

Each server has a coverage-area defined as the subnets and cells in the underlying transport network (each subnet may consist of one or more cells). In general, each subnet (and cell) is in the coverage area of the closest server at any given tier, but multiple servers may serve subnets at coverage-area boundaries. Servers at higher tiers have larger coverage areas and therefore, multiple tiers cover every subnet. We use the term *server zone* to refer to the coverage area of lowest tier servers.

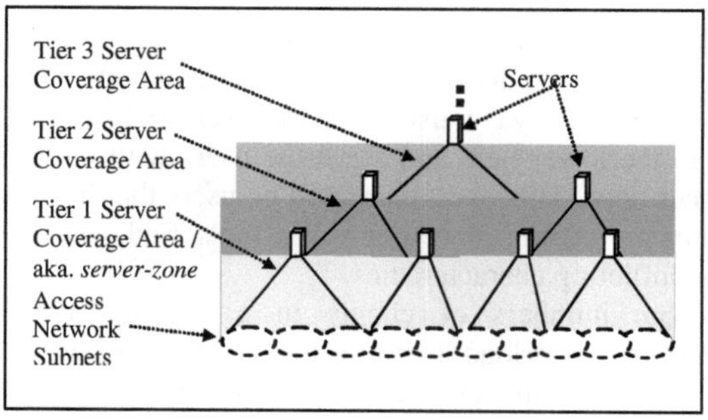

Figure 1. Tiered Topology of Servers in CDN

As in [16], a server handoff is performed whenever a client moves from the coverage area of one server to that of another server. However, with the multi-tier arrangement of servers, there is an opportunity to trade-off QoS with the number of server handoffs, by carefully choosing the tier of the server assigned after handoff. Due to the topological arrangement of the servers, it is likely that the data delivered from the servers in lower tiers will suffer less delay (and probably less jitter) than the data delivered from the servers in higher tiers; however, assignment to servers in lower tiers will result in more server handoffs and may increase the overhead for the network.

We assume that clients have different categories of QoS requirements in terms of delay and jitter. We further assume that each of these categories of QoS can be roughly characterized as the property of a particular server tier. For example, we can assume that highest QoS class, with strict delay and jitter requirement (QoS class 1) maps to the lowest tier server. A fixed mapping is of this type is, however, not necessary. Other QoS metrics, such as, available bandwidth, server response-time, and media quality may be used for server classification. Several dynamic QoS probing and measurement methods, such as in [5, 10] can be used to dynamically classify servers based on different QoS metrics.

We assume that there is a Request Routing function (RR) in each server zone; recall a server zone is the coverage area of lowest-tier servers. We recommend a proxy-based RR approach because it provides greater control and flexibility but other request redirection techniques, such as those discussed in [1], may also be used. In case of proxy-based RR, the function can co-reside with the lowest tier servers.

As the client moves from one server zone to another, it is assigned to a new serving RR. The new RR selects servers for any new sessions that the client may request and also initiates and coordinates server handoffs for the client's existing sessions, if needed. We assume that the RR has sufficient knowledge about the CDN topology and the client mobility information to select appropriate servers and route the requests correctly. We describe the information collection and server selection process in the following.

2.2 Considerations in mobility based server selection

We must consider the following in mobility based server selection.

Using lower tier servers reduces delay and jitter, but increases the number of server handoffs. Therefore, we must serve the users with high QoS requirement from the servers in the lower tiers and serve the users with high mobility from the servers in the higher tiers. However, there will be users that have high mobility and high QoS requirements at the same time. We must find way to assign server to clients with such conflicting characteristics.

If disproportionate numbers of clients in a region have similar mobility characteristics, they may overload a certain tier. Therefore, we must perform load balancing among tiers. As an example, consider the mobility patterns in commute hours. Since many users are mobile, the mobility rate is high for many clients. Similarly, in a region that spans high-speed freeways, the average user mobility would be higher than the regions that cover residential areas. Reduction in handoffs in high mobility rate regions requires that we serve users from servers in higher tiers. However, if we do not consider server load, servers in the higher tier servers may get overloaded.

When the client requests belong to various differentiated QoS classes, we must maintain a desired differentiation of the mean QoS for sessions of each QoS class.

2.3 Measurement of mobility rate and server residence time estimation

We use client's expected *residence time* as a measure of its mobility and as a parameter in the server selection process. The *residence time* is defined as the time that elapses between when a client enters a region and when it leaves that region.

Each RR maintains the residence time value r_i for every client i in its server zone. This value is computed by the clients themselves as a running average of the residence time over their k most recent movements, and reported to the serving RR upon each movement. This movement can be a movement across a cell region, a subnet region or any other lower layer region boundary, but the information based on movements across subnet boundaries can be easily collected and reported using protocols, such as, Mobile IP [6] binding update messages. Each RR also maintains the average subnet residence time $\overline{r}$ over all the clients in its server zone. In addition, RR also maintains the mean server residence time $\overline{R}_t$ over all servers in each tier t. For server residence time, we consider the server's coverage area as the region.

After every successful handoff, the new RR send a server handoff notification message to the old RR. This message includes the client identifier, session identifier, and the identifiers of the new and the old servers involved in the handoff. Using the information contained in this message, the old RR determines the server residence time as the difference of the time at which the client was originally assigned to the old server and the time of server handoff. Information about the timing of original assignment can either be contained in the notification message, or can be assumed to be maintained by the old RR since it is in path of session control signaling traffic. The server tier can be determined using the server identifier. The old RR updates the mean server residence time for the corresponding tier after each handoff notification message.

Equipped with measurements of individual and aggregate residence times, an RR can estimate residence time of a client i with a server at tier t as:

$$E_{i,t} = \frac{r_i \overline{R}_t}{\overline{r}} \qquad (1)$$

In a refinement to mobility information measurement, every RR maintains the average subnet residence time $\overline{r}_s$ and server tier residence time $\overline{R}_{t,s}$ separately for clients in each subnet s in its coverage-area and uses these values for estimating residence time of a client in subnet s. This improves the accuracy of estimation because the extent of a clients mobility can be more accurately determined by comparing it with other clients in its close geographical proximity (in this case, a subnet), than by comparing it with clients over a larger area. Expected residence time of a client i in subnet s, with a server in tier t with this refinement is given as:

$$E_{i,t} = \frac{r_i \overline{R}_{t,s}}{\overline{r}_s} \qquad (2)$$

In the following text, we refer to equation (1) and equation (2) as *low granularity residence time estimation* and *high granularity residence time estimation* respectively.

Several mobility prediction schemes have been proposed (e.g., [2], see [3] for a survey), which can be used by our algorithm with minor modifications. These schemes will likely provide better accuracy but only at the expense of extensive state maintenance and large overhead.

2.4 Server load and QoS information collection

Each server periodically sends information to all the RR in its coverage area about the number of sessions that it can accept from the RR. The servers observe the load from each RR (i.e., the number of sessions emanating from the RR's server-zone) and allocate its remaining capacity to each RR proportionately to the load observed from that RR; we call this the *server load allowance*. The sum of load allowance reported to an RR by all servers at tier t is denoted as L_t. With our hierarchical topology of CDN, the number of RR in coverage area of each server is likely to be low, thus dispersal of such information is feasible.

Each RR obtains information about the nominal QoS obtained from each tier. This can be done using QoS probing techniques, such as in [5, 10], or QoS feedback information, such as through RTCP [14] or by other means. For the following discussion, we assume delay as the primary QoS metric, but other QoS metrics can be used in a similar manner with our algorithm. The nominal delay information for a tier t is represented as D_t. Each RR also maintains the number of client requests of each QoS class q that it sends to tier t as $N_{q,t}$.

2.5 Server selection algorithm

Listing 1 shows the pseudo-code for the server selection process. When an RR needs to select a server for a request to establish new session of QoS class q, it starts by selecting the highest possible server tier t, whose reported nominal delay D_t meets or exceeds the requirements of QoS class q. If the request is for a server handoff of an existing session then the RR sets t to the tier of the currently assigned server.

The RR then uses a heuristic (lines 4-7 in listing 1) to determine whether the client should be assigned to tier $t+1$. It finds the load allowance L_{t+1} for tier $t+1$, and determines whether the requesting client i is among the fastest clients. The reasoning behind the intuition here is that since the load allowance is limited and only a limited number of clients can be served from higher tiers, the reduction in number of server handoffs can be maximized if only the fastest clients are served from the higher tiers.

Instead of relying on any absolute value for residence time or speed that would qualify a client for service from higher tiers, we simply use the client's relative mobility with respect to the other clients in its region and see if the client is fast enough. The heuristic used to determine whether a client i is "fast enough" is to determine whether the client is among L_{t+1} fastest clients in the server zone. This heuristic is based on the assumption that the residence time distributions of the clients in the recent past are similar to those of the immediate future. The RR sums the load due to all the clients whose residence times are less than the client i, and determines whether it is less than the load allowance L_{t+1}; if so, the client i is a candidate to be moved to tier $t+1$.

The RR now determines whether moving the client i to tier $t+1$ will maintain the desired separation among the adjacent QoS classes. The comparison is performed using the weighted mean QoS of client sessions of each QoS class (line 12, 23-25 in listing 1). If the QoS separation is maintained, the client is assigned to a server at tier $t+1$.

If either of the above conditions is not satisfied, i.e., if client is not "fast enough" or that the QoS differentiation is violated, the RR checks whether the client should be moved to the tier $t-1$ (lines 10-16 in listing 1). For this, the load allowance of tier $t-1$ must be positive. Additionally, RR must ensure that moving the client to a lower tier would not increase the number of handoffs. This is only possible if the client is relatively slow moving. This is done by calculating whether the estimated residence time for this client in a lower tier, $E_{i,t-1}$, exceeds the average residence time of current tier t (line 11 in listing 1). If $E_{i,t-1}$ is greater than the average residence time of current tier, it is an indication that the client i is slow moving, and thus assigning it to a lower tier will not increase the number of server handoffs.

Listing 1: Mobility Based Server Selection Algorithm

```
1.  proc selectServer (i: Client, q: QoS, t: serverTier,
                       mode: String)
2.     selectedTier ← t;
3.     L_{t+1} ← findLoadAllowance (t+1)
4.     if clientInFastest (L_{t+1}, i)
5.        if (qosSeparationMaintained(q))
6.           selectedTier ← t+1;
7.        endif
8.     elseif(mode = "eager" ||
          (mode = "lazy" & LoadAllowanceLow (L_{t+1}, t+1))
9.        L_{t-1} = findLoadAllowance (t-1)
10.       if (L_{t-1} > 0)
11.          if (E_{(i,t-1)} > R̄_t )
12.             if (qosSeparationMaintained(q))
13.                selectedTier ← t-1;
14.             endif
15.          endif
16.       endif
17.    endif
18.    serverIdentifier ← findServer(selectedTier)
19.    return serverIdentifier;
20. endproc
21.
22. proc qosSeparationMaintained (q: QoSClass): bool
23.    return (meanDelay(q+1)-meanDelay(q)> δ_{q,q+1} ) &
24.            (meanDelay(q)-meanDelay(q-1)> δ_{q-1,q} )
         %% δ is desired delay separation of QoS Classes
25. endproc
26.
27. proc meanDelay (q: QosClass): QoS
28.    T ← number of server tiers that cover server-zone
    of the RR
```

29. $\bar{D}_q \leftarrow \frac{\sum_{t=1}^{T} N_{q,t} D_t}{\sum_{t=1}^{T} N_{q,t}}$ `%% weighted mean QoS of class q`

30. `return` $\bar{D}_q$

31. `endproc`

32. `proc clientInFastest (Allowance L, Client i): bool`

33. `C` $\leftarrow$ `total number of clients of RR`

34. `return` $\sum_{j:r_j<r_i}^{C} U_j$ `< L;`

`%%` U_j `is the number of sessions of client j`

35. `endproc`

```
36. proc LoadAllowanceLow (L: Allowance,
                           t: serverTier): bool
37.     return (L < minThreshold * MaxAllowance(t));
38. endproc
```

In order to study the impact of accuracy of residence time estimation, we perform experiments using both the low granularity residence time estimation and the higher granularity residence time estimation techniques for $E_{i,t-1}$.

Note that if the residence time is under-estimated, moving a client to lower tiers will likely increase the number of server handoffs. To study this effect, we have introduced two modes, *lazy* and *eager*, in our algorithm. In the eager mode, the RR takes the risk and selects tier $t-1$ if it the client's estimated residence time with tier $t-1$ exceeds the average residence time. In the lazy mode, however, RR acts more cautiously; it additionally checks whether there are indications that the higher tier servers are starting to be overloaded. Specifically, when the load allowance reported by higher tier servers has falls below a certain threshold (lines 08, 36-38 in listing 1), only then the lazy mode considers assigning slow moving clients to the servers in lower tiers.

We compare the results of these several possibilities through simulations in the next section.

3. Simulation Setup

We have simulated behavior various variations of our algorithm. The following subsections describe the system layout, the simulation scenarios and the results obtained.

3.1 Mobility simulation

To get realistic mobility information, we have built a custom mobility simulator. This tool simulates movement of mobile clients in a digitized geographical map comprising roads, streets, and railways. For this paper, we have used a map of San Francisco Bay Area spanning 3575 sq. miles.

Using this simulator, we can specify populations at different locations in the map. The simulator chooses the origin and destination points of the mobile clients such that the number of points chosen from an area is proportional to its population. Any desired flow of client movement in the network can be specified by creating appropriate set of population distributions.

Once the origin and destination points are chosen, the simulator uses Dijkstra's shortest path algorithm to compute the path for the hosts. The simulator is capable of using different metrics for computing shortest path. These include distance, estimated travel time, or a preference of type of transport. Presently we consider clients with distance and time based short paths only, and in equal proportions.

The simulator supports four mediums of transport: on-foot, car, train, and metropolitan bus service. On-foot mode is used over very short distances, less than 0.25 miles, and clients traveling in this mode do not necessarily abide by the road layouts and conditions. Clients in cars and buses and trains follow the roads directions and conditions. There is, however, an important difference between these modes; the clients in cars move asynchronously, whereas, the clients on buses must travel together according to route and schedules. It is also the case with trains except that they have specialized tracks. The simulator allows us to specify the capacity and schedules for buses and trains. In the current simulator, only one client occupies each car.

Once the path is calculated, each client host travels along its path and varies its speed according to speed limits and congestion conditions on the roads and highways and by schedules of trains and buses; clients slow down in congested segments and wait on designated stations for the next bus or train to arrive. Clients on roads can re-route in response to traffic congestion by calculating a new path that avoids congested segments.

The geographic map is overlaid with 187 cells of varying radii. The radius of the cell varies between 0.2 miles to 5 miles depending on population distribution and the terrain; this is in line with typical 3G network planning [22]. The radio network is further overlaid with 59 network layer (IP) subnets. Each subnet has a router that serves 2 to 4 radio cells. We have simulated movement of 2500 clients. We log events as these clients move from one cell to another. These events are then converted into network subnet level movement events and can be directly consumed by our server selection algorithm.

Figure 2 shows a screen shot of the mobility simulator showing bay area map, radio network base-stations covering the area and population distributions (the rectangular boxes).

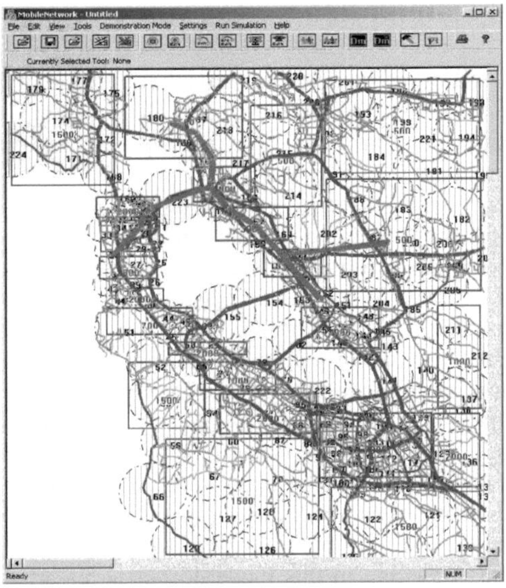

Figure 2. Mobility Simulator

3.2 CDN layout

Network topology

In the simulation scenario, the tiered CDN topology comprises 34 surrogate servers arranged in 3 tiers, thus $t \in \{1, 2, 3\}$. We have 21 servers in tier 1 and thus 21 server zones, each containing an RR. There are 8 servers in tier 2 and 4 in tier 3. The CDN has 4 origin servers that place content on the surrogate servers if it is already not there. The model also contains background traffic servers to create configurable amount of background traffic to load the network. We have modeled the server selection algorithm and the network using a combination of Matlab [9] and OPNET [12].

For our current simulation, we assume that when there is no load on the network, there is 80ms end-to-end delay from a client in the access network to the nearest tier 1 server. Similarly, there is 160ms end-to-end delay from nearest tier 2 servers and 240ms delay from nearest tier 3 servers, thus D_1 =80ms, D_2 =160ms and D_3 =240ms. In general, any real-time delay estimation process, such as in [5, 9], can be used to obtain these values, but for simplicity we do not simulate these processes here. We also assume the desired delay separation between QoS classes is 20 ms, thus $\delta_{1,2}$ = $\delta_{2,3}$ = 20ms.

Server characteristics

The behavior of surrogate servers is characterized by:

- The time to process a session request; for the current simulation it is 100ms.
- The time to process mobility notification message (e.g., binding update message of Mobile IP [6]), currently it is fixed at 10ms.
- The maximum media packet delivery rate, current value is 10,000 packets per second.
- The cache hit-ratio i.e., the probability of finding the content locally. In case of a cache miss, the server sends a content placement request to the origin server to retrieve the content.

Origin servers have same characteristics as surrogate servers, except that the content is always assumed present at origin servers. Surrogate servers act as clients to origin servers. The data-rate of content placement is higher than the actual streaming data-rate from the surrogate server to the client; in our simulation, it is 1Mbps. Servers process all requests (session and mobility notification) on first-come first-served basis. Because of high packet delivery rate, the number of sessions that can be served from a server is usually limited by the network capacity.

Client characteristics

We assume that client requests belong to one of the three QoS classes, high (1), medium (2) and low (3). We assume that the tier 1 servers are appropriate for high QoS class. Tier 2 servers are appropriate for medium QoS class and tier 3 servers are appropriate for low QoS class.

The number of client requests in the three QoS classes is proportional to number of servers in three tiers, thus the ratio is 21:8:4. This ensures that if there is no mobility based server selection (simulation scenario 5, explained in next subsection), the clients are evenly distributed across the tiers. This is important for obtaining a fair assessment of benefits of our scheme. In our present simulation, we assume that data rate for all three QoS classes is 64kbps and 20 packets per-second and end-to-end delay is the main QoS determinant. We assume that each client has only one active session at a time, thus U_j is 1.

3.3 Simulation scenarios

We have simulated five scenarios.

1. Lazy server selection mode using high granularity residence time estimate (using equation 2).
2. Eager server selection mode using high granularity residence time estimate (using equation 2).
3. Lazy server selection mode using low granularity residence time estimate (using equation 1).
4. Eager server selection mode using low granularity residence time estimate (using equation 1).

5. Non-adaptive server selection. In this scenario we do not attempt to reduce the number of handoffs. A client request is always sent to the highest tier that can meet or exceed the QoS requirements.

We have simulated each scenario for different simulation parameters. These are explained in next section. We use the scenario 5 as a baseline for comparison.

Simulation Parameters

- Session duration: The session durations vary between 75 seconds and 1500 seconds. Each client waits for a random time between 0 and 20 seconds before the start of the first session and between subsequent sessions.
- Server capacity: We have also obtained results for different server capacities. In particular, we have simulated the scenarios for server capacities varying between 50 to 300 simultaneous sessions.
- Load allowance threshold for lazy mode: For the lazy mode, the minimum load allowance threshold is 10% of maximum reported load allowance.
- Network delay: We have also performed simulations for different network delays to confirm the independence of algorithm from absolute QoS values.

Table 1 summarizes different simulation parameters.

Table 1. Summary of Simulation Parameters

Parameter	**Values/Range**
Mobility Related Parameters	
Number of users	2500
Simulation area	3575 sq. miles
Simulation duration	7200 seconds
Number of base stations and subnets	187 base stations, 59 subnets
Server Selection Parameters	
t: tier identifier	{1,2,3}
q: QoS class identifier	{1,2,3}
D_t: delay to the tier t in millisecond, $t \in \{1,2,3\}$	{{80, 160, 240}, {80, 120, 160}}
$\delta_{q,q+1}$: Desired delay separation between QoS class q and q+1.	$\delta_{1,2} = \delta_{2,3} = 20\text{ms}$
Total servers	34
Servers at each tier $t \in \{1,2,3\}$	{21, 8, 4}
Session duration (seconds)	{50, 100, 200, 500, 1000, 1500}
Server capacity (number of simultaneous sessions)	{50, 75, 100, 200, 300}
Server overload threshold for Lazy mode.	10% of maximum reported load allowance
Content Distribution Parameters	
Server's session request processing time	100ms
Server's new address binding update processing time	10ms

4. Results

Figure 3 shows the reduction in the number of server handoffs as compared to the non-adaptive server selection scenario. Detailed results for { D_1 =80ms, D_2 =160 ms and D_3 = 240ms}, { $\delta_{1,2}$ = $\delta_{2,3}$ = 20ms} are included here. Later in the section, we will summarize how varying network delays impacts our server selection algorithm.

Impact of session duration

Both lazy and eager approaches achieve better performance for longer session durations because the algorithm gets more time to adapt to behavior of the mobile client. In fact, for very small sessions, there is a slight increase in number of handoffs, because the algorithm does not get a chance to correct a wrong estimate before the session is ended.

Impact of server capacity:

Reduction in number of handoffs is similar in both lazy and eager approach for smaller server capacity, since the minimum load allowance threshold is reached more quickly. However, as server capacity increases, lazy approach provides significant improvement in reduction of number of handoffs. We found that for eager approach the reduction in number of handoffs was often limited by the desired delay separation whereas for lazy approach the limiting factor is the load allowance of higher tiers. The reduction in number of handovers in figure 3c and 3d for longer duration and low load allowance is not well behaved. This is because with lower load allowance, the criterion for fastest users is strict, and an error is more likely.

Impact of granularity of residence time estimation

Both lazy and eager approaches give better performance when high granularity (subnet-specific) residence time estimation is used because higher accuracy reduces the probability of assigning a fast moving user to a lower tier.

Impact of network delay

We have performed simulations for different values of D_t and $\delta_{q,q+1}$,and we found that the performance of the algorithm is largely independent of absolute value of QoS metric (D_t) and depends more on the desired separation $\delta_{q,q+1}$ For example we kept the desired separation at the same level, that is, $\delta_{1,2}$ = $\delta_{2,3}$ = 20ms but reduced the server delays such that D_1 = 80ms, D_2 = 120ms and D_3 =160ms.

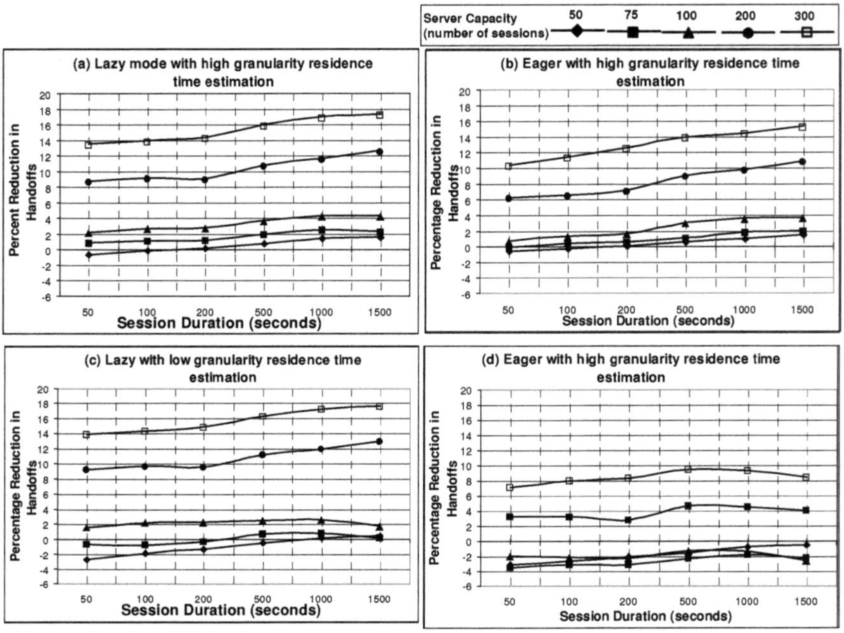

Figure 3. Percentage reduction in server handoffs for various scenarios

We observed that the performance of our scheme does not degrade significantly. The mean deterioration in reduction in number of handoffs is 0.9%. This is in comparison with the results in figure 3, where D_1=80ms, D_2=160ms and D_3= 240ms. The higher deterioration (maximum deterioration observed is 2.5%, which still corresponds to 15% reduction in number of handoffs) is only for longer session durations and higher server load allowances, but since we already have significantly larger reduction in number of handoffs for these parameters, this deterioration is not worrisome.

We also observed that for eager mode, lower network delay actually slightly improves the reduction in the number of handoffs. Because in this case the delay separation limit is reached more quickly that stops the algorithm for making further erroneous server assignment.

Impact on delay and jitter

Figure 4 shows the impact on the mean end-to-end delay for the three QoS classes. Both lazy and eager approaches maintain delay separation among the QoS classes; however, the overall delay for each QoS class is more with lazy approach. Observe that as the load allowance increases, the delay separation between the different QoS classes tends to decrease in all scenarios. The reason is that under lower server capacity, the load allowance diminishes more quickly, resulting in fewer transitions from one tier to another and thus maintenance of higher delay separation.

We have also obtained results for end-to-end jitter; these are omitted here but are quite similar in nature.

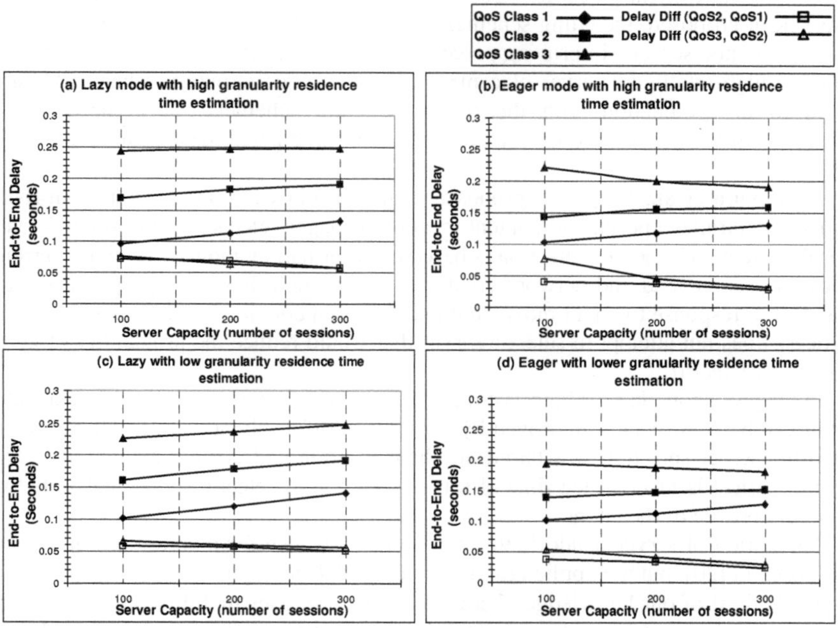

Figure 4. Impact on End-to-End Delay

Comparing lazy and eager approaches

Although we are able to achieve significant reduction in number of handoffs for each of the four simulation scenarios, the results highlight several comparative strengths and weaknesses of the two approaches. Lazy mode achieves more reduction in number of handoffs than the eager mode (comparing figure 3a and 3c with figure 3b and 3d). Lazy mode is also less susceptible to errors in residence time estimation (comparing figure 3c and 3d). Lazy mode, with its cautious approach, moves fewer clients to lower tiers thus reducing the cases where a wrong client is assigned to a lower tier server. Assigning users to higher tiers also has some disadvantages. Comparing figures 4a and 4c with figures 4b and 4d respectively we can see that there is that clients incur overall higher delays with lazy mode compared with the eager mode.

At the network level, assigning clients to higher tiers translates to lesser traffic localization and higher utilization of network resources. This can be a significant factor if network bandwidth is more expensive than placing servers at lower tiers.

5. Related Work

There has been a great deal of work on web performance enhancements. For CDN, several content placement, server placement, and server selection techniques have been proposed, including schemes geared towards load balancing among local and distributed servers, improving perceived QoS such as [5], and reducing network cost through intelligent server placement. Several request routing schemes [1] are used to in conjunction with the server selection schemes. A comprehensive overview of various techniques used for enhancing web performance can be found in [8].

Yoshimura et al., [17] have proposed a multimedia CDN architecture. In their work, a centralized portal and content location manager selects an appropriate server for the client. The server selection is based on information about the client's current location, server load and network conditions. Although they have proposed server handoff in response to client movement under certain conditions, the server selection process does not attempt to select a server that would reduce the probability of need for server handoff.

In [16] we show that for long-lasting streaming multimedia sessions, the delay and jitter can be significantly improved by server-handoff as clients move. However, in [16] the server selection was based on client location and content availability. No explicit effort is made at the time of server selection to select a server that suits the mobility profile of the client.

The notion of server handoffs to improve performance for mobile users has been proposed in [20], but not applied to multimedia or CDN systems.

To the best of our knowledge, there is no work that uses mobility information for server selection in CDN. In estimating residence time and dealing with mobility information, our focus has been to enable server selection in a fully distributed manner. Additionally, we try to minimize the state information that should be maintained for the purpose of tracking client movement or making prediction, because maintaining such information for a large number of clients can consume significant resources. Mobility information gathering, as it is currently defined in our scheme, can be easily collected using existing protocols such as Mobile IP [6] binding update signaling messages. It is likely that mobility prediction algorithms, such as used for resource reservation and call admission control [19] in mobile networks can be used.

6. Conclusions

We have presented a mobility aware server selection scheme that can significantly reduce the number of server-handoffs necessary for streaming multimedia content delivery via a high density CDN. In our server selection algorithm, we use mobility information about the clientele, along with traditional server selection criteria, such as expected QoS from the server, traffic localization, and server load. The effectiveness of our server-selection algorithm depends on the accuracy of the clients residence time estimate. While the estimation schemes that we have proposed are decentralized, stateless and simple, they are not very accurate. Nonetheless, our simulations show that our server selection algorithm can reduce the server handoffs by up to 10-15% at the expense of small increase in mean end-to-end delay, while maintaining delay separation among different QoS classes. More elaborate mobility information prediction method can improve the performance of our algorithm at the expense of increased complexity.

References

[1] A. Barbir, B. Cain, R. Nair, and O. Spatscheck. Known CN request-routing mechanisms. IETF Internet Draft, work in progress. Apr. 2003.

[2] C. Cheng, R. Jain, and E. van den Berg. Location prediction algorithms for mobile wireless systems, in *Handbook of Wireless Internet*. M. Illyas and B. Furht (eds.), CRC Press, Dec. 2002.

[3] I.R. Chen and N. Verma, Simulation study of a class of autonomous host-centric mobility prediction algorithms for cellular and ad hoc networks. In *Proc. of the 36th Annual Simulation Symposium*, Orlando, FL, USA, Mar 2003.

[4] Cisco's Mobile CDN. http://www.cisco.com/warp/public/784/packet/apr02/p49-cover.html.

[5] Z. Fei, S. Bhattacharjee, E. Zegura, and M. Ammar. A novel server selection technique for improving the response time of a replicated service. In *Proc. IEEE INFOCOM*, Apr 1998. San Francisco, California.

[6] D. Johnson, C. Perkins, and J. Arrko. Mobility Support in IPv6. IETF Internet Draft, work in progress. May 2003.

[7] Intelligent Content Distribution Service, AT&T. http://www.research.att.com/news/2002/September/ICDS.html

[8] A. Iyengar, E. Nahum, A. Shaikh, and R. Tewari. Enhancing Web performance. In *Proc. of the 2002 IFIP World Computer Congress*. Aug 2002. Montreal, Canada.

[9] The Mathwork Inc. http://www.mathworks.com/

[10] J. Matta, A. Takeshita. End-to-end voice over IP quality of service estimation through router queuing delay monitoring. In *Proc. of IEEE Globecom*, Nov 17-22, 2002. Taipei, Taiwan.

[11] NTT DoCoMo. FOMA subscriber growth. http://www.nttdocomo.com/

[12] OPNET Technologies. http://www.opnet.com

[13] M. Rabinovich and A. Aggarwal. Radar: A scalable architecture for global Web hosting service. In *Proc. of the Eighth Int'l World Wide Web Conf.*, May 1999.

[14] H. Schulzrinne, S. Casner, R. Frederick and V. Jacobson. RTP: A transport protocol for real-time applications. IETF RFC-1889. January 1996.

[15] S. Sen, J. Rexford, D. Towsley. Proxy prefix caching for multimedia streams. In *Proc. of IEEE INFOCOM*. April 1999.

[16] M. Tariq, A. Takeshita. Management of cacheable streaming multimedia content in networks with mobile hosts. In *Proc. of IEEE Globecom*, Nov 2002. Taipei, Taiwan.

[17] T. Yoshimura, Y. Yonemoto, T. Ohya, M. Etoh, S. Wee. Mobile streaming media CDN enabled by dynamic SMIL. In *Proc. of WWW Conference*, May 2002, Honolulu, Hawaii.

[18] H. Yumiba, K. Imai, and M. Yabusaki, IP-based IMT network platform. *IEEE Personal Communications*, Oct 2001, pp 18-23.

[19] T. Zhang. E. Berg, J. Chennikara, P. Agrawal, J. Chen, T. Kodama. Local predictive resource reservation for handoff in multimedia wireless IP networks. *IEEE Journal on Selected Areas in Communications*, 19(10):1931-1941, Oct 2001.

[20] R. Jain and N. Krishnakumar, Network support for personal information services to PCS users. *IEEE Networks for Personal Communications (NPC)*, Jan. 1994.

[21] A. K. Elmagarmid, J. Jing, and O. Bukhres. An efficient and reliable reservation algorithm for mobile transactions. In *Proc. of the 4th International Conference on Information and Knowledge Management (CIKM'95).*

[22] Nokia Network Planning White Paper: http://www.nokia.com/downloads /aboutnokia/press/pdf/MWR_Planning_A4.pdf

PERFORMANCE OF PEPS IN CELLULAR WIRELESS NETWORKS

Pablo Rodriguez and Vitali Fridman
Microsoft Research, Cambridge

Abstract In this paper we study the main performance problems experienced by Web applications in Cellular Wireless networks. To this extend we identify the particular problems at each protocol layer (transport, session and application layer). We present our practical experience results, and investigate how a Wireless Performance Enhancing Proxy can improve the performance over these networks.

1. Introduction

GSM networks all over the world are being upgraded to support the General Packet Radio Service (GPRS). GPRS provides wireless data services that enable ubiquitous mobile access to IP-based applications. In addition to providing new services for today's mobile user, GPRS is important as a migration step toward third-generation (3G) networks. GPRS allows network operators to implement new IP-based mobile data applications, which will continue to be used and expanded for 3G services. In addition, GPRS provides operators with a testbed where they learn about the challenges and technical problems of deploying such a network.

Over the last several years important advances have been made in the lower layers of the wireless protocol stack. As a result, new modulation protocols, smarter schedulers, and optimized error correction techniques were introduced. Despite the advances in the wireless data link and MAC layers, little attention has been paid to evaluation of the performance of the higher layers of the protocol stack, e.g., TCP, HTTP, etc. Optimized wireless networking is one of the major hurdles that mobile computing must solve if it is to enable ubiquitous access to networking resources. However, current data networking protocols have been optimized primarily for wired networks and do not work well in wireless networks. GPRS wireless environments have very different characteristics in terms of latency, jitter, and error rate as compared to wired networks. As a result applications experience unnecessary losses, bursty behavior, slow response times, and sudden losses of connectivity. This mismatch between traditional wireline protocols and the wireless bearer can significantly reduce the benefits provided by link layer protocols. Accordingly, traditional protocols have to be adjusted and tuned to this medium.

F. Douglis and B.D. Davison (eds.), Web Content Caching and Distribution, 19-38.

Some of the performance problems observed in GPRS networks have also been observed to some extent in other long-thin or long-fat networks (e.g., Satellite, WLANs, Metricon Ricochet). However, little in depth analysis has been done about the real underlying problems that impact data performance over GPRS. In this paper we investigate the performance of layers 4-7 of the protocol stack over GPRS networks. We first analyze the characteristics of a GPRS network and then study the performance issues of TCP, DNS and HTTP.

To overcome the problems posed by wireless links we introduce a **Wireless Performance Enhancing Proxy** (W-PEP) architecture that attempts to handle the underlying characteristics of GPRS wireless links. The W-PEP sits at the edge of the wireline network, facing the wireless link, and monitors, modifies, and shapes traffic going to the wireless interface to accommodate the wireless link peculiarities. W-PEP implements a number of optimizations at the transport, session and application layers that overall provide a significant improvement to the end user experience. It can also includes caching, logging and billing subsystems. Some of these W-PEP proxies have been recently deployed in wireless commercial networks primarily to enhance remote Web access for business users using their laptops or PDAs. However, little data has been published about the real performance improvement provided by these W-PEPs. In this paper we present one of the first studies of the real benefits of these proxies in 2.5G GPRS networks.

The rest of the paper is organized as follows. The next section discusses related work. In Section 3 we present a brief overview of cellular wireless networks. Section 4 discusses the architecture of the Wireless Performance Enhancing Proxy. Section 5 describes the latency components of a typical Web transfer in a wireless link. Section 6 presents transport and session level optimizations. Section 7 discusses application level optimizations and Section 8 compares then with transport and session level optimizations. Finally, we conclude the paper in Section 9 with some future work discussion.

2. Related Work

Previous solutions to improve data delivery in wireless data networks have been proposed at the physical, link and MAC layers, at the transport layer (TCP optimizations) as well as at the application layer (data compression).

In the literature, physical/link/MAC layer enhancements have been proposed that aim to provide improved scheduling algorithms over wireless links to increase the total system throughput, provide fairness or priorities between the different users, assure minimum transmission rates to each user and incorporate forward error correction on the link to reduce retransmissions. The scheduling algorithms aim to control the system or user throughput at the physical layer. For data applications, it is equally important to consider the data performance at higher layers in the protocol stack, especially at the transport (TCP) layer. Techniques such as the ACK Regulator ([7]) has been proposed to monitor and control the flow of acknowledgment packets in the uplink channel and therefore regulate the traffic flow in the downlink channel of a wireless link. This solution avoids buffer overflow and the resulting congestion avoidance mechanism of TCP. Similarly ([5]) proposes to avoid slow-start and congestion avoidance all together by clamping the TCP window to an static estimate of the Bandwidth Delay product of the link. At the application layer several data compression techniques have been proposed ([8, 3]) to increase the effective throughput of wireless links. Examples include degrading the quality of an image, reducing the number of colors, compressing texts, etc.

Several proxy based protocols have been proposed to fix TCP problems over wireless links, e.g. Snoop ([2]), I-TCP ([1]), and W-TCP ([10]). More recent papers suggest the use of a small proxy at the mobile host combined with a proxy on the wireline network to implement a new transport protocol that replaces TCP ([4]).

Despite the large amount of work in this area, most of the research in wireless performance enhancing proxies has focused on solving the impact of highly lossy wireless networks on TCP. However, as we will see in this paper, current 2.5G and 3G networks provide very strong link-level reliability mechanisms that hide losses from the higher protocol layers. This suggests the need for revisiting the use of performance enhancing proxies in next generation wireless networks. In this paper we consider the impact of Wireless Performance Enhancing Proxies in current 2.5G and 3G wireless networks. We consider the range of optimizations that can be implemented at the W-PEP at different layer of the stacks, e.g., transport, session, and application. We first discuss the existing problems and then we propose possible solutions that we validate by using data from real wireless deployments.

3. Overview of Cellular Networks

GPRS networks represent an evolution of GSM networks and are considered to be an intermediate step towards 3G networks, e.g., UMTS. That is why GPRS networks are given the name of 2.5G. While GSM networks are mostly a European phenomenon, similar efforts are under way in the USA and in Asia. For instance, CDMA 1xRTT is the equivalent of GPRS in the USA market. CDMA 1xRTT is supposed to evolve to the equivalent 3G standard, CDMA 1xEDVD, at about the same time when GPRS is expected to evolve to UMTS. Despite the difference in the cellular wireless standards in different countries, all these networks share a similar set of problems. To better understand the problems posed by these networks we next show the result of a set of pings through a live commercial GPRS network in Europe. Similar results were obtained in other networks both in Europe and the USA.

```
Pinging www.microsoft.com [192.11.229.2] (32 bytes):
Reply from 192.11.229.2:  bytes=32 time=885ms TTL=109
Reply from 192.11.229.2:  bytes=32 time=908ms TTL=109
Reply from 192.11.229.2:  bytes=32 time=4154ms TTL=109
Reply from 192.11.229.2:  bytes=32 time=870ms TTL=109
Reply from 192.11.229.2:  bytes=32 time=795ms TTL=109
Ping statistics for 192.11.229.2:
Packets:  Sent = 80, Received = 80, Lost = 0 (0% loss),
Approximate round trip times in milli-seconds:  Min = 761ms, Max = 4154ms, Avrg = 1001ms
```

As we can see the average RTT delay measured by a ping request is equal to one second. Moreover, the variability in the delay is quite extreme, ranging from 761 msec to 4 sec. An important point to note is that there were *zero losses*. A deeper analysis using tcpdump also revealed *zero out of order* packets. The main reason for this is the strong link-layer reliability implemented by cellular wireless networks. Cellular wireless networks implement ARQ retransmissions, FEC, and other related techniques to ensure that packets are not lost or misordered by the air interface. As an example, link-layer GPRS protocols provide 10^{-9} packet loss rates and similar values for out-of-order packet reception. The fact that reliability is implemented at the link layer ensures that higher layers see practically no losses or packets out of order. However, the drawback is that the variability and value of the wireless RTTs increases drastically. This is an inherent problem of cellular wireless links that will not go away with the arrival of 3G wireless networks.

Another problem is the fact that these wireless networks have a very low throughput. For instance GPRS networks provide a throughput of 15-25 Kbps while CDMA 1xRTT provide a throughput of 50-70 kbps for HTTP traffic. These problems require an in-depth study of their impact on the higher layer protocols. As we will see in this paper, a careful study of these problems and a set of intelligent optimizations techniques can overcome many of the problems that plague GPRS links.

4. Wireless PEP

Wireless Performance Enhancing Proxies (W-PEP) are required to minimize the big performance mismatch between terrestrial and wireless links. By splitting the connection between the terrestrial and the wireless side into two different connections, W-PEPs can significantly improve end-to-end protocol performance. Previous work in wireless performance enhancing proxies assumed high losses in the wireless links, however, 2.5 and 3G links provide reliable and orderly delivery by implementing link-layer reliability. As a result we consider it crucial to revisit the concept of W-PEPs, taking into account the peculiarities of 2.5 and 3G wireless networks.

The W-PEP can be located in multiple places in the network. The most natural place is to collocate the W-PEP with the GGSN since this the first IP node where all IP traffic is anchored from the Base Station System (BSS) (Figure 1). Other locations could also provide interesting benefits and a more distributed architecture (e.g., at the SGSN, at the Base Station). However, deploying the W-PEP at these locations would need substantial work since it requires accommodating a layer 3-7 proxy between layer-2 nodes. The W-PEP can be deployed using an explicit proxy configuration or a transparent proxy configuration plus a layer-4 switch. W-PEPs provide a number of enhancements that improve the overall end-to-end user experience. These optimizations cover transport layer optimizations, session level optimizations and application level optimizations.

One important characteristic of W-PEP is that it does not require any modifications to the client or server stacks and performs all the optimizations transparently. This makes it very easy to deploy and avoids the hassle of having to provide and maintain new purpose build client software. However, we point out those optimizations that would benefit more from having a special client-software.

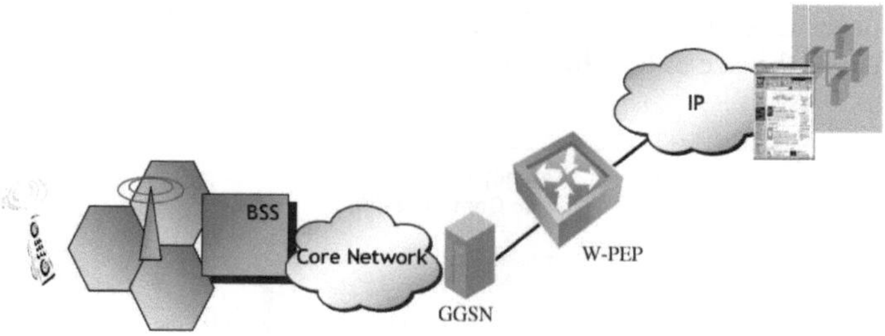

Figure 1. W-PEP network architecture

In the rest of the paper we will show the performance benefits of a W-PEP implementation. This implementation runs on Solaris 8 and enables TCP and session optimizations, application-level optimizations, and caching. The W-PEP was tested on a GPRS cell where several cell parameters could be configured. The primary cell parameters available for configuration were the number of time-slots available for GPRS traffic, the signal to noise ratio, the amount of background traffic, and the duration of the periods with no signal. For most experiments we considered a typical average loaded GPRS cell with the following parameters: 18 dB of SNR, phone with four downlink slots and 2 uplink slots, background traffic of voice calls with 60 seconds of active time and 15 seconds of inter-arrival time exponentially distributed, 10 background users, and no data slots being reserved exclusively for GPRS traffic.

To remove dependency on the real Internet and its unpredictability, we downloaded and hosted all necessary objects on our own web server. We also hosted our own DNS server with all necessary records to reproduce the exact setup of the target Web pages.

5. Latency Components

To have a better understanding of the different factors that determine the overall latency to download a document, we have disaggregated the total download time of a document into various components. To do this we use tcpdump traces. We try to determine the impact of: a) the time to resolve DNS names; b) the time to establish TCP connections; c) the idle times between the end of the reception of an embedded object and the beginning of the reception of the next one; and d) the transmission time of the data payload. We do not explicitly consider the processing time of the server since we assume that in a wireless link, the server processing time is negligible compared to the other latency components. Figure 2 shows the GPRS latency components for the top-level page of the 10 most popular web sites. These Top 10 pages are described in table 4. From this Figure we see that time to deliver the payload accounts for most of the latency (about 65%). This is an expected result since the throughput of the GPRS link is quite low. The idle RTTs in between GET requests for embedded objects account for the second largest latency component, especially in pages with many objects. These idle times represent a significant overhead since the RTT through wireless links can be quite high. HTTP headers also account for a large overhead since many objects in Web pages tend to be quite small and the Header sizes could potentially be on the same order as the size of the object itself.

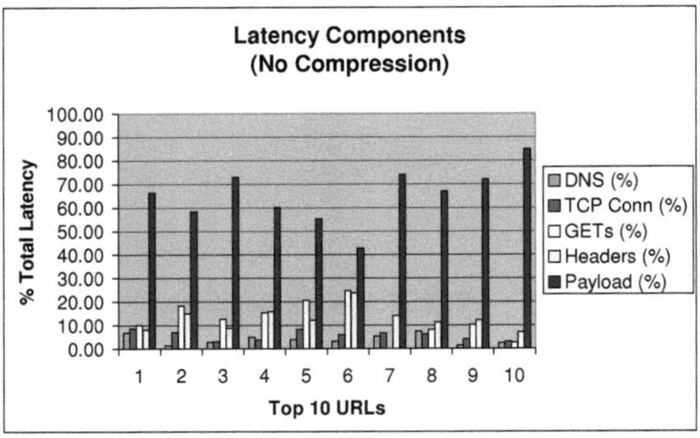

Figure 2. Latency Distribution (No compression)

At the transport and session layer, DNS queries and TCP connection setups account for the smallest portion of the overhead. Most of these overheads are barely noticeable in terrestrial links with quite small RTTs. However, in GPRS links where RTTs are in the order of seconds, these overheads can significantly affect the end-user experience. In the next sections we describe in more detail these overheads and identify solutions for some of these problems.

6. Transport/Session Optimizations

TCP faces several problems in GPRS links. Given the Radio Link Protocol (RLP) link-layer retransmissions, TCP sees very large and variable delays. On top of this, the throughput of the wireless link is very small. The two possible options to deal with these problems are to tune/alter the TCP/IP networking stack at an intermediate node only (i.e., W-PEP), or to change/replace the networking stack at the mobile host and intermediate node. The latter approach may be based on a completely new transport protocol (e.g., based on UDP) that handles the particularities of wireless links and replaces TCP. This new transport protocol should provide reliability, congestion control, flow control, and fairness. However, TCP has proven to be a very flexible and robust protocol that can be adapted to many types of networks. Creating a completely new transport protocol may not prove to be the most efficient solution since one may end up re-implementing most of the TCP features. Therefore, in this paper we propose solutions that do not replace TCP but instead optimize it for cellular wireless links.

These type of solutions can be implemented in an intermediate proxy such as the W-PEP and do not require any modifications to TCP/IP stack on either end of connection (servers or mobile clients).

Before describing in detail the problems and solutions for TCP performance in wireless links, we would like to have a rough estimate of how much improvement we should expect from TCP optimizations. From Figure 2 we can estimate the best case scenario for TCP and session-level optimizations. By eliminating TCP setup, and removing DNS lookups, the best improvement possible is about 18%. As we will see later, this number goes up to 30% once the content is compressed since the relative impact of TCP setup and DNS queries increases (See Figure 5). Next we will consider how to deal with specific TCP problems in more detail.

6.1 TCP Tuning

To better understand the problems of the TCP performance in wireless links we run the following experiment. We first calculate the FTP throughput obtained when downloading a large file. Then we compare it with the throughput obtained when downloading a Web page of the same size using HTTP (the page had 10 embedded objects). The results can be seen in table 1.

Table 1 shows that the throughput provided by an FTP transfer of a large file is quite close to the maximum TCP throughput than can be achieved through the wireless link. The main reason for this is that FTP uses a single connection and therefore TCP connection setup overhead is very small. This result also shows that the long-term throughput provided by the tested Solaris TCP implementation is very good. We tested other OS stacks (e.g., Linux) and the performance was slightly worse, however, it became comparable when tuning several important TCP parameters in the manner that we will describe in this section.

When we compare the FTP throughput of a large file with the HTTP throughput of a size-equivalent Web page, we noticed see that the HTTP throughput drastically drops from $37-39$ Kbps to $21-28$ Kbps. This is due to the large number of TCP connections opened by the browser and the DNS lookups needed to resolve the domain names of embedded objects. In addition to the TCP and DNS problems, HTTP also suffers from the fact that there are multiple idle RTTs in between object requests since pipelining is not enabled. We will later discuss in more depth the impact of pipelining.

Table 1. TCP Throughput

	Throughput (Kbps)
Maximum Airlink GPRS Rate	53.6
Maximum TCP Throughput Rate	43-45
FTP Rate	37-39
HTTP Rate	20-28

As discussed earlier in the paper, we will use a standard TCP/IP stack and tune it to optimize TCP performance in GPRS networks. Different stacks require different levels of tuning, however, the optimal combination of TCP parameters should be the quite similar for different OS implementations. In order to tune a given TCP stack we consider critical parameters that could potentially impact the TCP wireless performance. These parameters include the maximum transfer unit size, the slow start window, and several timers that determined the retransmission timeouts calculation:

- a) A careful selection of the MTU allows for efficient link utilization. Selecting a small MTU increases the chance of a successful transmission through a lossy link; however, it also increases the overhead of header to data. Selecting a large MTU, on the other hand, increases the probability of error and the delay to transmit one segment. However, wireless links such as GPRS and CDMA provide low-level techniques that strongly reduce the probability of loss. Having large MTU has several benefits, such as a smaller ratio of header overhead to data and a rapid increase of TCP's congestion window (TCP's congestion window increases in units of segments). As a result, a large MTU (tcp_mss_def about 1460 bytes) may prove to be quite beneficial to optimize TCP in GPRS networks.

- b) Traditional slow start, with an initial window of one segment, is too slow over wireless networks. Slow start is particularly harmful for short data transmissions ([9]), which is commonly the case for web traffic. Increasing the initial slow start window (tcp_slow_start) to three to four segments may improve the transfer time significantly and therefore it is highly recommended as a possible TCP optimization.

- c) A correct estimation of a connection timeout (RTO) is very important since it may lead to unnecessary segment retransmissions and a sharp decrease in TCP throughput. To estimate RTO, the sender uses periodic measurements of the RTT between the client and the server. However, it takes several RTTs before the sender can efficiently estimate the actual delay of the connection. In between, many packets can be retransmitted inefficiently. One way to minimize spurious retransmission due to incorrect RTT estimation is to set the initial values of the RTT closer to its real values in the wireless link. Thus, parameters such as $tcp\_rtx\_interval_initial$ and tcp_rtx_min can play an important role in optimizing TCP performance.

To have a better understanding of the impact of a wrong estimation of the retransmission timer, in Figure 3(a) we show a TCP trace for a page download from a standard Web server which parameters have not been tuned for GPRS links. We focus on the number of retransmissions that were received at the client (marked by an R). The results show that at the beginning of the download the Web server has a completely wrong (too short) estimation of the RTT, thus, it generates many unnecessary retransmissions. As the download progresses the retransmissions disappears since the RTT estimation improves. However, it takes about 30 seconds for the Web server to properly estimate the parameters of the wireless connection and stop sending unnecessary retransmissions. This overhead can be catastrophic for short Web transfers. We obtained similar results with other pages and found that in many situations the number of retransmissions accounted for up to 50% of the packets. As we will see later, W-PEP solves this problem.

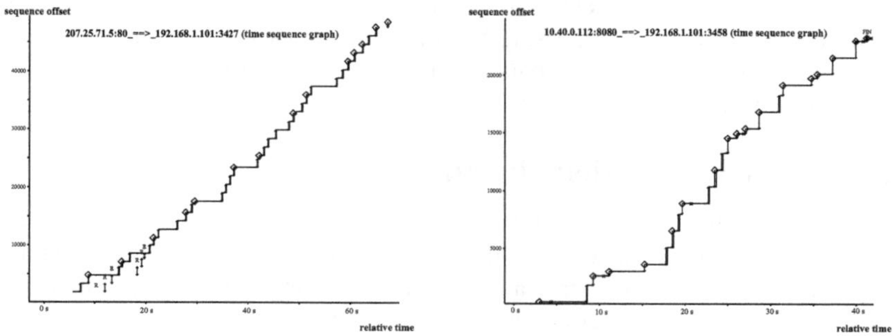

(a) W-PEP disabled. (b) W-PEP enabled.

Figure 3. TCP Behavior.

TCP Parameter Selection. To determine the optimal TCP parameter mixture for GPRS links we run an experiment where a 50 KB file is downloaded multiple times with a different combination of the parameters described in the previous section (tcp_mss_def, tcp_slow_start, $tcp\_rtx\_interval$, and tcp_rtx_min). We used all possible combinations of these parameters with the following input values: tcp_mss_def = (Default, 1500, 1400, 1200) Bytes, $tcp\_slow\_start\_initial, tcp\_slow\_start\_idle$ = (Default, 4, 3, 2) segments, $tcp\_rtx\_interval\_initial$ = (Default, 7000, 5000, 3000) msec, and tcp_rtx_min = (400, 3000, 5000) msec, where DEFAULT are the default TCP settings of the Solaris TCP stack. The default TCP settings for Solaris 8.2 are: MSS = 536, Slow start initial =4, Slow start after idle= 4, Retransmission initial =3000, Retransmission min =400. We repeated each download 20 times, averaged the results, and compared them with the results obtained using the default TCP parameter setup. Due to space limitations we are not showing here the complete set of results for all combinations.

The measurements showed that most combinations provided little benefit versus the default configuration, however, some of them provided an improvement of up to 17% improvement over the default configuration. The optimal tuning configuration consisted on: MSS = 1400, Slow start initial = 4, Slow start after idle =4, Retransmission initial = 7000 ms, Retransmission min = 3000ms. These results indicates that large MTUs, large values of the slow start window, and values of the initial and minimum RTT that are closer to those in the wireless links, can significantly improved the overall performance of TCP. In addition to reducing latency, having a better tuned TCP stack also improves bandwidth usage since it decreases the number of retransmissions. To illustrate this point, we now repeat a similar experiment than the one presented in Figure 3(a) using W-PEP with a tuned TCP stack. The results are presented in Figure 3(b). We can note that with W-PEP there are no retransmissions, even at the beginning of the connection. We repeated the same experiment with many other Web sites and we always saw a significant reduction in the number of retransmissions. Without W-PEP, the number of retransmissions for certain Web sites sometimes accounted for half of the data delivered, therefore, significantly reducing the available bandwidth and download rates. Using W-PEP, on the other hand, the number of retransmissions due to spurious timeouts was negligible. As a result, a well-tuned TCP stack can drastically improve the wireless link efficiency.

6.2 TCP Connection Sharing

In standard TCP/IP implementations each new TCP connection independently estimates the connection parameters to match an end to end bandwidth and round trip time of the network. The two critical connection parameters that require estimation are RTT and congestion window size. For each connection the estimation process normally starts with the same default values, set in the stack configuration. This estimation process takes several RTTs to converge. In wireless links with very large and highly variable RTTs, estimation process converges very slowly. For short connections typical for Web access, it is common to have most of the traffic flow over connections with sub optimal parameters, therefore underutilizing the available capacity. It can be observed however that RTT and congestion parameters for connections to the same mobile host are likely to have very similar values, primarily determined by device and network capabilities (e.g., number of uplink and downlink time slots), and by current network load and conditions (e.g., current link error rate).

To minimize the amount of time required for TCP to converge to the right connection parameters, we instrumented W-PEP TCP stack to cache RTT and congestion window parameters of currently running or recently expired connections, and to reuse them as a starting values for the new connections *to the same mobile host*. A second advantage of this approach is that connection parameters are effectively estimated over considerably longer period of time, spanning several connections to the same mobile host. This provides for much better estimates of link parameters. We set a timeout before the cached values expire to 2 minutes. If there are no new connections to the same mobile host over this period of time, the cached values are cleared. We found this value to be a good heuristic timeout.

Whenever a new connection is initiated, W-PEP re-uses previously cached connection parameters. The new connection immediately starts sending data at the same rate at which a previous connection had been sending just before it finished. Therefore, we virtually eliminate the overhead of the slow start phase of most connections, the convergence process happens much faster, and estimated connection parameters are more precise.

One drawback of this approach is that at the beginning of a given connection W-PEP may produce bursts of data that are higher than it would otherwise be the case; however, current GPRS networks provide large buffers that can easily absorb it.

To evaluate the performance of this TCP connection state sharing technique we downloaded the Top 10 pages with and without this feature enabled, and averaged the results over multiple rounds. We used HTTP1.1 persistent connections. The results showed an average improvement of 12% over all pages. This is a significant improvement that can clearly have an effect on end-user experience, and it does not require any modification of the client or server TCP stacks.

6.3 Number of TCP connections

One of the main factors that impacts the performance of TCP through GPRS networks is the number connections opened and closed. Each time a new connection is opened, there is a corresponding connection setup overhead which includes several idle RTTs before TCP is able to fully utilize the available link capacity. Next we study how standard browsers such as Netscape or IE use/reuse TCP connections and what their impact is on GPRS networks. To study this, we considered a page with 25 embedded objects and 142 KB of data that was downloaded by the browser. While the download was happening we collected TCP traces using tcpdump to study the dynamics of TCP. The traces where captured on a Windows 2000 laptop. The address of the laptop was 192.168.1.101. Version of IE used was 5.50. The IE was set to use an explicit HTTP proxy at the address 10.40.0.112 port 8080. "Use HTTP1.1 through proxy connection" (in IE Options) was turned on. The collected traces cover the period time between the initial page request and when all its embedded objects are received in full.

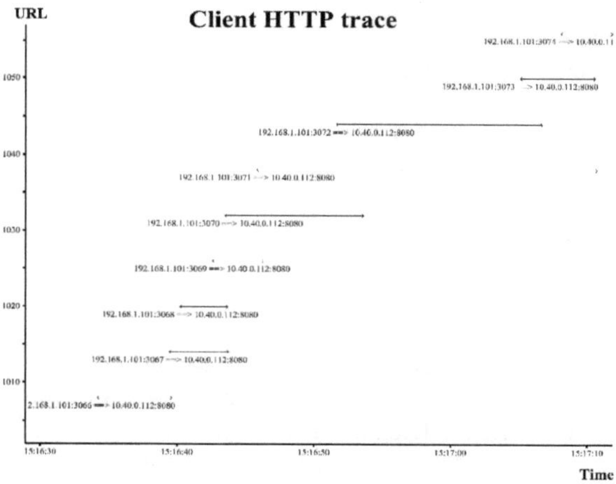

Figure 4. TCP Connection Behavior

Figure 4 shows a time graph of the TCP connections opened and closed by the browser. From this figure we see that the total number of connections used to retrieve the page by the browser is equal to 10 connections (the first connection coincide with the beginning of X axis on the graph, and can only be seen with appropriate magnification of this corner). All connections were closed by the browser and not by the proxy or the server. The number of connections used is very large. According to HTTP1.1 specifications, browsers are supposed to use only two connections to a proxy. In the same Figure we can also note that IE never uses more then 2 connections in parallel, however, it routinely opens new connections, just to close an "older" connection in lieu of newly opened one.

By examining the breakdown of number of URLs served per connection, we noticed that the bulk of objects is retrieved using 2 or 3 connections, however the rest of the connections are only used to retrieve 1 or 2 objects per connection. In fact, in this experiment, 6 out of 10 connections were used to retrieve only 1 object. We have conducted the same experiment with other Web sites and we observed the same behavior.

To better understand this behavior we analyzed the source code of TCP connection scheduler of a popular browser. Next we describe the algorithm used when a new object request is received. We assume that the browser is explicitly connected to the proxy, thus, all connections opened to the proxy can be potentially reused to retrieve any object from any Web site.

```
if (num_conn_open < max_conn){
   if (idle_conn & !conn_expired()){
       reuse_conn();
    }
   elseif{
      open_new_conn();
      close_expired_conn();
    }
}
elseif{
   open_new_conn();
   close_oldest_idle_conn();
}
```

The main motivation behind this algorithm is the following. Browsers open multiple connections in parallel to fetch embedded objects as fast as possible. The browser first opens a TCP connection to fetch the home page (i.e., index.html file). If embedded objects need to be fetched, the browser will try to re-use any existing connections that have finished previous downloads (i.e., idle connections) and that have not yet expired (as determined by the keep-alive timeout). If no connections are idle, the browser will open new connections until it reaches the maximum number of connections allowed. For an HTTP 1.1 browser in explicit proxy mode this limit is usually two. Once the browser has reached its maximum number of connections, if a new object request arrives, the browser still creates a new connection. At this point, the browser is exceeding its maximum connection allowance. To balance the number of connections, the browser closes the first connection that becomes idle or any connection that has been active for more than the maximum allowed delay.

This is an aggressive behavior that prevents browsers from getting slowed down by connections that are stalled. In a wireless network, however, connections take a long time to complete. The browser frequently thinks that the connection is stalled and opens a new connection to try to get the download going as fast as possible. At the same time it closes other pending active connections that according to the browser's algorithm are not progressing at a fast enough rate. This can lead to a behavior where many connections are frequently opened and closed and a single object can take a very long time to complete download since it never gets to progress. This is counter productive in wireless networks since opening and closing many connections creates and extremely high overhead that significantly increases the download times. One way to overcome these problems is to ensure that browsers for wireless networks try to keep a constant number of connections that they re-use them as much as possible, and use connection timers that are in accordance with the delays/throughputs of wireless links. These modifications require slight changes or re-configuration of the client browsers.

Even though the number of connections should be kept constant as much as possible, the number of connections should not be very small to prevent stalled connections from delaying the overall download. In the next experiment we try provide some intuition of why using a very small number of connections may not be very efficient. To this end, we calculate the performance of a file download using one single connection vs. four connections. We consider a large 1 MB file. When using four connections, the file is divided into four equal pieces and all of them are fetched in parallel. The results show that the throughput obtained with four connections is about 11% higher than with a single connection. One reason for this higher throughput is that multiple asynchronous connections can better utilize the wireless link capacity since some connections are able to grab additional bandwidth while others are idle or slow.

6.4 Temporal Block Flow Release

Wireless Base Station Controllers (BSC) and SGSNs allocate wireless link resources to a particular mobile host only for a certain period of time. After a mobile host is idle for more than a preconfigured period of time, the wireless link resources are released. The logical resource in question is called TBF (temporary block flow) and we will refer to this behavior as TBF release. This improves the utilization of a given GPRS channel since the period of channel inactivity is limited. However, when the mobile host comes back and starts requesting data again it goes through the acquisition of a new TBF and associated time slot. Acquiring a new TBF and time slot is an expensive process that adds an initial latency before data transfer can be initiated. In the next experiment we try to determine the TBF release timeout, i.e., the idle period of time before the mobile host's GPRS channel is released. To this end we periodically ping the W-PEP from the mobile host with increasing pinging intervals. The following table shows that there is a jump in the value of RTT when the time between pings increases from 5 seconds to 6 seconds. This indicates a TBF release timeout of around 6 seconds. This empirical experiment to determine the TBF value was later confirmed by a number of GPRS vendors.

A TBF equal to 6 sec can create very frequent wireless channel releases and acquisitions, especially given that user think times in between page downloads are frequently higher than 6 seconds. As a result when the think time between requests is higher than the TBF value, the mobile host suffers an extra delay to acquire a new GRPS channel before any data can be transmitted. In order to determine the impact of the TBF release in a page download, we consider the following experiment. We downloaded the Top 10 pages repeatedly, with a new page request happening 10 seconds after the end of the previous page download finished. We then repeated the same experiment while having a background ping from the W-PEP to the mobile host every 5 seconds. This background ping happens before the TBF is released, thus, the mobile host gets to keep the channel and does not need to re-acquire on each page. Only one ping request is required per mobile host to keep the wireless GPRS channel active. The results show that keeping the GPRS channel and not releasing it in between page requests provides a 15% latency improvement. In a real implementation this background ping should stop after the mobile host inactivity period goes over a given threshold, e.g., 20 sec, to avoid flooding the wireless link with unnecessary packets and to release the GPRS channel.

6.5 Session-level overheads: DNS

DNS is a session-level protocol that is used to resolve the server names associated with all objects in a Web page. However, DNS queries can have a significant impact in the GPRS performance. For example, we observed that www.britannica.com has 14 different domain names used by objects embedded on its home page, or www.cnn.com has 6 different embedded domain names. Performing DNS queries through the wireless interface can drastically increase the overall download time of this and similar pages. The way of overcoming DNS delays is by caching DNS responses and re-using them for a certain Time-to-live without having to re-contact the DNS server each time a given domain is accessed. However, popular domains names are frequently served by content distribution networks, which set very small TTLs in their DNS responses. When TTL response is very small, the browser has to repeat the same DNS query over

Table 2. Ping times for different inter-ping intervals

Ping Interval (sec)	Avrg. Ping Time (msec)
1	1667
2	1726
3	1664
4	1778
5	1726
6	2110
7	2551
8	2345
9	2123
10	2314

and over again to resolve a given domain name. Performing repeated DNS queries through terrestrial links may not have a significant performance impact; however, in wireless links this can be a source of major overhead.

In order to estimate the impact of DNS queries we conducted the following experiment. We download the main CNN page with all its embedded objects. First we download it ensuring that all DNS lookups required to resolve the embedded domain names happen trough the GPRS link. Then we repeat the same experiment with all DNS lookups being satisfied from the local DNS cache, thus avoiding the GPRS link. To make sure that in the first experiment DNS lookups were not satisfied from a local client DNS cache, the DNS Time To Live key value was set to 0 in the Windows registry, and IE was restarted after every run (IE keeps its own DNS cache). The results obtained show that avoiding DNS lookups over the wireless interface reduces the response time by 16%. This is a significant time reduction that requires special attention. There are several ways to fix this problem by having the proxy do the lookups over a terrestrial link in a transparent way. However, due to its lengthy considerations we prefer to make this the subject of another paper ([11]).

7. Application Level Optimizations

Application level optimizations are intended to minimize the overhead of the application level protocol (e.g., HTTP) and to minimize the time to delivery the payload. Figure 2 showed that payload transmission time accounts for the major portion of the delay since the bandwidth of the GPRS link is quite low. The idle time in between object requests (GETs) as well as the HTTP headers also account for a large portion of the total delay, however, its significance depends on the number of objects in a given page. Pages with a large number of small objects have a higher number of idle RTTs and higher proportion of HTTP headers to content, while pages with a low number of large objects barely experience these kinds of overhead.

Given that the time to deliver the payload accounts for most of the transmission time, minimizing the amount of data delivered will significantly reduce the total download time. To this end we have instrumented the W-PEP to intercept all Web requests and process the responses before passing them to the mobile host. Once W-PEP downloads the Web document requested, it performs the following actions: a) *transform* the page format into a suitable page format readable in the mobile host. If the mobile host is a laptop then there is no page transformation; b) *lossless content compression*, e.g., compresses text/html files; c) *lossy compression* of images. Regarding lossy compression, W-PEP has several adjustable levels of compression which can be configured by the W-PEP administrator, the content provider, or the end-user. The number of parameters that can be adjusted include: number of colors in a GIF, level of quality degradation in JPEGs, and whether animated GIFs should be converted into static ones or not.

7.1 Compression Results

Next, we present the compression factors attained by W-PEP on different types of Web pages. We focus on optimizing the downlink channel since it carries most of the data in WEB applications. We assume that the page is already pre-formatted to fit the device screen, and therefore, we do not consider the impact of content transformation. Instead, we measure the compression achieved by W-PEP through lossy or lossless compression. For lossless compression, W-PEP used gzip on all possible content-types. Some servers already support gzipping of text/html content, however, we found that most pages were uncompressed, and therefore, W-PEP had to compress them. Recent versions of most browsers support the reception of compressed content, which is uncompressed on the fly. For lossy compression we selected 16 colors for GIFs, no animated GIFs (in banners), and a level of JPEG quality degradation that was barely perceived by the human eye. W-PEP caches all transformed content, thus, it only needs to compress it or transcode it once and then multiple requests for the same object can be served from the cache, substantially increasing scalability.

We considered an experiment using the Top 100 pages to understand how W-PEP would behave in a real scenario. The average compression factor for the Top 100 pages is 2.83, however, some specific pages with large portions of text and highly compressible images can be compressed by a factor close to 6.

Table 3 presents the compression factor for each content type individually as well as the percentage of a page corresponding to each type. We can see that GIFs as well as HTML content amount for a large portion of all files and can be highly compressed, which helps achieving a high overall compression factor.

Table 3. Compression Factors by Content Type.

Content Type	% of Content	Compression Factor
Octetstream	0.45	x2.3
Xjavascript	4.50	x2.73
Xpotplus	0.60	x2.5
Xshockwaveflash	0.68	x3.1
GIF	77.33	x2.44
JPEG	4.80	x1.93
PNG	0.38	x1.92
CSS	0.23	x4.94
HTML	8.41	x3.84
Text/PLA	0.60	x3.45

7.2 Acceleration Results

Given the compression factors calculated in the previous section we would like to determine how this data compression factors translate into a latency reduction experienced by the end user. To this end we calculated the compression factors and the latency reduction seen by the end user for the Top 10 Web pages (see Table 4). The latency reduction is calculated as the ratio between the total page download time with W-PEP and without W-PEP. We can see that the average latency reduction for these

Table 4. Top 10 Web sites. Impact of Compression, Acceleration, and Pipelining.

Name	Objects	Domains	Size (KB)	Compress.	Speedup	Pipelining
1-Altavista	6.00	3	25.4	2.15	1.51	1.87
2-Chek	14.00	1	37.7	2.77	1.67	2.42
3-CNN	36.00	8	185	2.72	2.8	4.42
4-Excite	16.00	5	55.4	3.46	2.43	4.13
5-Fortunecity	36.00	7	142	3.46	1.97	3.57
6-Go	16.00	2	33.7	3.66	1.72	2.96
7-Google	2.00	1	10.3	3.96	1.76	2.00
8-Lycos	5.00	3	30.3	4.09	1.76	2.36
9-MS	16.00	2	87	3.95	2.37	3.25
10-Yahoo	4.00	2	37.4	3.53	3.18	3.59

pages is 2.12, and the average compression factor is 3.18. The average ratio between latency/compression is 0.64, thus, the compression factor for most pages is higher than the latency reduction achieved. This indicates that there is not a perfect direct translation between compressing data and reducing latency, though, the correlation is quite high. This is caused by other overheads that are not affected by compressing data (e.g., idle RTTs, TCP connection setup, DNS lookups, etc.) and that still accounts for a significant portion of the total download latency as we will see later.

To better understand how different factors affect the overall latency after compression has been applied, we re-calculated the results of Figure 2. The new latency distribution with compressed data is presented in Figure 5. From this Figure we see that payload's transmission time has been reduced substantially and accounts for a much smaller portion (about 40%) of the total latency. This causes other latency factors such as idle RTTs, or HTTP headers to account for a much higher portion of the delay, sometimes equal or more than the actual payload delivery.

7.3 Impact of Pipelining

We will now consider the impact of pipelining. Pipelining is required to avoid the idle RTTs that occur when browsers wait until an embedded object is fully received before requesting the next object. This overhead can be quite important for GPRS links since the RTT delays through a GPRS link are usually over one second. One way to avoid these idle RTTs is by requesting a given embedded object before the previous one has been fully downloaded. This is known as *request pipelining* and its benefits have been well studied in the past on wireline and satellite networks. To estimate the benefits associated with pipelining in a cellular network, we compare the latency obtained when downloading the Top 10 Web pages with and without request pipelining. The results obtained are presented in table 4. This table shows that the speedup factor obtained when pipelining is turned on is 3.06, which is a 42% improvement versus not having pipelining (table 4). Similar results were obtained in ([6]).

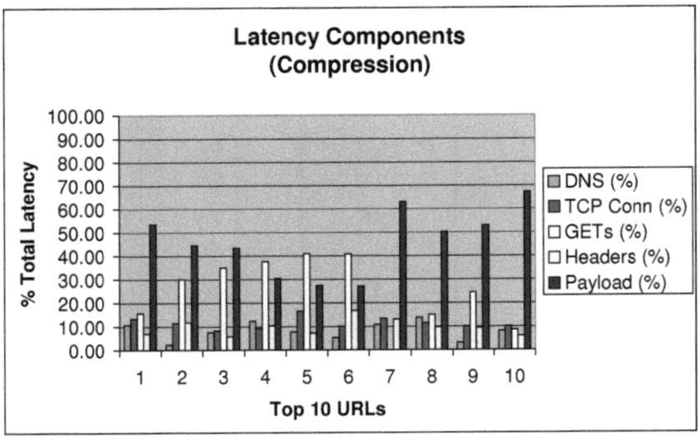

Figure 5. Latency Distribution (Compression).

Assuming that pipelining was turn on, we now attempt to determine whether the relationship between compression and latency reduction improved. To this end we compare the latency reduction obtained with pipelining vs. the compression factors. (We do not show the individual results for each page.) The results obtained show a new average ratio between latency reduction and compression of 0.90 versus 0.64 without request pipelining. This indicates that when request pipelining is employed, gains achieved by data compression translate almost fully into a latency reduction factor.

Despite the advantages of pipelining, especially in wireless networks, most servers still do not support it. Browsers, on the other hand, frequently support request pipelining. In order to enable request pipelining through the wireless link, W-PEP supports request pipelining. Pipelining is not always supported in the wireline connection to the origin servers. But having W-PEP implement pipelining over the wireless link provides most of the benefit since the wireless link dominates the end-to-end latency.

8. Comparison

In this section we try to determine the relative impact of the different optimizations presented in this paper. We compare transport-level optimizations with application-level optimizations. To do this, we measured the latency of downloading certain Web objects when W-PEP performs only transport-level optimizations, and when W-PEP performs both transport-level and application-level optimizations. The application-level optimizations included pipelining and compression of Web objects (lossless and lossy). The transport-level optimizations considered were TCP tuning, TCP connection sharing, TBF release avoidance, use of two persistent TCP connections, and no DNS lookups.

We first compared both sets of optimizations for a large object, a 400 KB image, and then we repeated the same experiment with a Web page. In Figure 6(a) we show the speedup obtained when downloading this large image for the cases of no W-PEP, W-PEP with transport-level optimizations only, and W-PEP with transport-level and application-level optimizations. From this figure we see that relative benefits obtained from optimizing long-term TCP behavior are quite small (about 6%). Compressing the large image, on the other hand, provides a compression factor over 800% with very small visual perception impact. It is clear from this example that application-level optimizations are much more important than transport-level optimizations when considering a large single object that is highly compressible.

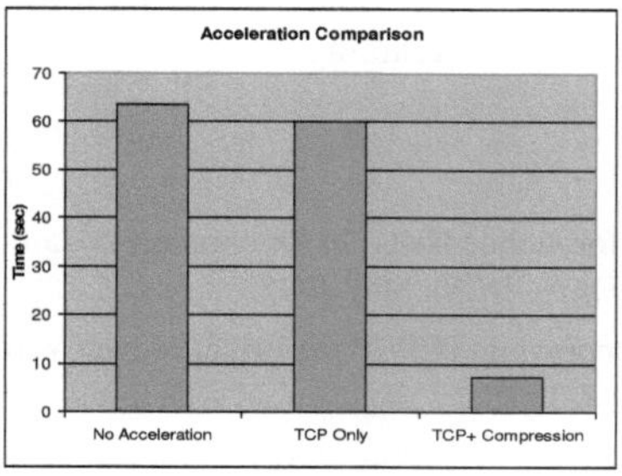

(a) 400 KB image.

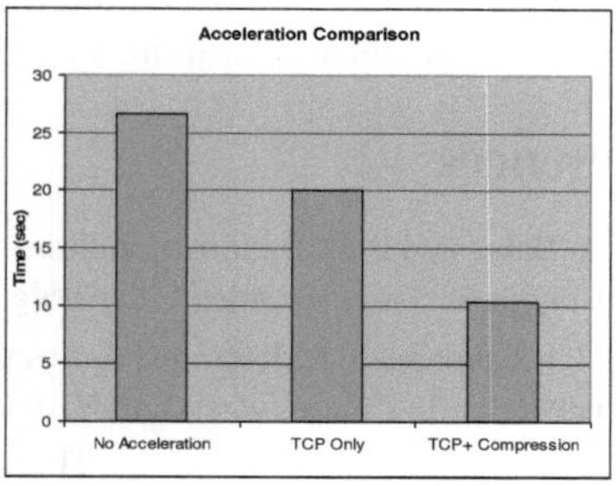

(b) 100 KB page. 8 embedded objects. 3 different domain names.

Figure 6. Comparison between application-level and transport-level optimizations.

In Figure 6(b) we repeated the same experiments for a typical Web page. Given that this page has multiple objects and it is hosted in several different domain names, the overhead of DNS lookups and TCP connection setup is much higher. Using the previously mentioned transport-level optimizations, the speedup obtained equals 30% instead 6% obtained with a single large object. Transport-level optimizations provide better results in this case since the complexity of the page is higher, with many objects hosted in multiple domains, which is the case of many popular Web sites. Regarding application-level optimizations, we note that they do not work that well with the tested Web page since the objects were quite small and not very compressible. The actual speedup obtained when compressing text and images on the tested Web page was close to 250%, which is a much lower number than the 800% obtained previously with a single large image. Still this factor is much higher number than the 30% provided by the transport-level optimizations alone. Therefore, application-level optimizations amount for a larger portion of the overall latency reduction, although their relative impact highly depends on the type of pages and their complexity.

9. Conclusions

In this paper we have identified the main problems experienced by Web applications in 2.5G and 3G wireless networks. We have considered the problems at each layer of protocol stack separately, e.g., transport, session, and application. Given the peculiarities of these wireless networks with very large and highly variable latencies and low throughputs, we introduced a Wireless Performance Enhancing Proxy (W-PEP). The W-PEP does not require any modification to the sender or the client's stack, and it can be deployed completely transparently at the border between the Wireline and the wireless network. We showed that the suite of optimizations implemented at the W-PEP can provide significant latency reduction to the end user, thus, fostering the utilization of such links to access Web content.

As possible future directions of this work we are considering use of W-PEP to handle temporary loss of connectivity, (e.g., when users go inside a tunnel) or how it can be used to minimize the impact when the mobile user switches access networks, e.g., GPRS to WiFi. Finally, other possible areas where W-PEP could improve performance are dynamic content acceleration, wireless gaming or streaming.

References

[1] A. Bakre and B. R. Badrinath. Indirect TCP for mobile hosts. In *Proceedings of 15th Int'l Conference on Distributed Computing Systems (ICDCS)*, May 1995.

[2] H. Balakrishnan, R. Katz, and S. Seshan. Improving TCP/IP performance over wireless networks. In *Proceedings of ACM MOBICOM*, 1995.

[3] Bytemobile Inc. The Macara Optimization Service Node. http://www.bytemobile.com/html/products.html.

[4] R. Chakravorty, A. Clark, and I. Pratt. GPRSWeb: Optimizing the Web for GPRS links. In *ACM/USENIX First International Conference on Mobile Systems, Applications and Services*, 2003.

[5] R. Chakravorty, S. Katti, J. Crowcroft, and I. Pratt. Flow aggregation for enhanced tcp over wide-area wireless. In *IEEE INFOCOM*, 2003.

[6] R. Chakravorty and I. Pratt. WWW performance over GPRS. In *IEEE MWCN*, 2002.

[7] M. C. Chan and R. Ramjee. TCP/IP performance over 3G wireless links with rate and delay variation. In *Proc. of ACM Mobicom*, Sept. 2002.

[8] Forelle Systems Inc. The Venturi Server. http://www.fourelle.com/pdfs/Venturi_V2.1_Brochure.pdf.

[9] V. Padmanabhan. *Addressing the challenges of web data transport*. PhD thesis, Computer Science Division, University of California at Berkeley, 1998.

[10] K. Ratnam and I. Matta. W-TCP: An efficient transmission control protocol for networks with wireless links. In *In Proceedings of Third IEEE Symposium on Computer and Communications*, 1998.

[11] P. Rodriguez, S. Mukherjee, and S. Rangarajan. Session level techniques for improving web browsing performance on wireless links. In *Bell-Labs Technical Report*, 2003.

EDGE CACHING FOR DIRECTORY BASED WEB APPLICATIONS

Algorithms and Performance

Apurva Kumar and Rajeev Gupta
IBM India Research Lab

Abstract In this paper, a dynamic content caching framework is proposed for deploying directory based applications at the edge of the network, closer to the client. The framework consists of a Lightweight Directory Access Protocol (LDAP) directory cache and the offloaded application running at a proxy. The LDAP directory cache is an enhanced LDAP proxy server which stores results and semantic information for search requests (*queries*) and answers incoming queries which are semantically contained in them. A simplified query containment approach based on the concept of LDAP *templates* is proposed. Caching algorithms have been proposed which take advantage of referential locality in the access pattern. A generic framework is used to offload the application at the edge and to support prefetching of LDAP queries based on application logic. A real enterprise directory application and real workloads are used to evaluate performance of the caching algorithms. The LDAP directory cache architecture, along with the proposed algorithms can be used to improve performance and scalability of directory based services

1. Introduction

There has been a growth of websites providing dynamic content on the web. Typically dynamic content is generated in response to a user request (*query*) evaluated against a database. Techniques used for traditional (static) content caching are, in general, not applicable for caching dynamic content.

Directories are specialized databases, which are capable of storing heterogeneous real world information in a single instance. The Lightweight Directory Access Protocol (LDAP) provides a means for accessing and managing remote and distributed directories [1,2]. LDAP directories are being used to store address books, contact information, customer profiles, network resource information, policies etc. Directories have assumed special significance in enterprises where a single directory instance containing employee and other organizational records is used by a wide variety of internet and intranet applications. The heterogeneous nature of information

F. Douglis and B.D. Davison (eds.), Web Content Caching and Distribution, 39-56.

stored in directories allows them to be used by a wide variety of applications. However, this also means that an overloaded directory could be a potential bottleneck in an enterprise infrastructure.

Database caching has been established as an effective means to improve performance of client server relational database systems [5-8]. In [5], an architecture for database caching at application servers is presented. In [6,9] an active query-caching framework for form-based proxy caching of database-backed websites is described and significant gains compared to passive query caching are reported.

While the above strongly motivate the need for a framework to be developed for caching/prefetching directory queries for web applications, the mechanisms for database query caching are not directly applicable because of the inherent differences in data models and query languages of directories and relational databases [10,11].

LDAP replication has widely been used for improving performance, scalability and availability of directory based web applications. However, since a typical directory can contain millions of records, the search performance of replicas suffers due to disk access latency [4]. Partial replication is useful only if applications need to access a part of the directory. An LDAP caching solution, which provides significant query hit ratio, while caching only a small fraction of the records in the directory is thus desirable.

There are two distinct problems associated with LDAP query caching: *(i)* Determining whether an LDAP query is contained in another query (*query containment*), *(ii)* Using *caching algorithms* which make efficient caching, prefetching, cache replacement decisions to maximize the fraction of queries answered from the cache. The complexity of *(i)* is the subject of [12]. The authors consider the general query containment problem for LDAP and show it to be NP-complete in the size of the query. The authors of [13] introduce the notion of generalized queries and propose algorithms for *(ii)*. They use a real directory but synthetic workloads and a single type of query for performance evaluation.

In our work we reduce the complexity of the query containment problem by introducing the concept of LDAP templates (Section 4). For *(ii)* we determine the caching algorithm to be used based on the type of query. The proposed directory cache is an LDAP proxy server extended for query caching. The performance of an LDAP cache implementation has been evaluated for an enterprise directory containing employee records of a large organization and real workloads for an employee white pages application.

Most directory enabled applications are web based and use directories to generate dynamic content in response to HTTP user requests. Resources that are deployed closer to the client are becoming under-utilized with increasing dominance of dynamic content in the web. To reduce client latency and to improve scalability of the origin application server and directory server, there is a need to offload some components of the application to a proxy server at edge of the network where the LDAP cache is located. We use a generic web publishing framework to offload the cacheable component of a directory based application to the edge of the network. The offloaded application can also be used to prefetch LDAP queries based on application logic to improve the cache performance as described in Section 7.2.

The paper is organized as follows: Section 2 describes notations and terminology used in the paper. Section 3 describes the proposed caching framework. Section 4 discusses the query containment problem and proposed solution. Section 5 discusses the caching algorithms. Section 6 describes the set of extensions required in typical directory servers to incorporate query caching. A generic XML based publishing framework and its use in application offload and prefetching is described in Section 7. In Section 8, performance of caching algorithms is evaluated for a real enterprise directory application. Contributions of the work are discussed in Section 9.

2. Notations

This section introduces briefly some relevant notations and terms used in the rest of the paper. LDAP assumes the existence of one or more directory servers jointly providing access to a *Directory Information Tree* (DIT), which is made of entries. An *entry* is defined as a set of *attribute* value pairs. Each entry has a distinguished name (DN) belonging to a hierarchical namespace. A *directory cache* (or *LDAP cache*) stores a subset of directory entries which correspond to results of cached queries obtained from a *master directory server.*

The functional model adopted by LDAP is one of clients performing protocol operations against servers. LDAP defines three types of operations: query operations, like *search*, *compare*, update operations like *add*, *modify*, *delete* and connect/disconnect operations like *bind*, *unbind*. The most common operation is *search*, which provides a flexible means of accessing information from the directory. The search operation also referred as a *query* consists of the following parameters which represent the semantic information (or *metadata*) associated with a query [1]:

base: A DN that defines the starting point of the search in the DIT.
scope: { BASE, SINGLE LEVEL, SUBTREE }
Specifies how deep within the DIT to search from the *base*.
filter: A boolean combination of *predicates* using the standard operators: AND `(&)`, OR `(|)` and NOT `(!)`, specifying the search criteria.
attributes: Set of required attributes from entries matching the filter.

LDAP filters are represented using the parentheses prefix notation of RFC 2254 [3], e.g.: `(&(sn=Doe)(givenName=John))`. Filters without any NOT operators are called *positive filters*. *Predicates* of the form (*name operator value*) where *operator* {=, ≤ , } are considered. *name* is an attribute name and *value* is its asserted value termed as *assertion value*. Example of predicates are: `(sn=Doe)`, `(age 30)`, `(sn=smith*)` where "`Doe`" "`30`" and "`smith*`" are assertion values representing equality, range and substring assertions, respectively.

3. LDAP Caching Framework

Figure 1 shows various components of the proposed LDAP caching framework for directory based web applications. The dashed arrows show the path of a client request in the absence of the framework. A remote client request is serviced by a web application which accesses a master directory to answer it.

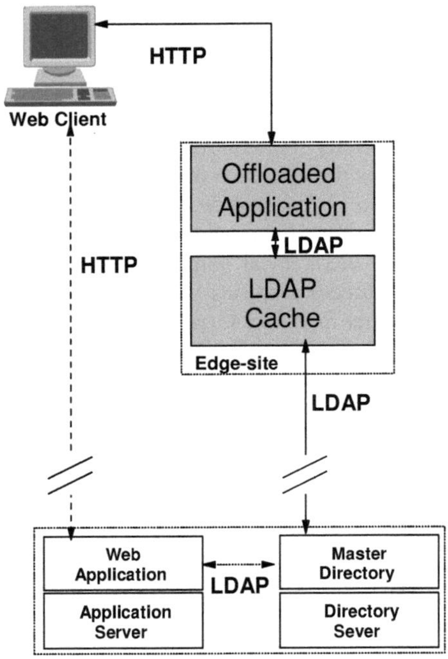

Figure 1. LDAP Caching Framework

Using the framework a remote client request is redirected, possibly using DNS redirection, to an edge site closer to the client. The edge site consists of an offloaded application which converts user requests to LDAP queries and sends them to the LDAP cache. Stored metadata for cached queries and the semantic information of an incoming query is used to determine whether it is contained in any of the cached queries. Contained queries are answered locally while other queries are forwarded to the master directory server. This improves performance for remote clients by reducing latency and makes the service more scalable by offloading web and LDAP requests to the edge site.

4. LDAP Query Containment

4.1 Query containment problem

An LDAP query Q, is said to be *contained* in another query Q', if all of the following are true:

C1: The *base* of Q is same as or descendant of the *base* of Q'.

C2: The *scope* of Q' is SUBTREE.

OR

(The *scope* of Q' is same as the *scope* of Q) AND (the *base* of Q' is same as the *base* of Q)

OR

(The *scope* of Q' is SINGLE LEVEL) AND (*scope* of Q is BASE) AND (*base* of Q is a direct descendant of the *base* of Q').

C3: *attributes* in Q is a subset of *attributes* in Q'.

C4: The *filter* of Q is semantically *contained* in the *filter* of Q'.

In general, a contained query is *answerable* (from the cache) if *attributes* of Q in C3 above is interpreted as the union of required *attributes* and the attributes appearing in its search filter. We neglect the impact of directory schema (e.g. mandatory attributes) on the cache answerability problem. Using schema information can provide a more complete cache answerability algorithm. However, as noted in [12] it makes the general problem NP complete in the size of the query.

4.2 General filter containment

An LDAP filter F_1 is *contained* in F_2 if it is not possible for an entry to satisfy F_1 but not F_2. This condition is formalized by Proposition 1.

Proposition 1:
An LDAP query filter F_1 is semantically contained in another query filter F_2 if and only if the expression $F_1 \wedge \neg F_2$ is *inconsistent.*

For the expression $F_1 \wedge \neg F_2$ to be inconsistent:

$$\neg\exists\, x_1, x_2..x_n \text{ such that } F_1 \wedge \neg F_2 \text{ is satisfied.}$$

where attribute set $\{x_1, x_2..x_n\}$ is the union of attribute sets appearing in the filters F_1 and F_2.

If $F_1 \wedge \neg F_2 = B_1 \vee B_2 \ldots\ldots \vee B_k$ where each B_i is a conjunct of simple predicates, then each B_i should be inconsistent, i.e. the following boolean expression should evaluate to TRUE.

$$(\neg\exists x_1, x_2..x_n(B_1)) \wedge (\neg\exists x_1, x_2..x_n(B_2)) \ldots \wedge (\neg\exists x_1, x_2..x_n(B_k)) \qquad (1)$$

where $\neg\exists x_1,x_2..x_n$ (B_i) represents the condition that B_i is not satisfiable for any values of attributes $x_1,x_2..x_n$ in their valid ranges.

Let the set A_{XY} represent the union of sets of assertion values in LDAP filters *X* and *Y*.

Proposition 2:
For positive LDAP filters F_1 and F_2 containing equality and range predicates, the condition for F_1 to be contained in F_2 can be expressed as a boolean expression in conjunctive normal form (CNF) with each simple predicate of the form:

$(a \quad b)$ where $a,b \quad A_{F1F2}$. □

Sketch of proof: The expression in *(1)* is a conjunct. The condition for each B_i being inconsistent requires that the predicates in B_i should impose an empty range for at least one of the attributes appearing in it. Thus the condition of each B_i being inconsistent is disjunctive and *(1)* can be written in CNF. It is easy to show that a possibly empty range for an attribute x_j imposed by the predicates of B_i is $(a_{xj}, b_{xj}]$, or $[a_{xj}, b_{xj})$ where $a_{xj}, b_{xj} \quad A_{F1F2}$. For this range to be empty $a_{xj} \quad b_{xj}$.

Example: Let F_1 be $(a \leq p) \wedge (b \quad q)$ and F_2 be $(a = x) \vee (b \quad y)$
The condition for F_1 to be contained in F_2 is easily seen to be $(q \quad y)$, but the example helps in illustrating proposition 2.
Here, $A_{F1F2} = \{p, q, x, y\}$.
F_1 is contained in F_2 if the following expression is inconsistent:

$F_1 \wedge \neg F_2 = ((a \leq p) \wedge (b \quad q) \wedge (a > x) \wedge (b < y)) \vee ((a \leq p) \wedge (b \quad q) \wedge (a < x) \wedge (b < y))$
$B_1 = ((a \leq p) \wedge (b \quad q) \wedge (a > x) \wedge (b < y))$.
$B_2 = ((a \leq p) \wedge (b \quad q) \wedge (a < x) \wedge (b < y))$.
For B_1 to be inconsistent: $(x \quad p) \vee (q \quad y)$
For B_2 to be inconsistent: $(q \quad y)$
Thus F_1 is contained in F_2 if:

$((x \quad p) \vee (q \quad y)) \wedge (q \quad y) \Rightarrow (q \quad y)$ □

In the worst case all *m* predicates in F_1 might have to be compared with all *n* predicates in F_2. Thus checking containment of an incoming filter with a cached filter requires *O(mn)* such comparisons.

4.3 Template based filter containment

To reduce the complexity of the problem, we introduce the concept of *LDAP templates*. Most query filters generated by applications are different from other filters of the same type in only their assertion values. A *template* is a filter with one or more assertion values unspecified. A template is also represented using the LDAP filter representation of [3] except that the missing assertion values are replaced by the "_" character. Examples of templates are: `(&(cn=_)(ou=research))`, `(uid=_)`, `(&(sn=_)(givenName=_))`. A substring template for generating queries with the initial string specified is `(sn=_*)`. Templates can be used to generate actual queries by supplying missing assertion values.

In template based query containment, queries belonging to only a specified set of templates, *T*, are cached and answered. The template based approach has several advantages. Firstly, the number of query comparisons are reduced since queries of a given template *A* can possibly be answered by only a subset of the templates $T_A \subseteq T$. E.g. a query of template `(&(sn=_)(ou=_))` can not answer a query of template `(sn=_)`. Secondly, conditions for a query *a* of template *A* to be contained in a query *b* of template *B* T_A can be represented as a predetermined generic boolean expression $C_{cnf}(A,B,A_{ab})$ (using a generalization of proposition 2). E.g. query `(age=X)` can be answered by query `(age Y)`, if (X Y). Thirdly, cache specific parameters used by the caching algorithms can be specified per template. E.g. different cacheability and consistency requirements can be specified for different templates.

The following observation about positive filters belonging to the same template can be made:

Proposition 3:
Let F_1 and F_2 be two positive LDAP query filters belonging to the same template. F_1 is contained in F_2 if each predicate in F_1 is contained in the corresponding predicate of F_2. □

Containment problem for filters of the same template having *n* predicates using Proposition 3 requires *O(n)* comparisons of assertion values.

4.4 Query containment algorithm

To evaluate whether a query *q*, with filter *f*, is contained in a cache having set of cacheable positive templates *T*, the following steps are performed:

1) Find template of the incoming query filter, say *a*. If $a \notin T$, return FALSE.
2) For all templates, *b* T_a
   For all queries q_t (with filter f_t) in *b*
     if all of *C1, C2, C3* are satisfied for *Q=q* and $Q'=q_t$,
       if *a=b*
         if all predicates of *f* are contained in corresponding predicates of f_t,
           return TRUE.
       else
         if $C_{cnf}(a,b,A_{fft}) = 1$, return TRUE.
   return FALSE.

An important aspect of LDAP query containment is that all comparisons of assertion values should be performed using the correct syntax and matching rules for the corresponding attribute. E.g. assertion values in the two filters `(telephoneNumber=2686-1100)` and `(telephoneNumber=26861100)` are equal according to the equality matching rule described for the commonly used `telephoneNumber` attribute.

The algorithms described in the section can be extended for substring assertions by interpreting substrings as range assertions.

5. LDAP Caching Algorithms

In this section, we consider improving performance of LDAP caching, for a given cache size, by making efficient caching, prefetching and cache replacement decisions. The directory cache is considered 'hit' by an incoming query if the query is semantically contained by any cached query. The cache can answer such queries without contacting the master server. For caching to be useful, the access pattern should demonstrate locality of reference. In the context of query caching, we define *spatial locality* as accessing of semantically or otherwise related queries in a short time span. Similarly, *temporal locality* is defined as accesses for the same query in a short time span. We extend this to include those accesses, which are contained in previous accesses. In the rest of the section we discuss the proposed caching algorithms.

Query based caching
Incoming user queries are cached if the number of entries in the result set is within specified limits. No prefetching is done. Cache replacement removes the least recently used (LRU) query. □

The limit on number of entries is to restrict the maximum size of a cached query.

The algorithm can not take advantage of spatial locality defined above. By caching user queries themselves, one can not answer queries which are related but not contained in them. Another problem with query based caching is that user queries typically return only a few entries. The number of such queries required to provide a reasonable hit ratio could be very large, thereby increasing query containment costs and storage requirements.

In [13], *super-query caching* (called template caching in their work) is considered and its performance for a real directory but synthetic workloads is evaluated. Generalized form of a user query (super-query) is prefetched when it is estimated to correspond to a hot region. An example of generalized query for the user query `(mail=foo@xyz.com)` is `(mail=*@xyz.com)`. Our version of the super-query caching algorithm which is simpler than [13], is described below:

Super-query caching
A *popularity index* (defined below) is maintained for each of several candidate super-queries. When this index crosses a *threshold*, all the entries corresponding to the super-query are fetched and cached. Cache replacement removes the least popular super-query (and its entries).

The *popularity index* is a measure of hotness of a super-query. It is defined as the expected normalized cost saving as a result of having a super-query (*Q*) in the cache:

$$PI(Q) = \frac{h(Q)c(Q)}{s(Q)} \tag{2}$$

where *s(Q)* is number of entries for that super-query, *h(Q)* is its hit frequency, and *c(Q)* is the retrieval cost of the super-query, which is assumed to be same for the super-queries belonging to same template. The candidate list is maintained as described below:

For each incoming query a corresponding super-query is added to the candidate list if it is not already present. Otherwise, its hit statistic *h(Q)* is updated. Super-queries which do not have a single hit for the last *N* incoming queries are removed from the candidate list. □

The *threshold* is set as $k.PI(Q_L)$ where Q_L is the least popular super-query actually cached. We use the values *k=2, N=500* for our experiments. The popularity index metric is similar *utility value* of greedy dual size popularity (GDSP) algorithm [16].

Compared to query caching, using super-query caching reduces the meta data to be maintained for the same cache size but increases overheads of maintaining statistics to estimate popularity of super-queries. Such prefetching is able to exploit any spatial locality arising out of semantic relation between user queries due to the presence of semantic hot regions.

However, for certain types of queries it is difficult to imagine accesses having hot-regions which can be semantically described by generalized queries. Using super-query caching for such types of queries yields inferior hit-ratio performance and significantly larger overheads compared to query caching. As an example consider a name search based on surnames generating queries `(sn=_*)`. Even if there is a semantic region (e.g. `sn=smith*`) with large number of accesses, the region might not be *hot* since a possibly large number of entries associated with the region brings down its popularity index.

The authors of [13] consider only one type of query, `(telephoneNumber=_)`, which requests an entry with the given telephone number. They generate workloads with between 70-90% queries uniformly distributed inside a hot region and show that super-query caching performs better than simple query caching.

The reason for query caching not performing well in this case can be easily seen. Firstly, generating workloads uniformly within the hot region rules out temporal locality. Secondly, since a maximum of one entry per query is returned, hits are more likely to be from repeat queries rather than contained queries when query caching is used.

From these observations, we infer that the type of query should be considered while making the decision of using query or super-query caching. Since a template specifies the type of query we propose the following template based caching algorithm.

Template based caching
Queries belonging to only a specified number of templates are cached. For each template either query or super-query caching is performed. The cache is dynamically partitioned such that the incremental hits provided by the least recently used queries (query caching) or least popular super-queries corresponding to the same incremental cache size is approximately equal for all the templates. □

The solution was proposed for the disk cache partitioning problem [17] and is applicable to the query cache partitioning problem as well.

6. Directory Server Extensions

A typical directory server architecture consists of a protocol *front-end* which receives a client request, uses the *backend* to perform the operation corresponding to the request and sends results back to the client. The backend abstracts the database from the front-end by performing the read and write operations corresponding to various LDAP operations.

Most directory servers, e.g. IBM Directory Server, SunOne (iPlanet)/Netscape directory servers support plugins to extend their functionality. Server plugins are libraries including custom functions which can be called before or after an LDAP operation is performed (pre and post operation plugins). Plugins can also be used to integrate a new database with the directory server. Plugins use the *SLAPI API* [21] for interacting with the front end and backend. LDAP query caching can be implemented using a pre-operation plugin for the search operation, called *pre-search* plugin and a database plugin called *proxy backend.* The pre-search plugin implements query containment and other caching algorithms (discussed in Sections 4,5). The proxy backend does not have an associated database and uses the LDAP client API to perform the requested operation on the master server. The cached entries are stored in the *default* database *backend* of the directory server.

The caching algorithms described in Section 5 implement query (or super-query) level cache replacement. When a query is removed from the cache, all the entries which were returned for only that query are removed. To support this a multi-valued attribute `query_id` should be present in each cached entry. The values for the attribute in an entry are the unique identifiers of the queries for which the entry was returned. When a query with identifier, ID, has to be removed, an internal *search* is performed with the filter `(query_id=ID)`. A resulting entry is removed (using an internal *delete* operation) if it has a single value for the attribute, otherwise it is modified (using an internal *modify* operation) to remove the value ID from the `query_id` attribute.

To support weak consistency, queries are assigned a time to live after which they expire and are removed (as described above) from the cache. The time to live can be specified on a per template basis depending on the nature of the queries in the template.

Since the cache consists of a subset of entries from the master server, the DIT at the cache is sparse. This requires the following to be allowed by the default backend:

(i) Adding an entry without a parent, *(ii)* Deleting an entry with a child, *(iii)* Searching the cache without the *base* in the cache. Also, since the cached entries in general consist of only the required *attributes* in the corresponding query, schema checking (for mandatory attributes) should be disabled while adding or modifying an entry.

Our reference implementation of the directory cache [18] for the OpenLDAP directory server uses the native *slapd* API and a callback mechanism instead of *SLAPI,* which is a recent addition to OpenLDAP [15].

7. Application Offload and Prefetching

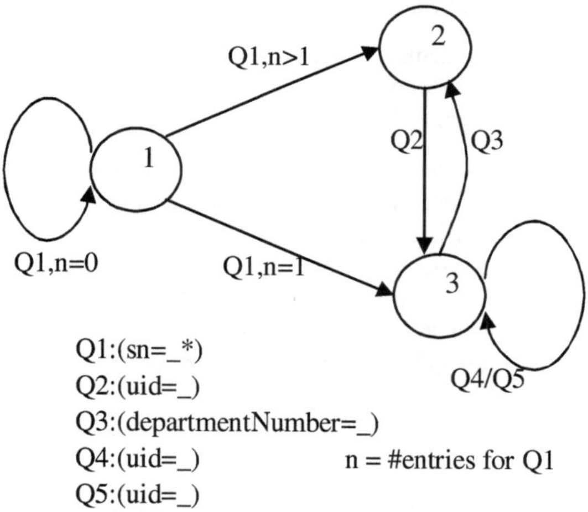

Figure 2: Application state diagram

Application offloading refers to pushing some of the processing from the backend server to a proxy server. For several reasons an offloaded application could be different from the original application. Firstly, it allows cache specific optimizations, e.g. displaying part of a page, which is available from the cache while fetching the rest of the information.

Secondly, it allows prefetching queries based on application logic. Such prefetching can take advantage of the relationship between queries in the same user session. Since these queries might not be semantically related, super-query caching can not exploit this.

Thirdly, in most cases it is not required to offload the complete application but only those components for which interaction with the origin server is not required.

The applications running at the edge are typically much simpler than the backend applications and it is possible to model them using a generic web publishing framework. Modeling applications using such a framework allows separation of content, logic and style and makes it easy to deploy and modify the application.

To model offloaded directory based applications, we use Apache Cocoon [19] which is an XML based publishing framework. An offloaded application is represented as a pipeline of XML-to-XML transformations. A directory application can be modeled in Cocoon using eXtensible Stylesheet Language Transformation (XSLT) and LDAP transformers. XSLT [20] is a language for transforming XML documents. Next we show how the framework can be used to model the offloaded version of a real enterprise directory application.

7.1 Directory application modeling: Example

We consider how a real enterprise web based directory application can be modeled using the Cocoon framework. The application has the following states:

State 1: A blank form is presented to the user.
State 2: A list of entries is displayed.
State 3: An individual record is displayed.

In state 1, the user is presented with a form for name search (based on surname). On submitting the input, if more than one entries are returned, a list with links for obtaining details of each entry is displayed (state 2). If a single entry is returned, an individual record is displayed (state 3), which has links to the *manager* and *secretary* entries for an employee and a link to obtain a list of employees in the same department. Clicking on links in state 3 can either result in a list (state 2) or an individual record (state 3). *Q1-Q5* are templates of the corresponding LDAP queries generated.

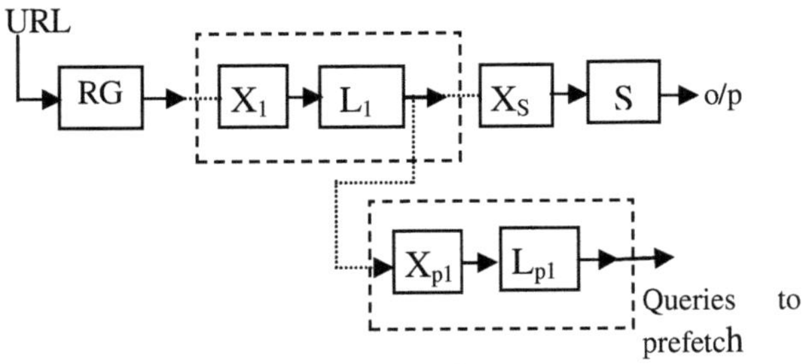

Figure 3. Processing of a user request

For each state the user interaction encoded in a URL is an input to a pipeline of transformers. E.g. in state 1 the user input (say `smith`) encoded in a URL is processed by a request generator (*R*) and converted into an XML representation. An XSLT transformer X_1 maps it to a Directory Services Markup Language (DSML) document representing corresponding LDAP query `(sn=smith*)`. The LDAP transformer (L_1) performs the corresponding LDAP search operation and converts the result to DSML. Depending upon whether one or more than one entries are returned the XSLT transformer (X_s) uses the results to create an HTML output corresponding to state 3 or 2 respectively which is returned to the client using a serializer (S).

In general the pipeline is of the form $R, X_1, L_1, X_2, L_2 \ldots X_m, L_m, X_s, S$ since a request might require a chain of LDAP queries with the result of ith stage providing input for generating LDAP queries for the $i+1^{th}$ stage. For the offloaded application each L_i points to the local LDAP cache.

7.2 Prefetching

The naming model of directory supports entries containing references to other entries using DN syntax attributes. Applications displaying such entries typically include a link to the referenced entries. Similarly web front ends of directory based applications contain links which are indicative of future accesses. Application logic based prefetching can be an important tool to improve hit ratio. Such prefetching can be easily performed using the application offload framework.

Continuing with the example application, if the search results in a single entry, and the accesses indicate a high probability of the *manager* link being accessed, then a new pipeline X_{p1}, L_{p1} (Figure 3) can be used to prefetch the manager entry. This ensures that if the manager link is accessed in the next click the LDAP cache will be able to answer the query. In general, a pipeline X_{pi}, L_{pi} could be used after each X_i, L_i segment to prefetch LDAP queries at the ith stage.

8. Performance of Caching Algorithms

We consider the IBM enterprise directory containing more than half a million employee and organizational records for measuring performance of the LDAP caching framework. Each employee entry is approximately 6KB in size. The web based employee white pages application described in Section 7 is the most popular application using the directory. The application is accessed by employees spread across more than 100 countries. The results in this section are based on a workload consisting of more than a million web accesses over a day. The distribution of query-types in the workloads is given in Table 1.

The directory uses standard LDAP schema for representing common entities. E.g. employee entries use the `inetOrgPerson` object class. The `uid` attribute of this object class is used as a unique identifier for each employee. Query containment for `(uid=_)` queries is equivalent to evaluating the query at the cache and returning TRUE if a matching entry is found. In general, query containment for queries of a template with known number of resulting entries does not require metadata. Thus it is possible to answer those `(uid=_)` queries for which the corresponding entry is cached as a result of some other type of query e.g. `(sn=_*)`. This requires complete entries to be fetched from the master server instead of just the required attributes for all cached queries. For all the algorithms only `(sn=_*)` and `(uid=_)` are considered as cacheable templates, i.e. queries or super-queries belonging to only these templates are cached. Cache size is measured in number of entries.

Table 1: Distribution of query types in the workload

Query type	*approx % contribution*
`(uid=_)`	51
`(sn=_*)`	28
`(departmentNumber=_)`	16
`Others`	5

Performance of query and super-query caching. Figure 4 shows hit ratio v/s percentage of master server entries (of class `inetOrgPerson`) cached, for the template `(sn=_*)`, when query and super-query caching are performed. Query caching provides higher hit ratio for a given number of cached entries. A hit ratio of 32% is achieved by caching 6% of the entries. This is due to the significant temporal locality in the access pattern for this template. Super-query caching does not perform well as anticipated in Section 5. Analyzing the hits in query caching reveals two main reasons. Repeat queries account for nearly half the hits. Query refinement (a broad search followed by a restricted one) by users is another important factor.

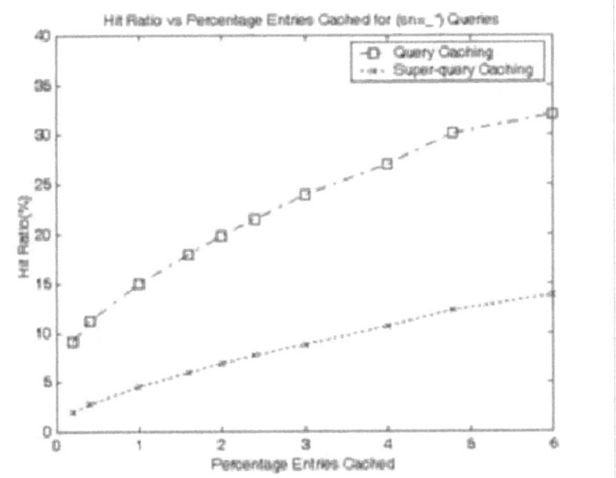

Figure 4. Hit ratio for (sn=_*)

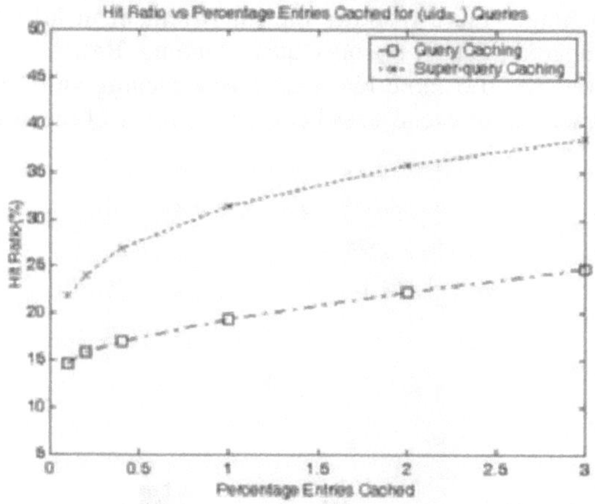

Figure 5. Hit ratio for (uid=_)

Figure 5 shows the comparison for `(uid=_)` template. In this case super-query caching performs better. The access pattern contains spatial locality which is well captured by super-queries. The hit ratio achieved is close to 40% when 3% of the entries are cached.

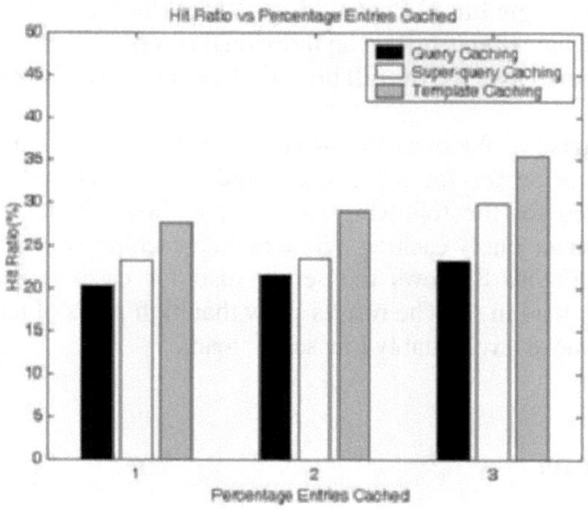

Figure 6. Performance of Template Caching

Template based caching. Next we present performance of template caching by simultaneously caching different templates with their corresponding best performing algorithms viz. query caching for `(sn=_*)` and super-query caching for `(uid=_)` queries. The performance of template based caching is compared against only query or super-query caching. Figure 6 shows the improvement in hit ratio for template based caching compared to query or super-query caching. Results show that template based caching gives 30-50% more hits than query caching and 15-20% more hits than super-query caching for cache sizes between 1.25-5 % of the directory size.

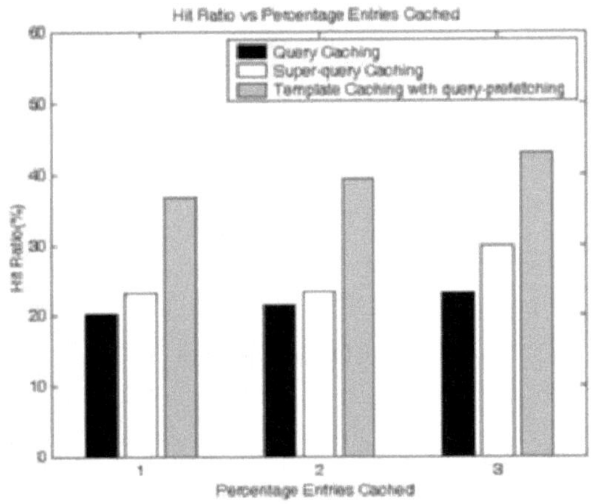

Figure 7. Hit ratio comparison

Effect of application logic based Prefetching. We perform application logic based prefetching of the *manager* entry when an individual record is displayed as described in Section 7.2. This increases the overall hit-ratio by approximately 10%.

Server load comparison. We measure the server load as the total number of entries served by the master server for the entire length of a workload execution. The workload is executed for the following cases: *(i)* without cache, *(ii)* only query caching, *(iii)* only super-query caching, *(iv)* template caching with application logic based prefetching. Figure 8 shows the server load for cases *(ii)*, *(iii)*, *(iv)* as a percentage of server load in *(i)*. The results show that high gains of template caching with prefetching come at a reasonably low server load.

9. Conclusions

We consider the problem of improving performance of directory based web applications by caching of LDAP queries. The concept of LDAP templates is defined and is used to reduce the complexity of the query containment problem for queries with n predicates to $O(n)$. Caching algorithms which make efficient decisions of caching, prefetching and removing LDAP queries are proposed to improve the cache performance. We have described directory server extensions required in commercial and opensource directory servers to incorporate LDAP caching. Since directories are typically accessed through a web application, we describe means of modeling the offloaded component of the backend application at an edge site. The offloaded application is also used to improve performance of the LDAP cache by prefetching queries based on application logic. The combination of application offload and LDAP caching reduces client latency and improves scalability of the origin directory and application servers.

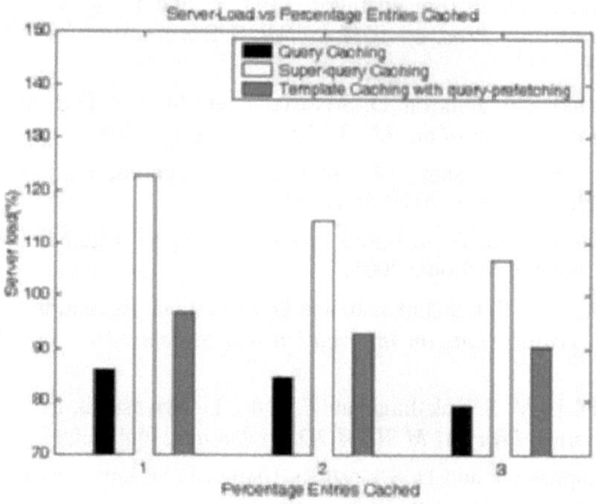

Figure 8. Server load comparison

We evaluate performance of the caching algorithms for a real enterprise directory application using real workloads. The results identify the types of queries which are candidates for query caching or generalized query caching. Results show that for a workload with multiple types of queries the proposed template based caching algorithm provides a 35% hit ratio for a cache size which is 3% of the backend directory and an improvement of 50% over query caching and 15% over super-query caching. We observe that using application logic based prefetching can significantly improve performance by providing up to 45% hit ratio while keeping the server load reasonably low. The algorithms along with the architecture, which supports consistency, security using standard protocols could provide the basis for improving performance and scalability of directory services.

References

[1] M. Wahl, T. Howes, and S. Kille, RFC 2251: Lightweight Directory Access Protocol (v3), www.ietf.org/rfc/rfc2251.txt.

[2] M. Wahl, A. Coulbeck, T. Howes, and S. Kille, RFC 2252: Lightweight Directory Access Protocol (v3): Attribute Syntax Definitions, www.ietf.org/rfc/rfc2252.txt.

[3] T. Howes, RFC 2254: The string representation of LDAP search filters, www.ietf.org/rfc/rfc2254.txt.

[4] X. Wang, H. Schulzrine, D. Kandlur, and D. Verma, Measurement and analysis of LDAP performance, *Proc. of ACM International Conference on Measurement and Modelling of Computer Systems (SIGMETRICS)*, 2000.

[5] Q. Luo, S. Krishnamurthy, C. Mohan, H. Pirahesh, H. Woo, B. Lindsay, and J. Naughton, Middle-Tier Database Caching for e-Business, *Proc. of ACM SIGMOD*, 2002.

[6] Q. Luo, J. F. Naughton, R. Krishnamurthy, P. Cao, and Y. Li, Active query caching for database Web servers, *Proc. of ACM SIGMOD Workshop on the Web and Databases (WebDB)*, 2000.

[7] S. Dar, M. Franklin, B. Jonsson, D. Srivastava, and M. Tan, Semantic data caching and replacement, *Proceedings of the 22nd VLDB Conference*, 1996.

[8] P. Deshpande, K. Ramasamy, A. Shukla, J. Naughton, Caching multidimentional queries using chunks, *Proc. ACM SIGMOD*, 1998.

[9] Q. Luo, J. F. Naughton, Form-Based Proxy Caching for Database-Backed Web Sites, *Proc. VLDB Conference*, Rome 2001.

[10] H. V. Jagadish, L. V. S. Lakshmanan, and D. Srivastava, Revisiting the hierarchical data model, *IEICE Transactions on Information and Systems*, Vol. E00-A, No. 1, January 1999.

[11] H. V. Jagadish, L. V. S. Lakshmanan, T. Milo, D. Srivastava, and D. Vista, Querying network directories, *Proc. ACM SIGMOD Conference*, Philadelphia, PA, June 1999.

[12] S. Cluet, O. Kapitskaia, and D. Srivastava, Using LDAP Directory Caches, *Proc. ACM Principles of Database Systems*, 1999.

[13] O. Kapitskaia, R. T. Ng and D. Srivastava, Evolution and revolutions in LDAP directory caches, *Proceedings of the International Conference on Extending Database Technology (EDBT)*, 202-216, 2000.

[14] P. A. Larson and H. Z. Yang, Computing queries from derived relations, *Proc. VLDB*, 1985.

[15] OpenLDAP Project, web page: (http://www.openldap.org).

[16] S. Jin and A. Bestavros, Popularity-aware greedy dual-size Web proxy caching algorithms, *Proc. 20th International Conference on Distributed Computing Systems (ICDCS)*, 2000.

[17] D. Thiebaut, Improving disk cache hit ratio through cache partitioning, *IEEE Transaction on Computers*, Vol.41, No.6, June1992.

[18] A. Kumar, The OpenLDAP Proxy Cache (http://www.openldap.org/pub/kapurva/proxycaching.pdf)

[19] Apache Cocoon, web page: http://xml.apache.org/cocoon

[20] XSL Transformations web page, (http:// www.w3.org/TR/xslt)

[21] Netscape Directory Server: Plug-in Programmer's Guide (http://enterprise.netscape.com/docs/directory/61/plugin/preface.htm)

COMPUTING ON THE EDGE: A PLATFORM FOR REPLICATING INTERNET APPLICATIONS

Michael Rabinovich[1], Zhen Xiao[1], and Amit Aggarwal[2]
[1]*AT&T Labs – Research,* [2]*Microsoft*

Abstract Content delivery networks (CDNs) improve the scalability of accessing static and, recently, streaming content. However, proxy caching can improve access to these types of content as well. A unique value of CDNs is therefore in improving performance of accesses to dynamic content and other computer applications. We describe an architecture, algorithms, and a preliminary performance study of a CDN for applications (ACDN). Our system includes novel algorithms for automatic redeployment of applications on networked servers as required by changing demand and for distributing client requests among application replicas based on their load and proximity. The system also incorporates a mechanism for keeping application replicas consistent in the presence of developer updates to the content. A prototype of the system has been implemented.

1. Introduction

Content delivery networks (CDNs) have become a popular method for providing scalable access to Web content. They currently provide access to static and streaming content. However, proxy caches can improve the delivery of these content types as well. In particular, as shown in [8], if proxies were deployed ubiquitously, the additional benefit of CDNs in delivering static content would be marginal.

A unique value of CDNs is in delivering dynamic content because this content cannot be cached by proxies. We refer to such a CDN as an Application CDN, or ACDN. An ACDN will allow a content provider (application provider in this case) to not worry about the amount of resources provisioned for its application. Instead, it can deploy the application on a single computer anywhere in the network, and then ACDN will replicate or migrate the application as needed by the observed demand.

One could implement an ACDN using general utility computing systems such as Ejacent [6] and vMatrix [1], which migrate the entire dynamic state of the running application from one server to another. These are complex systems that have to address all process migration issues that have been a subject of many years of research. Our key observation is that, because Web service applications have well-defined boundaries between processing individual requests, and that usually the server that starts processing a request is the one required to complete it, these applications do not have to be migrated at an arbitrary time. We exploit this specificity of our target application

F. Douglis and B.D. Davison (eds.), Web Content Caching and Distribution, 57-77.

class by allowing applications to migrate or replicate only at the request boundaries, when the dynamic state needed to be transferred is minimal. In a sense, instead of *migrating* an application to the new server, we simply *deploy* the application at the new server. Once the new server is up and running, the system can optionally decommission the application at the old server. Automatic deployment of an application is a much simpler task than the migration of a running application used in utility computing. Indeed, the latter is at the same time more fine-grained, in that the migration can occur at any time, and more heavy-weight since the transferred state must include the entire memory footprint of the application at the time of the transfer. We extend a typical approach used by software distribution systems to implement automatic deployment.

Another simplification is that we currently maintain replica consistency only for updates to the application by the content provider. For updates that occur as a result of user accesses, we either assume they can be merged periodically off-line (which is the case for commutative updates such as access logs) or that these updates are done on a shared back-end database and hence they do not violate replica consistency.

This paper presents a system design of an ACDN that uses the above approach and proposes the algorithms for deciding when and where to replicate or migrate an application, and how to distribute incoming requests among available replicas. We also report the results of a preliminary study of the performance of our approach. A functional prototype of our ACDN has been implemented and its demo has been presented at SIGMOD'2002 [12].

2. Issues

An ACDN has a fundamental difference with a traditional CDN that delivers static content. The latter uses caches as CDN servers. Each CDN server is willing to process any request for any content from the subscriber Web site. The CDN will either satisfy the request from its cache or obtain the response from the origin server, send it to the requesting client, and store it in its cache for future use. In contrast, to be able to process a request for an application locally, an ACDN server must possess a deployed application, including executables, underlying data, and the computing environment. Deploying an application at the time of the request is impractical; thus the ACDN must ensure that requests are distributed only among the servers that currently have a replica of the application; at the same time, the applications are redeployed among ACDN servers asynchronously with requests.

Thus, ACDN must provide solutions for the following issues that traditional CDNs do not face:

- Application distribution framework: ACDN needs a mechanism to dynamically deploy an application replica, and to keep the replica consistent. The latter issue is complicated by the fact that an application typically contains multiple components whose versions must be mutually consistent for the application to be able to function properly.

- Content placement algorithm: the system must decide which applications to deploy where and when. Content placement is solved trivially in traditional CDNs by cache replacement algorithms. However, an ACDN must make explicit decisions to replicate or migrate an application because it is too costly to replicate an application, especially at the time of the request.

- Request distribution algorithm: in addition to load and proximity factors that traditional CDNs must consider in their request distribution decisions, the request distribution mechanism in ACDN must be aware of where in the system different applications are currently deployed. Indeed, while a traditional CDN can process a request from any of its caches, an ACDN must assign the request to one of the servers that possess the application.

- System stability: under steady demand, the system behavior should reach a steady state with respect to request distribution and replica placement. Highly oscillating request distribution necessarily implies periods of highly suboptimal distribution; endless redeployment of application replicas only consumes bandwidth and adds load on the servers.

- Bandwidth overhead: creating a remote application replica consumes bandwidth for sending the application from the source to the target server. Thus, one has to be cautious and create replicas only when there is a reason to believe that the benefits will overweight this overhead.

In the following sections we describe our solutions to these problems.

3. Architecture Overview

The general architecture is straight-forward and depicted in Figure 1. Its main goal is to rely completely on the HTTP protocol and Web servers without any modifications. Not only does this simplify the adoption of the system, but it also allows easy firewall and NAT traversal and hence a deployment of the system over public Internet.

Each ACDN server is a standard Web server that also contains a *replicator*, which implements ACDN-specific functionality. In our initial design, we assume homogeneous servers to simplify the comparison of their loads. The replicators are implemented as a set of CGI scripts and so are a pure add-on to any standard Web server[1]. There is also a global *central replicator* that mainly keeps track of application replicas in the system. Although the central replicator is theoretically a bottleneck, this is not a concern in practice since the amount of processing it does is minimal; it can in fact be physically co-located with the DNS server. The central replicator is also a single point of failure. However, it is not involved in processing user requests. Thus, its failure only leads to a stop in application migrations or replications and does not affect the processing of requests by the existing replicas. Furthermore, the central replicator only contains soft state that can be reconstructed upon its recovery or replacement.

The server replicator contains the following CGI scripts: the start-up script, the load reporter, the replica target script, the replica source script, and the updater script.

[1]All scripts in our prototype are implemented as FastCGI for scalability. Using servlets in place of CGI scripts is also possible.

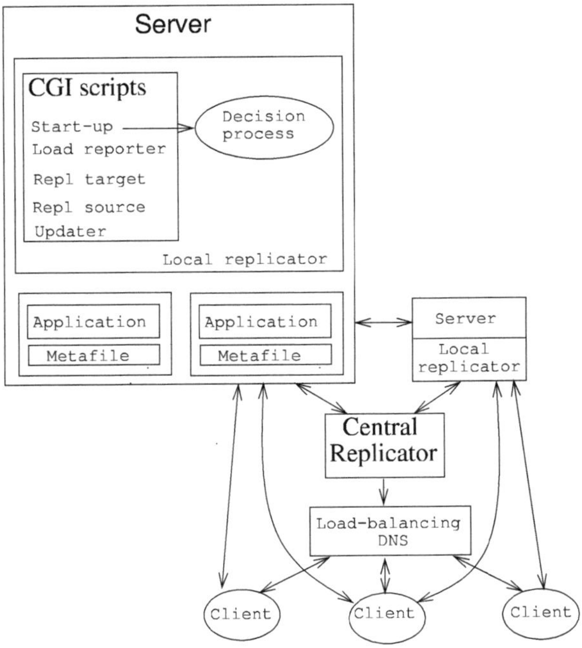

Figure 1. ACDN prototype architecture

The *start-up script* is invoked by the system administrator when (s)he brings a new ACDN server on-line. The script forks a *decision process* that periodically examines every application on the server and decides if any of them must be replicated or deleted. The *load reporter* is a script periodically invoked by the central replicator. The script returns the load of the server, which we measure as an output of a Unix *uptime* command.

The rest of the scripts implement the application distribution framework and are considered below.

4. Application Distribution Framework

Our implementation of the framework is based on the concept of a metafile, inspired by ideas from the area of software distribution such as the technology by Marimba Corp. [9]. The next subsection describes the metafile and the subsequent subsections show how it is used to implement the application distribution framework.

4.1 The metafile

Conceptually, the metafile consists of two parts: the list of all files comprising the application along with their last-modified dates; and the *initialization script* that the recipient server must run before accepting any requests. Figure 2 provides an example

```
FILE /home/apps/maps/query-engine.cgi 1999.apr.14.08:46:12
FILE /home/apps/maps/map-database 2000.oct.15.13:15:59
FILE /home/apps/maps/user-preferences 2001.jan.30.18:00:05
SCRIPT
  mkdir /home/apps/mapping/access-stats
  setenv ACCESS_DIRECTORY /home/apps/maps/access-stats
ENDSCRIPT
```

Figure 2. An example of a metafile

of a metafile[2] that represents a map-drawing application consisting of three files: an executable file that is invoked on access to this application and two data files used by this executable to generate responses. The metafile also contains the initialization script that creates a directory where the application collects usage statistics and sets the corresponding environment variable used by the executable. The initialization script can be an arbitrary shell script. When the initialization script is large, the metafile can include just a URL of the file containing the script.

The metafile is treated as any other static Web page and has its own URL. Using the metafile, the entire application distribution framework can be implemented entirely over standard HTTP. Indeed, the operations that the framework must support include replica creation, replica deletion, and replica consistency control. Migration of an application is accomplished by replica creation followed by the deletion of the original replica. Let us consider these operations in turn.

4.2 Replica creation

The replicator on each ACDN server contains replica target and replica source CGI scripts. The process of replica creation is initiated by the decision process on an ACDN server with an existing replica (called the source server below) and entails the following steps.

1 If the reason for replication is the overload of the source server, the source server queries the central replicator for the least-loaded server in the system, which will be asked to create a new application replica (referred to as the target server). A subsequent negotiation between the source and target servers prevents a herd effect when many servers attempt to replicate their applications to the the same target server at once. When the reason for replication is improving proximity to demand, the source server identifies the target server locally from its replica usage (see Section 1.5 for details).

[2]Our prototype uses an XML markup for encoding metafiles, which is stripped from the example in Figure 2 for clarity.

2 The source server invokes the replica target CGI script on the target server, giving the URL of the application metafile as a parameter. This URL also serves as a logical application ID inside the system.

3 The replica target script at the target server invokes the replica source script at the source server, with the metafile URL as a parameter.

4 The source server responds with the tarball of the application, which the replica target script unpacks and installs. The replica target script also executes the initialization script from the metafile.

5 Upon the execution of the initialization script, the replica target script informs the central replicator about the new application replica. The central replicator sends this update to the DNS server, which recomputes its request distribution policy based on the new replica set (see Section 1.5).

All the above interactions occur over standard HTTP, and require only a few CGI scripts.

4.3 Replica deletion

Replica deletion is initiated by the decision process on a server with the application replica and involves the following steps.

1 The server sends to the central replicator a request for permission to delete its replica.

2 If this is not the last replica in the system, the central replicator sends the deletion update to the DNS server, which recomputes the request distribution policy that excludes this replica.

3 Once the DNS server confirms the adoption of the new request distribution policy, the central replicator responds to the ACDN server with the permission to delete the replica. The permission response contains the DNS time-to-live (TTL) associated with the domain name of the application.

4 The ACDN server marks the replica as "to be deleted" and actually deletes it after the TTL time provided by the central replicator. This delay is required because residual requests for the application might still arrive due to earlier DNS responses still cached by the clients.

4.4 Consistency maintenance

In general, an application may change either because of the developer updates, defined as any modification to the application by the application authors or maintainers, or user updates, which occur as a result of user accesses. Developer updates can affect code (i.e., application upgrades) or underlying data (i.e., product pricing), while user updates involve data only (i.e., e-commerce transactions). We so far have implemented a solution to the developer updates, so our current prototype is suitable for applications that are read-only from the user perspective, such as informational sites. For updates that occur as a result of user accesses, we either assume they can be merged periodically off-line (which is the case for commutative updates such as access logs) or that these updates are done on a shared back-end database and hence they do not violate replica consistency.

There are three related issues in handling developer updates: replica divergence, replica staleness, and replica coherency. Replica divergence occurs when several replicas receive conflicting updates independently at the same time. Replica can become stale if it missed some updates. Finally, a replica can become incoherent if it acquired updates to some of its files but not others and so there is version mismatch between individual files.

Our system avoids replica divergence by allowing developer updates only to the *primary* application replica, so that all the updates can be properly serialized. The primary replica is appointed and kept track of by the central replicator. In particular, the central replicator selects a new primary before giving permission to delete the old primary replica.

The metafile provides an effective solution to the replica staleness and coherency problems. With the metafile, whenever some objects in the application change, the application's primary server updates the metafile accordingly. Whenever other Web servers detect that their cached copies of the metafile are not valid, they download the new metafile and then copy all modified objects together as prescribed in the metafile.

Thus, the metafile reduces the application staleness and coherency problems to cache consistency of an individual static page (the metafile). Once a replica detects that its cached metafile is stale it always obtains the coherent new version of the application. Any existing cache consistency mechanism for static pages can be used to maintain cache consistency of the metafile, including various validation and invalidation techniques (see [18], Chapter 10). The only difference is that updating an application must be asynchronous with request arrivals since doing it at the request time may create a prohibitive delay to the user latency. In our current prototype, the central replicator periodically invokes the updater script on all ACDN servers with a replica of the application; the updater script then validates the local metafile by issuing a GET If-Modified-Since request to the primary server, and acquires necessary updates for the corresponding application. In the meantime, the server keeps using the old version until the new version is ready. The details of the interaction between the server with a stale application and the application's primary server should be straightforward given the description of the replica creation procedure in Section 1.4.2 and omitted for brevity.

5. Algorithms

Two types of algorithms are inherent in any ACDN: an algorithm for content placement, which decides on the number and location of the application replicas, and an algorithm for request distribution, which chooses a replica for a given user request. We consider these algorithms in the next two subsections.

5.1 Content placement algorithm

The algorithm is executed periodically by an ACDN server, which makes a local decision on deleting, replicating, or migrating its applications. Allowing each server to decide autonomously ensures the scalability of our approach. The algorithm for a given application is shown in Figure 3. The algorithm utilizes three parameters, the *deletion threshold*, the *redeployment* threshold, and the *migration* threshold. The deletion threshold D characterizes the lowest demand that still justifies having a replica. In our experiments, it is expressed as the total number of bytes served by the server since the previous run of the of the algorithm[3]. The choice of the deletion threshold depends on the characteristics of the application as well as the underlying system and network.

The redeployment threshold reflects the amount of demand from the vicinity of another server that would warrant the deployment of an application replica at that server. Consider the decision to replicate the application at server i. Let B_i be the total number of bytes the current server served to clients from the vicinity of server i since the previous run of the algorithm. Let A be the total size of the application tarball (possibly compressed), and U be the total size of updates received in the same period. If $B_i/(A+U) > 1$ then the amount of bandwidth that would be saved by serving these requests from the nearby server i would exceed the amount of bandwidth consumed by shipping the application to server i and by keeping the new replica fresh. Hence, the bandwidth overhead of replication would be fully compensated by the benefits from the new replica within one time period until the next execution of the placement algorithm provided the demand patterns remain stable during this time. In fact, we might choose to replicate even if this ratio is less than one (e.g., if we are willing to trade bandwidth for latency), or only allow replication when this ratio exceeds a threshold that is greater than one (e.g., to safeguard against a risk that a fast-changing demand might not allow enough time to compensate for the replication overhead). Hence, the algorithm provides the redeployment threshold R, and replicates the application on server i if $B_i/(A+U) > R$.

One caveat is the possibility of a vicious cycle of creating a new replica without enough demand, which will then be deleted because of the deletion threshold. Thus, we factor in the deletion threshold into the replication decision and arrive at the following final replication criterion: To create a replica on server i, the demand from i's vicinity, B_i, should be such that $B_i/(A+U) > R$ and $B_i > D$.

The migration threshold M governs the migration decision. The application is migrated only if the fraction of demand from the target server exceeds M, $B_i/B_{total} > M$, and if the bandwidth benefit would be sufficiently high relative to the overhead, $B_i/A > R$. Note that the latter condition does not include the updates bytes because migration does not increase the number of replicas. To avoid endless migration back and forth between two servers, we require that the migration threshold be over 50%; we set it at 60% in our experiments.

[3]An alternative is to express it as the request rate. The implications and analysis of these two ways to express the deletion threshold are left for future work.

```
DecidePlacement():
/* Executed by server s */
  if load_s > HW, offloading = Yes;
  if load_s < LW, offloading = No;
  for each application app
    if B_total <= D
      delete app unless this is the sole replica
    elseif B_i/(A + U) > R AND B_i > D for some server i
      replicate app on server i
    elseif B_i/B_total > M AND B_i/A > R for some server i
      if server i accepts app migrate app to server i
    endif
  endfor
  if no application was deleted, replicated or migrated
  AND offloading = Yes
    find out the least loaded server t from the central replicator
    while load_s > LW AND not all applications have been examined
      let app be the unexamined application with the highest
      ratio of non-local demand, B_total/B_s;
      if B_s > D and B_t > D
        replicate app on t
        load_s = load_s - load_s(app, t)
      else
        if server t accepts app
          migrate app to t
          load_s = load_s - load_s(app)
        endif
      endif
    endwhile
  endif
end
```

Figure 3. Replica placement algorithm. $Load_s$ denotes load on server s, $load_s(app, t)$ denotes load on server s due to demand for application *app* coming from clients in server t's area, and $load_s(app)$ denotes load on server s due to application *app*.

The above considerations hold when the server wants to improve proximity of servers to client requests. Another reason for replication is when the current server is overloaded. In this case, it might decide to replicate or migrate some applications regardless of their proximity or their demand characteristics. So, if the server is overloaded, it queries the central replicator for the least-loaded server and, if the application cannot be replicated there (because the new replica might be deleted again), migrates the application to that server unconditionally. To add the stability to the system, we use a standard watermarking technique. There are two load watermarks, high watermark HW and low watermark LW. The server considers itself overloaded if its load reaches the high watermark; once this happens, the server continues considering itself overloaded until its load drops below the low watermark.

Again, one must avoid vicious cycles. The danger here is that after migrating an application to server i, the current server's load drops to the normal range, and the application would then migrate right back to the current server (because its initial migration worsened the client proximity). To prevent this, the target server only accepts the migration request if its projected load after receiving the application will remain acceptable (that is, below low watermark). This resolves the above problem because the current server will not accept the application back. This also prevents a herd effect when many servers try to offload to an underloaded server at once. To allow load predictions, the source server must apportion its total load to the application in question, $load_s(app)$, and to the requests that come from the target server's area, $load_s(app, t)$. As a crude estimate, we can apportion the total load in proportion to the number of relevant requests.

5.2 Request distribution algorithm

The goal of the request distribution algorithm is to direct requests to the nearest non-overloaded server with a replica of the application. However, an intuitive algorithm that examines the servers in the order of increasing distance from the client and selects the first non-overloaded server for the request (similar to the algorithm described in [2]) can cause severe load oscillations due to a herd effect [4]. Furthermore, our simulations show that randomization introduced by DNS caching may not be sufficient to eliminate the herd effect. The algorithm used in the RaDaR system [17] does not suffer from the herd effect but often chooses distant servers even when closer servers with low load are available. Thus, our main challenge was to find an algorithm that never skip the nearest non-overloaded server and yet reduce oscillations in request distribution.

An additional challenge was to make the algorithm compatible with our system environment. We use iDNS as our load-balancing DNS server [2]. For a given application[4] iDNS expects a *request distribution policy* in the form of tuples $(R, Prob(1), \ldots, Prob(N))$, where R is a region and $Prob(i)$ is the probability of selecting server i for a request from this region. Regions can be defined in a variety of ways and can be geographical regions (e.g, countries) or network regions (e.g., autonomous systems or BGP prefixes). For the purpose of this paper, we assume that each server i in the system is assigned a region R_i (represented as a set of IP addresses) for which this server is the closest. We also assume some distance metric between a region R_i and all servers that allows one to rank all servers according to their distance to a given region. The question of what kind of a distance metric is the most appropriate is a topic of active research in its own right; different CDNs use proprietary techniques to derive these rankings, as well as to divide client IP addresses into regions.

[4] We assume that every application uses a distinct (sub)domain name.

```
for each region R do
  for every server j in the system
    Prob(j) = 0;
  endfor
  for each server i_k with a replica of the application do
    if load(i_k) < LW
      Prob(i_k) = 1;
    elseif load(i_k) > HW
      Prob(i_k) = 0
    else
      Prob(i_k) = (HW − load(i_k))/(HW − LW)
    endif
  endfor
  residue = 1.0
  Loop through the servers with a replica of the application
  in the order of increasing distance from region R
  for each such server i_k do
    Prob(i_k) = residue * Prob(i_k)
    residue = residue − Prob(i_k)
  endfor
  let total be the sum of the Prob array computed above
  if total > 0
    for each server i_k with a replica of the application
      Prob(i_k) = Prob(i_k)/total
    endfor
  else
    for each server i_k with a replica of the application
      Prob(i_k) = 1/n, where n is the number of replicas
    endfor
  endif
  output (R, Prob)
endfor
```

Figure 4. Algorithm for computing request distribution policy for a given application.

In our ACDN, the central replicator computes the request distribution policy in the above format and sends it to iDNS. The policy is computed periodically based on the load reports from ACDN servers (obtained by accessing load reporter scripts as discussed in Section 1.3), and also whenever the set of replicas for the application changes (i.e., after a replica deletion, migration or creation).[5] The computation uses the algorithm shown in Figure 4.

[5]Technically, the policy is computed by a control module within iDNS that we modified; however, because this module can run on a different host from the component that actually answers DNS queries, we chose to consider the control module to be logically part of the central replicator.

Let the system contain servers $1, \ldots, N$, out of which servers $i_1, \ldots, i_n$ contain a replica of the application. The algorithm, again, uses low watermark LW and high watermark HW, and operates in three passes over the servers. The first pass assigns a probability to each server based on its load. Any server with load above HW gets zero weight. If the load of a server is below low watermark, the server receives unity weight. Otherwise, the algorithm assigns each examined server a weight between zero and unity depending on where the server load falls between the high and low watermarks. In the second pass, the algorithm examines all servers with a replica of the application in the order of the increasing distance from the region. It computes the probabilities in the request distribution policy to favor the selection of nearby servers. The third pass simply normalizes the probabilities of these servers so that they sum up to one. If all servers are overloaded, the algorithm assigns the load evenly among them.

6. Performance

In this section, we evaluate the performance of ACDN using a set of simulation experiments. We first evaluate the effectiveness of the request distribution algorithm in achieving a good balance of load and proximity. We then study the effectiveness of our content placement algorithm in reducing bandwidth consumption and user latency.

6.1 Request distribution

A request distribution algorithm would ideally direct client requests to their nearby replicas in order to reduce latency. At the same time, it must avoid overloading a replica in a popular region with too many requests. The simulation was conducted in a system with three replicas with decreasing proximity to the clients: replica 1 is closest to the clients, replica 2 is the next closest, and replica 3 is the farthest. The high watermark and the low watermark for the replicas are 1000 and 200 requests per second, respectively. Initially, there are 10 clients in the system. Each client starts at a randomized time and sends 5 requests per second. Then we gradually increase the number of clients to study the behavior of the algorithm when the replicas are overloaded. After that, we gradually decrease the number of clients to the original level to simulate a situation where the "flash crowd" is gone. To simulate the effect of DNS caching, each client caches the result of replica selection for 100 seconds. The results are shown in Figure 5. The top figure plots the number of requests served by each replica in sequential 100-second intervals. The bar graph on the bottom plots the request distribution every 10 minutes.

As can be seen from the figures, when the number of clients is small, the closest replica absorbs all the requests – the request distribution is determined by proximity. As the number of clients increases, load balancing kicks in and all replicas begin to share the load. Proximity still plays a role, however: note that replica 1 serves more requests than replica 2, which in turn serves more requests than replica 3. Also note that the load of none of the replicas ever exceeds the high watermark, which is usually set to reflect the processing capacity of the underlying server. When the number of clients decreases, the load in the three replicas decreases accordingly. Consequently, proximity starts playing an increasingly important role in the request distribution algorithm. When the load falls back to the original level, replica 1 again absorbs all the requests. The results indicate that our algorithm is efficient in utilizing proximity information while avoid overloading the replicas.

As targets for comparison, we also simulated two other algorithms: a pure random algorithm and the algorithm previously used in CDN brokering [2]. The pure random algorithm distributes requests randomly among the replicas regardless of the proximity. As can be seen from Figure 6, it achieves perfect load balancing among the three replicas. However, the clients might suffer unnecessarily from high latency due to requests directed to remote replicas even during low load.

The request redirection algorithm in the previous CDN brokering paper [2] works as follows:

- Select the closest replica whose load is less than 80% of its capacity. In our simulation, we set the capacity of all replicas to 1000 requests per second, the same as the high watermark used in ACDN.
- If no such replica exists, distribute the load evenly across all replicas.

The results are shown in Figure 7. When the load is low, replica 1 absorbs all the requests. When the load increases, the algorithm exhibits oscillations between the first and the second replicas, while the third replica remains unutilized. Moreover, note that the load on replica 1 can go substantially above the high watermark. These results show that, although this algorithm also takes into account both load and proximity, it does not perform as well overall as our ACDN algorithm. This is because it relies on a single load threshold (i.e. 80%) to decide whether a replica can be selected, which makes it susceptible to the herd effect. Although residual requests due to DNS caching add randomization to load balancing, they were not sufficient to dampen the herd effect. In contrast, our algorithm uses a pair of high watermark and low watermark to gradually adjust the probability of server selection based on the load of the server.

6.2 Content placement

The goal of the content placement algorithm is to detect hot regions based on observed user demands, and replicate the application to those regions to reduce network bandwidth and client perceived latency. The simulation was conducted using the old UUNET topology with 53 nodes which we used for our previous study [17]. 10% of the nodes represent "hot" regions: they generate 90% of the requests in the system. The rest 90% of the nodes are "cold" regions and generate 10% of the requests. In this simulation, the request rates from each hot region and each cold region are 810 and 10 requests per second, respectively. The simulation starts with an 100 second warm-up period during which clients in each region start at randomized times. To simulate the effect of changing demand patterns, the set of hot regions changes every 400 seconds. The application size used in the simulation is 10M bytes. The size of an application response message is 20K bytes. Every minute we introduce an update into the system that needs to be propagated to all the replicas. The size of the update is 5% of the application size. The redeployment threshold used in the simulation is 4. The deletion

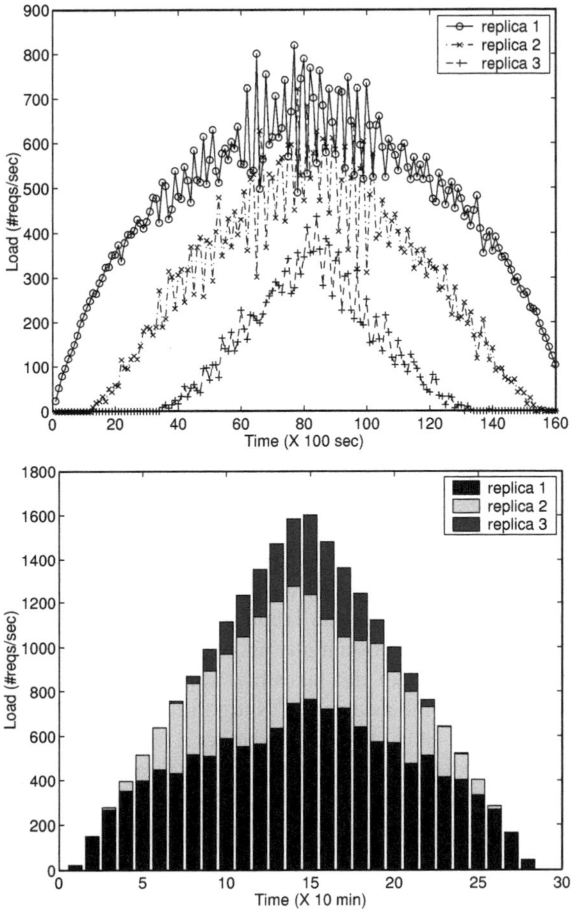

Figure 5. Request distribution in ACDN.

threshold is set to half the redeployment threshold times the size of the application. The algorithm makes a decision whether it needs to replicate or migrate every 100 seconds.

We compare our approach with two other algorithms: a static algorithm and an ideal algorithm. In the static algorithm, a replica is created when the simulation starts and is fixed throughout the simulation. In the ideal algorithm, we assume that the algorithm can get instantaneous knowledge as which regions are hot or cold and then replicates or deletes applications accordingly. It represents the optimal case which cannot be implemented in practice. The results of the simulation are shown in Figure 8. The top figure shows the amount of network bandwidth consumed in the simulation per second. This is measured as the product between the number of bytes sent and

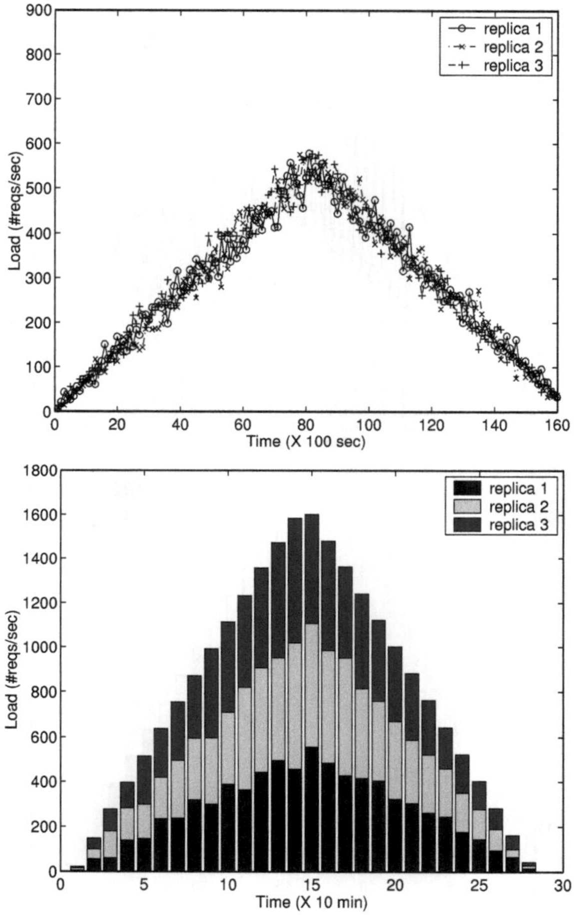

Figure 6. Request distribution in the random algorithm.

the number of hops they travel. For example, if a replica sends 1000 bytes to a client which is 3 hops away, the amount of network bandwidth consumed is 3000 bytes. The bottom figure shows the average response latency among all clients in the system. We assume that the latency on each link in the topology is 10ms. For this preliminary study, we also assume that the processing overhead at the replicas is negligible. Both figures indicate that our ACDN algorithm can quickly adapt to the set of hot regions and significantly reduce network bandwidth and response latency. The spikes in the top figure are caused by the bandwidth incurred during the application replication

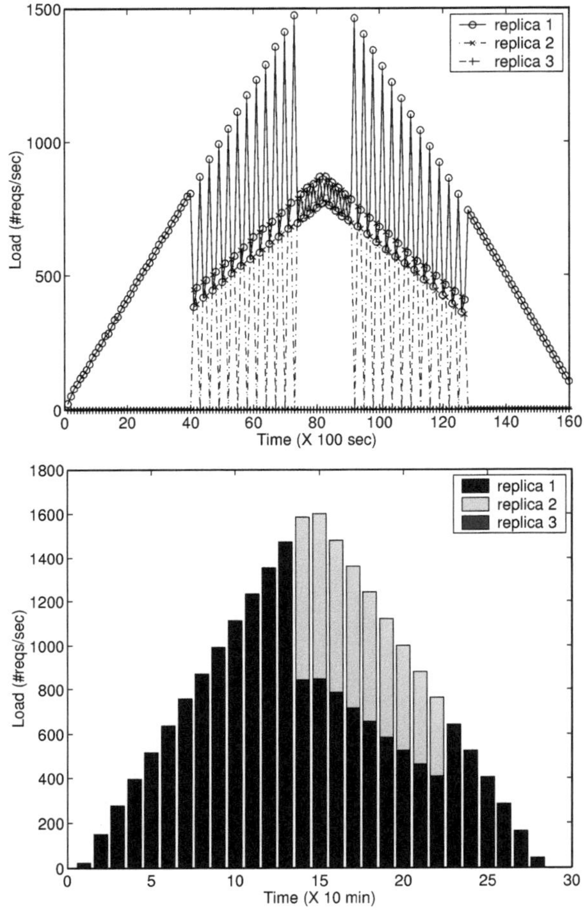

Figure 7. Request distribution in the brokering system.

process.[6] The migration algorithm was never triggered in this simulation because no region contributed enough traffic.

6.3 Redeployment threshold

Choosing an appropriate value for the redeployment threshold is essential for achieving good performance of the protocol. With a low threshold, more replicas will be created in the system. This allows more requests to be served efficiently by a

[6]Note that in some part of the curve the ACDN algorithm appears to perform slightly better than the ideal algorithm. This is due to random fluctuation in the simulation.

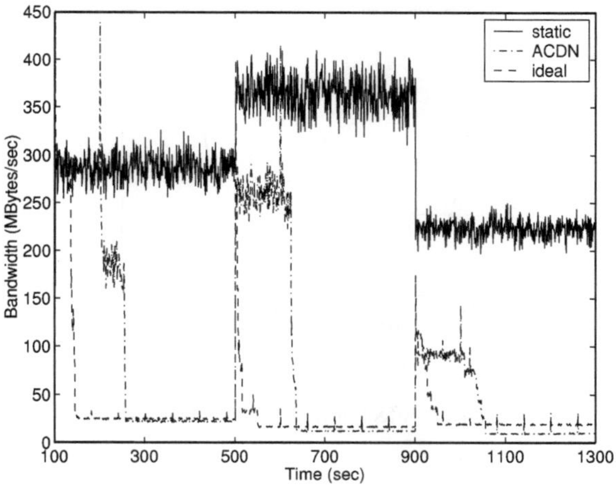

(a) Network Bandwidth Consumption

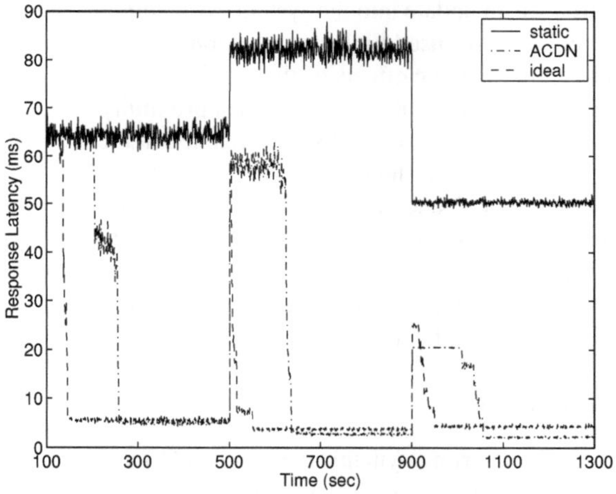

(b) Response Latency

Figure 8. Effectiveness of dynamic replication in ACDN.

nearby server, but increases the overhead for application replication and update propagation to all the replicas. On the other hand, a high redeployment threshold will result in fewer replicas, with less overhead due to application replication or updates, but also with less efficient processing of application requests.

Finding the right threshold is not trivial as it depends on many factors: the size of the application, the sizes and frequency of its updates, the traffic pattern of user requests, etc. We did a preliminary experiment to explore this trade-off for a 100M bytes application in the UUNET topology. In this simulation, the request rates from each hot region and each cold region are 81 and 1 requests per second, respectively.

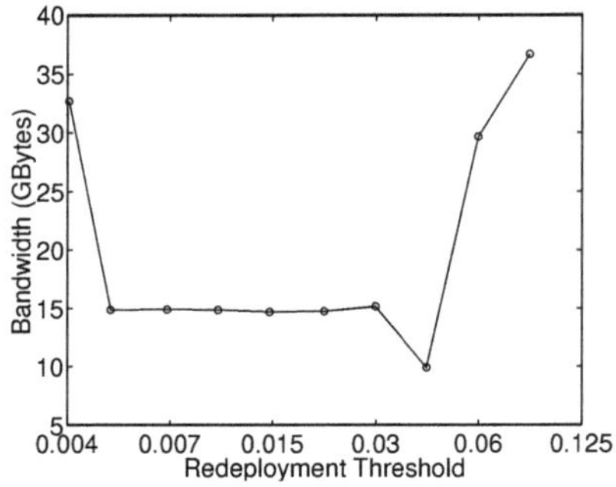

Figure 9. Effect of Redeployment Threshold

As before, we introduce an update into the system every minute that needs to be propagated to all the replicas. We used higher application size and lower request rates in this experiment to emphasize the effects of the overhead of creating and maintaining extra replicas relative to the benefits from increased proximity of replicas to requests. We vary the redeployment threshold and see its impact on the total amount of traffic on the network. The results are shown in Figure 9. The x-axis is the redeployment threshold used in the protocol, and the y-axis the total amount of bandwidth consumed in the entire simulation. The figure indicates that there is a "plateau" of good threshold values for this application: thresholds that are either too high or too low result in increased bandwidth consumption. As future work, we plan to investigate algorithms for adjusting the threshold automatically to optimize the performance of an application.

7. Related Work

The importance of supporting dynamic content in a CDN has been recognized and several proposals have been described that address this problem to various extents. Most of this work concentrates on caching dynamic responses and on mechanisms of timely invalidation of the cached copies, or on assembling a response at the edge from static and dynamic components [5, 14, 19, 10]. The fundamental difference between ACDN and these approaches is that the former replicates the computation as well as the underlying data used by the application while the latter handles only responses and leaves the computation to the origin server.

The Globule system [16] uses an object-oriented approach to content replication. It encapsulates content into special Globule objects, which include replication functionality and can be used to distribute static or dynamically generated content. Compared to our ACDN, Globule gives each object a flexibility to use its own policy for distribution and consistency maintenance while ACDN applies its policies to all hosted applications. On the other hand, Globule uses its own protocols to implement distributed objects and it requires compiling applications into Globule objects as well as modifying the Web server to be able to use Globule objects. Our ACDN is built entirely over HTTP, which simplifies firewall traversal, and works with existing applications and unmodified Web servers.

As discussed in the introduction, one can run an ACDN on top of a general process migration system such as Ejasent and vMatrix [6, 1]. Finally, one can also run ACDN servers on top of a distributed file system where each server acts as a client of the global file system and where each file is replicated among CDN servers through caching within the file system. This approach replicates computation but is limited to only very simple applications since it does not replicate the environment, such as resident processes. Also, ensuring that different components of the application are always mutually consistent becomes difficult since consistency is maintained for each file individually.

Turning to algorithms, ACDN involves two main algorithms - an algorithm for application placement and an algorithm for request distribution. Request distribution algorithms are closely related to load balancing and job scheduling algorithms. In particular, the issue of load oscillation that we faced has been well-studied in the context of load balancing (see. e.g., [4] and references therein). However, ACDN has to address the same issue in a new environment that takes into account client proximity in addition to server load. The algorithms by Fei et al. [7] and by Sayal et al. [20] use client-observed latency as the metric for server selection and thus implicitly account for both load and client proximity factors. Both algorithms, however, target client-based server selection, which does not apply to a CDN.

Many algorithms have also been proposed for content or server placement. However, most of them assume static placement so that they can afford to solve a mathematical optimization problem to find an "optimal" placement (see, e.g, [3, 21] and references therein). Even with empirical pruning, this approach is not feasible if content were to dynamically follow the demand. Some server placement algorithms use a greedy approach to place a given number of servers into the network. These algorithms still require a central decision point and are mostly suitable for static server placement. Our ACDN placement algorithm is incremental and distributed. Among the few distributed placement algorithms, the approach by Leff et al. [13] targets the remote caching context and does not apply to our environment where requests are directed specifically to servers that already have the application replica. The strategies considered by Kangasharji et al. [11] assume a homogeneous request pattern across all regions. Our algorithm can react to different demands in different regions and migrate applications accordingly. Finally, the strategy mentioned by Pierre et al. [15] places a fixed number of object replicas in the regions with the highest demand. Our algorithm allows a variable number of replicas depending on the demand and takes into account the server load in addition to client proximity in its placement decisions.

Our ACDN content placement algorithm is an extension of our earlier RaDaR algorithm [17]. However, because ACDN replicates entire applications, its placement algorithm is different from RaDaR in that it takes into account the size of the application and the amount of application updates in content placement decisions.

8. Conclusions

This paper describes an ACDN - a middleware platform for providing scalable access to Web applications. Accelerating applications is extremely important to CDNs because it represents CDNs' unique value that cannot be offered by client-side caching platforms. ACDN relieves the content provider from guessing the demand when provisioning the resources for the application and deciding on the location for those resources. The application can be deployed anywhere on one server, and then ACDN will replicate or migrate it as needed using shared infrastructure to gain the economy of scale.

We presented a system design and algorithms for request distribution and replica placement. The main challenge for the algorithms is to avoid a variety of "vicious cycles" such as endless creation and deletion of a replica, or migration of a replica back and force, or oscillations in request distribution, and yet to avoid too much deviation from optimal decisions in a given instance. Our preliminary simulation study indicated that our algorithms achieve promising results.

To date, we only experimented with one read-only application as a testbed for ACDN [12]. In the future, we would like to gain more experience with the system by deploying a variety of applications on it. In particular, we would like to explore various ways to support user updates to the application data, from relying on a shared back-end database to possibly replicating these updates among application replicas in a consistent way.

Acknowledgments

We would like to acknowledge Pradnya Karbhari who implemented the first version of the ACDN prototype and Fred Douglis for his involvement in the first prototype. We also thank the anonymous reviewers for comments on an early draft of the paper.

References

[1] A. A. Awadallah and M. Rosenblum. The vMatrix: A network of virtual machine monitors for dynamic content distribution. In *7th Int. Workshop on Web Content Caching and Distribution (WCW 2002)*, Aug. 2002.

[2] A. Biliris, C. Cranor, F. Douglis, M. Rabinovich, S. Sibal, O. Spatscheck, and W. Sturm. CDN brokering. In *6th Int. Workshop on Web Caching and Content Distribution*, June 2001.

[3] I. Cidon, S. Kutten, and R. Soffer. Optimal allocation of electronic content. In *Proceedings of IEEE INFOCOM*, pages 1773–1780, Los Alamitos, CA, Apr. 22–26 2001. IEEE Computer Society.

[4] M. Dahlin. Interpreting stale load information. *IEEE Transactions on Parallel and Distributed Systems*, 11(10):1033–1047, Oct. 2000.

[5] F. Douglis, A. Haro, and M. Rabinovich. HPP: HTML macro-preprocessing to support dynamic document caching. In *Proceedings of the Symposium on Internet Technologies and Systems*, pages 83–94. USENIX, Dec. 1997.

[6] Ejasent, Inc. Ejasent web site. http://www.ejasent.com/, 2003.

[7] Z. Fei, S. Bhattacharjee, E. W. Zegura, and M. H. Ammar. A novel server selection technique for improving the response time of a replicated service. In *INFOCOM*, pages 783–791, 1998.

[8] S. Gadde, J. Chase, and M. Rabinovich. Web caching and content distribution: A view from the interior. In *5th Int. Web Caching and Content Delivery Workshop (WCW5)*, 2000.

[9] A. V. Hoff, J. Payne, and S. Shaio. Method for the distribution of code and data updates. U.S. Patent Number 5,919,247, July 6 1999.

[10] A. Iyengar and J. Challenger. Improving Web server performance by caching dynamic data. In *Proceedings of the USENIX Symposium on Internet Technologies and Systems*, pages 49–60, Berkeley, Dec. 8–11 1997.

[11] J. Kangasharju, J. W. Roberts, and K. W. Ross. Object replication strategires in content distribution networks. In *Proceedings of the Sixth Int Workshop on Web Caching and Content Distribution (WCW)*, 2001.

[12] P. Karbhari, M. Rabinovich, Z. Xiao, and F. Douglis. ACDN: a content delivery network for applications (project demo). In *Proceedings of ACM SIGMOD*, pages 619–619, June 2002.

[13] A. Leff, J. L. Wolf, and P. S. Yu. Replication algorithms in a remote caching architecture. *IEEE Transactions on Parallel and Distributed Systems*, 4(11):1185–1204, Nov. 1993.

[14] Oracle Corporation and Akamai Technologies, Inc. ESI - accelerating e-business applications. `http://www.esi.org/`, 2001.

[15] G. Pierre, I. Kuz, M. van Steen, and A. S. Tanenbaum. Differentiated strategies for replicating Web documents. *Computer Communications*, 24(2):232–240, Feb. 2001.

[16] G. Pierre and M. van Steen. Globule: a platform for self-replicating Web documents. In *6th Int. Conference on Protocols for Multimedia Systems*, pages 1–11, Oct. 2001.

[17] M. Rabinovich, I. Rabinovich, R. Rajaraman, and A. Aggarwal. A dynamic object replication and migration protocol for an Internet hosting service. In *19th IEEE International Conference on Distributed Computing Systems (ICDCS '99)*, pages 101–113. IEEE, May 1999.

[18] M. Rabinovich and O. Spatscheck. *Web Caching and Replication.* Addison-Wesley, 2001.

[19] M. Rabinovich, Z. Xiao, F. Douglis, and C. Kalmanek. Moving edge-side includes to the real edge—the clients. In *Proceedings of the 4th USENIX Symposium on Internet Technologies and Systems*, Mar. 2003.

[20] M. Sayal, Y. Breitbart, P. Scheuermann, and R. Vingralek. Selection algorithms for replicated web servers. In *Workshop on Internet Server Performance*, June 1998.

[21] A. Wierzbicki. Models for Internet cache location. In *The 7th Int'l Workshop on Web Content Caching and Distribution (WCW)*, 2002.

SCALABLE CONSISTENCY MAINTENANCE FOR EDGE QUERY CACHES

Exploiting templates in Web applications

Khalil Amiri[1], Sara Sprenkle[2], Renu Tewari[3], and Sriram Padmanabhan[4]
[1]*Imperial College London,* [2]*Duke University,* [3]*IBM Almaden Research Center,*
[4]*IBM Santa Teresa Lab*

Abstract Semantic database caching is a self-managing approach to dynamic materialization of "semantic" slices of back-end databases on servers at the edge of the network. It can be used to enhance the performance of distributed Web servers, information integration applications, and Web applications of oaded to edge servers. Such semantic caches often rely on update propagation protocols to maintain consistency with the back-end database system. However, the scalability of such update propagation protocols continues to be a major challenge. In this paper, we focus on the scalability of update propagation from back-end databases to the edge server caches. In particular, we propose a publish-subscribe like scheme for aggregating cache subscriptions at the back-end site to enhance the scalability of the ltering step required to route updates to the target caches. Our proposal exploits the template-rich nature of Web applications and promises signi cantly better scalability. In this paper, we describe our approach, discuss the tradeoffs that arise in its implementation, and estimate its scalability compared to naive update propagation schemes.

1. Introduction

The performance and scalability of Web applications continues to be a critical requirement for content providers. Traditionally, static caching of HTML pages on edge servers has been used to help meet this requirement. However, with a growing fraction of the content becoming dynamic and requiring access to the back-end database, static caching is by-passed as all the dynamically generated pages are marked uncacheable by the server.

Dynamic data is typically served using a 3-tiered architecture consisting of a web server, an application server and a database; data is stored in the database and is accessed on-demand by the application server components and formatted and delivered to the client by the web server. In more recent architectures, the edge server (which includes client-side proxies, server-side reverse proxies, or caches within a content

F. Douglis and B.D. Davison (eds.), Web Content Caching and Distribution, 79-90.

distribution network(CDN) [2]) acts as an application server proxy by offloading application components (e.g., JSPs, servlets, EJBeans) to the edge [12, 7]. Database accesses by these edge application components, however, are still retrieved from the back-end server over the wide area network.

To accelerate edge applications by eliminating wide-area network transfers, we have recently proposed and implemented DBProxy, a database cache that dynamically and adaptively stores structured data at the edge [4]. The cache in this scenario is a persistent edge cache containing a large number of changing and overlapping "materialized views" stored in common tables.

The scalability of such a scheme depends on how efficiently we can maintain consistency with the back-end database without undue processing at the back-end or network bandwidth overheads. Consistency in semantic caches has been traditionally achieved through two approaches. The first is timeout-based. Data in the cache is expired after a specified timeout period, regardless of whether it has been updated at the back-end or not. The second approach is update-based consistency which relies on propagating the relevant changed data to the caches where it is locally applied.

Timeout-based consistency suffers from the disadvantage of increased latency and network message overheads as the cache needs to validate the data with the back-end or fetch any modified data. Update-based consistency propagates all the new data to the cache when any change happens, but suffers from serious scalability limitations both as the number of caches grow, and as the size of an individual cache, in terms of the number of cached views, increases. The scalability problem arises from the back-end overhead of "figuring out", when a row changes, which of the target caches it must be forwarded to.

One approach that was used initially in DBProxy was to propagate all changes at the back-end and apply them to the local cache. This increases the network bandwidth overhead and increases the size of the local cache. In this paper, we propose *template-based filtering* that efficiently aggregates cache subscriptions at the back-end by exploiting the similarity of "views" in edge query caches. Similar aggregations have been proposed for event matching (against client subscriptions) in publish subscribe systems.

We discuss the specifics of our approach, describe how we handle dynamic changes in cache subscription, and highlight the challenges and trade-offs that arise in this space.

The rest of the paper is organized as follows. We review the design of our prototype dynamic edge data cache in Section 2, and describe its consistency maintenance framework in Section 3. We review naive filtering schemes in Section 4 and describe our approach in Section 5. We discuss related work in Section 6 and summarize the paper in Section 7.

2. Semantic caching over the Web

2.1 DBProxy overview

We designed and implemented an edge data cache, called DBProxy [4], as a JDBC driver which transparently intercepts the SQL calls issued by application components (e.g., servlets) executed on the edge and determines if they can be satisfied from the local cache (shown in Figure 1). To make DBProxy as self-managing as possible, while leveraging the performance capabilities of mature database management systems, we chose to design DBProxy to be: (i) persistent, so that results are cached across instantiations and crashes of the edge server; (ii) DBMS-based, utilizing a stand-alone database for storage to eliminate redundancy using common tables and to allow for the efficient execution of complex local queries; (iii) dynamically populated, populating the cache based on the application query stream without the need for pre-defined administrator views; and (iv) dynamically pruned, adjusting the set of cached queries based on available space and relative benefits of cached queries.

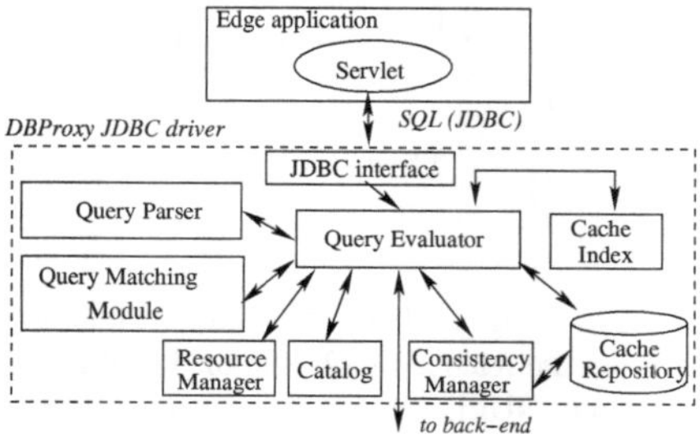

Figure 1. DBProxy key components. The query evaluator intercepts queries and parses them. The cache index is invoked to identify previously cached queries that operated on the same table(s) and columns. A query matching module establishes whether the new query's results are contained in the union of the data retrieved by previously cached queries. A local database is used to store the cached data.

2.2 Common Local Store

Data in a DBProxy edge cache is stored persistently in a *local stand-alone database*. The contents of the edge cache are described by a cache index containing the list of queries. To achieve space efficiency, data is stored in *common-schema tables* whenever possible such that multiple query results share the same physical storage. Queries over the same base table are stored in a single, usually partially populated, cached copy of the base table at the origin server. Join queries with the same "join condition" and over the same base table list are also stored in the same local table. This scheme not only achieves space efficiency but also simplifies the task of consistency maintenance, as discussed below. When a query is "worth caching", a local result table is created (if it does not already exist) with as many columns as selected by the query. The column type and metadata information are retrieved from the back-end server and cached in a local catalog cache. For example, Figure 2 shows an example local table cached at the edge. The local 'item' table is created just before inserting the three rows retrieved by query Q_1 with the primary key column (id) and the two columns requested by the query ($cost$ and $msrp$). All queries are rewritten to retrieve the primary key so that identical rows in the cached table are identified. Later, and to insert the three rows retrieved by Q_2, the table is *altered* if necessary to add any new columns not already created. Next, new rows fetched by Q_2 are inserted ($id = 450, 620$) and existing rows ($id = 340$) are updated. Note also that since Q_2 did not select the $cost$ column, a NULL value is inserted for that column.

Cached queries in DBProxy are organized according to a multi-level index of schemas, tables and clauses for efficient matching.

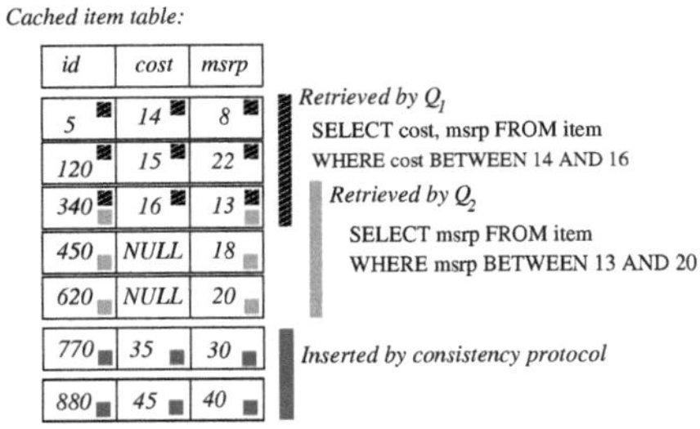

Figure 2. Local storage. The local $item$ table entries after the queries Q_1 and Q_2 are inserted in the cache. The rst three rows are fetched by Q_1 and the last three are fetched by Q_2. Since Q_2 did not fetch the $cost$ column, NULL values are inserted. The bottom two rows were not added to the table as a part of query result insertion, but by the update propagation protocol which re ects UDIs performed on the origin table. In the original design, changes are forwarded from the origin to the edge cache whether or not they match cached query predicates.

3. Consistency Management

Read-only queries received by DBProxy are satisfied from the local cache, whenever possible. Updates are always passed to the back-end database directly, bypassing the cache. The effects of updates trickle back to the edge cache, through a push or a pull approach.

While many Web applications can tolerate slightly stale data in the edge cache, they are nevertheless interested in reasonable consistency guarantees. For example, applications usually require that they observe the effects of their own updates on an immediate subsequent query. Since a query following an update can hit locally in a stale cache, updates not yet reflected in the cache would seem to have been lost, resulting in strange application behavior. Specifically we assume that three consistency properties hold. First, the cache guarantees *lag consistency* with respect to the origin. This means that the values of the tuples in the local cache equal those at the back-end at an earlier time, although the cache may have less tuples. Second, DBProxy exhibits *monotonic state transitions*, i.e, it does not allow an application to see a relatively current database state, then see a previous state corresponding to a point earlier in time. Finally, DBProxy supports *immediate visibility of local updates* where the result of a later read by an application will show the effect of the local update it had committed earlier.

To achieve these properties, DBProxy relies on protocols that force the cache to pull updates from the server on certain events, and also may decide to bypass the cache in others. The details of the consistency protocols and criteria are not the focus of this paper, but are more fully described in [3]. Suffice it to say, here, that regardless of the particular criteria, changed data must be efficiently propagated from the back-end database to the edge caches. In this section, we discuss two alternative approaches to update propagation, and describe a server-side filtering architecture which can be used to propagate changes in back-end tables only to the caches that require them.

3.1 Update propagation approaches

To maintain consistency, DBProxy relies on a data propagator, which captures all UDIs (updates, deletes, inserts) to the tables at the origin and forwards them to the edge caches either in their entirety or after filtering.

Zero-filtering propagation. This approach, which we initially implemented in DBProxy, propagates all changes to the edge caches, regardless of whether or not they match cached predicates. This places no filtering load on the server. Data changes are propagated to the edges tagged by their transaction identifiers and applied to the edge cache in transaction commit order. Since cached data is maintained as partially populated copies of back-end tables, changes committed to the base tables at the origin can be applied "as is" to the cached versions, without the need to re-execute the queries. Future queries that will execute over the cache will retrieve from these newly propagated changes any matching tuples. This solution presumes slowly changing data with few updates which is typical of some web environments.

Server-side filtering. The problem with propagation based consistency with zero filtering is that all changes to the back-end are propagated regardless of whether they are required by the cache or not. This not only wastes network bandwidth but also increases the size of the local cache database. Periodic garbage collection is required to clean the unaccessed tuples. An alternative approach is to filter the tuples that change in the back-end database and forward to each cache only the "relevant" ones. This filtering step can place, however, high load on the back-end site as the number of caches and the number of views per cache increase. The rest of the paper will describe filtering in more detail, and suggest an approach to make it more scalable.

4. Basic filtering

Notation. Before we discuss the details of filtering, we first define a few notational conventions to simplify the rest of the discussion. For each query in the cache we have a subscription at the back-end. We express a subscription S as a 3-tuple $S = (T, A, P)$, where T is the table name(s) accessed by the query. In case of a join, T may contain two or more tables. A is the set of attributes or columns projected by the query, and P is the search predicate. P is assumed to be in AND-OR normal form. When expressed in its AND-OR normal form, we denote by C_i the ith *disjunct* in that expression. Each disjunct contains several *predicate terms* (e.g., cost $<$ 15) ANDed together. These predicate terms are atomic conditions, such as equality or inequality predicates over columns.

Let N be the number of edge caches. For cache i, the set of subscriptions are denoted by S_i, and the number of subscriptions are therefore equal to $|S_i|$. The jth subscription of cache i is, therefore, $S_{ij} = (T_{ij}, A_{ij}, P_{ij})$.

Overview of Filtering. In this scheme, caches "subscribe" to a number of "logical" update streams. In particular, each cached view or query corresponds to one subscription. A filtering server at the back-end site manages subscriptions for the edge caches. Subscriptions are dynamic, that is they change as new views are added or evicted.

The filtering server, therefore, has to test each newly changed tuple (or row in a table) against all the $\sum_{i=1..N} |S_i|$ subscriptions. This results in a linear search overhead as the number of caches, N, increases, and as the number of subscriptions, $|S_i|$, per cache increases.

Precisely, for each tuple t_k let us assume that t_k^{old} is the old value and t_k^{new} is the value after a change (i.e., a UDI). In case of a newly inserted tuple t_k^{old} is NULL, similarly, when a tuple is deleted t_k^{new} is NULL. Assuming there is a single cached query denoted by S_{ij} with predicate P_{ij}, then the filtering algorithm decides to route tuple t_k to the target cache if either of the following two conditions hold:

- $t_k^{new} \in P_{ij}$
- $t_k^{old} \in P_{ij}$ AND $t_k^{new} \notin P_{ij}$

In the first case the tuple is inserted or updated in the cache. In the second case the tuple is either deleted or updated. The pseudo-code of the filtering algorithm is shown below.

```
filter (TUPLE Told, TUPLE Tnew, CACHE i)
begin
    for every SUBSCRIPTION Sij from CACHE i {
       for every DISJUNCT C in Pij {
         if( C.matches(Tnew) )
            return TUPLE_MATCH;
         else if ( C.matches(Told) )
            return TUPLE_MATCH;
       }
    }
    return TUPLE_NO_MATCH;
end
```

The filtering procedure above is invoked once for each cache, for each tuple updated by a transaction in the back-end server. For each subscription predicate $P_{ij} = C_1 \vee C_2 \vee ... \vee C_m$, the tuple is tested against each of the C_i's, and is forwarded to the edge cache if it matches any of them. Note that each of the C_i's is expressed as a bunch of atomic attribute tests (of the form $attr$ $\{=, <, ...\}$ val) ANDed together.

The complexity of the matching function grows with the size or complexity of the subscription. If the average number of disjuncts per subscription is m and the number of terms per disjunct is t, and if the average cache size is S, then assuming that there are N caches, the total complexity can be expressed as $O(N \times S \times m \times t)$. To simplify the analysis, when estimating filtering cost, we will assume, for the rest of the discussion, that each subscription has a single disjunct with t terms, in which the filtering complexity can be summarized as $O(N \times S \times t)$.

5. Template-based filtering

In order to improve the scalability of the basic filtering approach we propose template-based filtering that exploits the similarity among the subscriptions corresponding to the cached queries. Such similarities exist because queries are generated by applications which often define parametrized query templates which are instantiated into actual queries at run-time by binding the variable parameters.

Templates. In Java-based Web applications, two query programming styles are observed: i) explicitly declared templates where queries are pre-declared as *prepared statements* and where parameters are set through explicit calls, or ii) undeclared templates where queries are composed by the *string concatenation* of a fixed part and a variable part. Such templates are often seen in the servlet programs running at the Web server which process the inputs from the front-end interface consisting of webpage forms. If the template is not explicitly declared, our caching driver has to infer implicitly that some submitted queries follow the same template.

When referring to a group of similar query predicates instantiated by the same template, we call the part of the predicate that changes across the group as the *variant part*, to distinguish it from the *fixed part*. For the rest of this discussion, we will focus on a single disjunct in the query's predicate expression. Consider the two predicate expressions:

$$P_1 = (cost < 15 \wedge msrp < 8 \wedge num\_reviews = 5)$$
$$P_2 = (cost < 15 \wedge msrp < 8 \wedge num\_reviews = 10)$$

A comparison of these two predicate expressions suggests that they are likely two instantiations of a template P_t of the form: $P_t = (cost < 15 \wedge msrp < 8 \wedge num\_reviews =?)$ In this example, the predicates P_1 and P_2 contain a single disjunct $(cost < 15 \wedge msrp < 8 \wedge num\_reviews =?)$. Within the disjunct, $(cost < 15 \wedge msrp < 8)$ is the fixed part and $(num\_reviews =?)$ is the variant part. The atomic test $(cost < 15)$ is called a predicate term.

Templates in real applications. We looked at the number of templates used in two e-commerce benchmarks, TPC-W, which emulates an on-line bookstore and ECD-W, which is much more complex and emulates a customizable on-line shopping mall. The number of templates in these realistic full-fledged Java applications were 18 and 81, respectively. Furthermore, the percentage of queries generated by the top-5 templates where 79% an 68% respectively.

We detect templates on-line by comparing the structure of search predicates in a query stream. Our template detection algorithms are described in [5].

5.1 Template-based filtering: Single cache case

The main intuition behind our approach is that the fixed part of the predicate should be tested only once for all the query subscriptions generated from the same template. This is similar to the motivation behind event matching schemes in publish-subscribe systems. However, because of the prevalence of templates in Web applications, we suggest using specialized data structures, to aggregate the parameter instantiations of the variant part of the predicate. This allows for fast matching of a new tuple against the collection of the variant parts. We call these indexed aggregate data structures, Merged Aggregated Predicates (MAPs).

In the above example, assume that many subscriptions are received which are all generated by the same template. In that case, the MAP can be a hash table aggregating the variant integer terms that appear in the $num\_reviews =?$ test. The filtering test can therefore be reduced to testing the $num\_reviews$ attribute of the tuple against existence in the hash table. The tests against the $cost$ and $msrp$ attributes would be performed only once. The complexity of testing a tuple against a hash-table MAP is constant. If we denote by K the number of templates, which is usually much smaller than the number of subscriptions, $K << S$, the total complexity is $O(K)$, which is independent of cache size, unlike the cost of naive filtering, $O(S \times t)$. Of course, not all MAPs are hash tables, and therefore the complexity of looking up a MAP is not always constant, but is at worst logarithmic in cache size, as we show below.

Example MAPs. We describe through a few examples, the various possible MAPs. Consider the following template: $P_t = (cost <? \wedge msrp < 8 \wedge num\_reviews = 5)$

In this case, the subscriptions can be aggregated into a single value corresponding to the maximum upper bound parameter received for the cost attribute (e.g., for cost < 10, cost < 20, cost < 50, we only store the value 50). Consider, alternatively, a slightly more complex expression:

$$P_t = (cost\ BETWEEN\ ?\ AND\ ? \wedge msrp < 8\ \wedge\ num\_reviews = 5)$$

Our approach in this case is to merge successive instantiations of a BETWEEN predicate into an interval set, where all overlapping intervals are merged together. For example, the intervals $cost\ BETWEEN\ 10\ AND\ 20$, $cost\ BETWEEN\ 15\ AND\ 25$ will be merged and represented as $[10, 25]$. Further, the intervals are sorted to enable efficient binary searching. Thus intervals [10, 20], [45, 60], [25, 35] will be sorted to be [10, 20], [25, 35], [45, 60]. If a new interval say [15, 30] is added that overlaps with the previous intervals they are merged together and the list becomes [10, 35], [45, 60]. More formally, the sorted interval array $SA = ([x_1, X_1], ..., [x_m, X_m])$ has the following non-overlapping property:

$$x_i \leq X_i < x_{i+1} \text{ for every } i = 1, ..., m-1.$$

A more complex MAP arises when the predicate contains parameters corresponding to tests that involve more than a single attribute. For example:

$$P_t = (cost <? \wedge msrp <? \wedge num\_reviews = 5)$$

In this the case, the aggregation of various instantiations of the template is not straightforward. Note that we can associate a MAP with each parameter (the $cost$ and the $msrp$ upper bounds). However, a tuple can "hit" in both MAPs, while not matching any subscription (because it matches the $cost$ test corresponding to one subscription, and the $msrp$ test corresponding to another).

More generally, consider the case where the variant part of the disjunct contains more than one variant predicate term. We denote the variable part of the disjunct, C_j, as $\hat{C}_j$. Note that the columns referred to in $\hat{C}_j$ could be different. For a disjunct with predicate terms conditioning over the same column, we can sometimes convert $\hat{C}_j$ to a single BETWEEN predicate (e.g., in case of a combination of $\leq$ and $\geq$). However, in the general case, $\hat{C}_j$ can have predicate terms that refer to several different columns, for example: $\hat{C}_j = (col_1 < \ ?) \wedge (col_2 = \ ?) \wedge (col_3 \ BETWEEN \ ? \ and \ ?)$. The MAP associated with $\hat{C}_j$ is a list of regions in a k-dimensional space. In this particular example, the region is upper-bounded by a single value along the first dimension (col_1), corresponds to a single value along the second dimension (col_2), and is upper and lower-bounded along the third dimension (col_3). In this 3-dimensional space, the region is described by a rectangular plane bounded on three sides and extending to infinity along the fourth side.

A single instantiation of the variant disjunct, $\hat{C}_j$, can therefore be thought to correspond to a region, or rectangle in k-dimensional space. The dimensionality of the region space is upper-bounded by the number of columns that appear in $\hat{C}_j$. We associate a MAP with $\hat{C}_j$ which maintains the set of rectangles corresponding to the variant disjunct of all cached queries. Overlapping or adjacent rectangles are merged if possible. In general, a MAP contains a set of overlapping and/or disjoint rectangles.

Testing if a tuple matches the aggregation of several subscriptions reduces to inspecting the corresponding MAP to verify whether the "point" corresponding to the tuple is contained in any of the k-dimensional rectangles aggregated in the MAP. To allow quick search of such composite MAPs, one approach is to organize the rectangles corresponding to cached queries using a *memory-resident multi-dimensional index* such as an R-tree [10]. The rectangles are organized hierarchically using their minimum bounding rectangles (MBR). Testing a tuple against the aggregated subscriptions translates into searching the multi-dimensional index to find out if there exists at least one rectangle at the leaf level which contains the tuple.

Note that in the TPC-W application trace, we observe that less than 5% of the templates have more than one variable parameter in a disjunct. So, in practice, the need for R-trees and multi-dimensional MAPs should be limited.

Complexity analysis. Testing a tuple against a subscription is logarithmic in the number of aggregated subscriptions. Assume there are K templates in the cache, with each template having a single disjunct. Thus, there are S/K subscriptions from the cache associated with a given template, and the complexity is $O(K \times log(S/K))$ compared to the $O(S \times t)$ of the naive approach.

Subscription management. When queries are added to or removed from the cache, the subscriptions and the corresponding MAPs need to be updated in the filtering server. However, because MAPs represent aggregations of a set of values, updating them is not straightforward when a value in the aggregated set is removed.

A few alternatives can be distinguished here. One is to perform subscription deletion by entire template. A cache cannot unsubscribe by removing an individual query (i.e., a template instantiation). Instead, caches are allowed only to unsubscribe by dropping all the queries in a particular template.

The other alternative is to allow query-based unsubscriptions. For this, the MAPs must be modified to support individual query removals. For hash tables, removing a value can be implemented relatively easily. However, if the MAP is a min-max aggregation or a sorted interval list, updating the MAP after removing an element is not straightforward. The problem is recomputing a max over a set of values, when a random element is removed requires maintaining all the elements in the set. The same observation applies for the merged interval list. Therefore, under this scheme, we need to maintain all the individual elements that have been aggregated into the MAP. This set is not accessed during tuple tests, but is used to "recompute" the MAP when an element is removed. Note that MAPS can be lazily updated when unsubscriptions occur because sending more data to the cache than necessary increases load but does not compromise consistency.

5.2 Template-based filtering: Multiple caches

In this section, we consider the case where multiple caches subscribe to update streams at the back-end. One option is to use the single-cache template-based matching algorithm for each cache. This approach results in $O(N \times K \times log(S/K))$ filtering complexity for N caches. In this case, efficiency is gained only in aggregating queries from a single cache.

We investigate whether this simple algorithm can be improved if the templates are shared among caches. This arises in practice because often multiple "instances" of the same Web application can be executing on many edge servers. In this case, the subscriptions coming from the various caches will also be similar, having been generated by the same set of templates. Suppose that two caches have identical templates:

$T_1 : p_1 \wedge p_2 \wedge p_v$ and $T_2 : p_1 \wedge p_2 \wedge p_v$, where p_i are predicate terms.

When testing a tuple against these two template subscriptions, we need to test the tuple against $p_1 \wedge p_2$ for both caches. Then, we need to test the tuple against the aggregations of the variant parts. Suppose the variant part p_v is a simple equality query $p_v :\ attr ==?$. Suppose that cache 1 has cached $attr$ values $7, 12, 15$ and cache 2 has cached $attr$ values $12, 15, 16$. Under the simple scheme, the values for each cache would be stored in separate hash-tables. During the matching process, the filter server would consider each cache's MAP separately. The filter server would look up the value of `tuple.attr` in each cache's MAP hash-table. Alternatively, the values corresponding to the MAPs of various caches can be merged into a single unified MAP, which is capable of remembering which cache(s) each value is associated with. This can be achieved as follows. First, consider inequality tests. Without loss of generality, assume the variant part is of the form $attr <=?$. While in the single cache case, the filter server maintains a single maximum value for all subscriptions that were generated by the same template, in the multi-cache case, the filter server maintains a sorted list of the maximum query parameter for each cache. The tuple matches queries for every cache whose maximum query parameter is greater than or equal to the value of the tuple's attribute. Merging MAPs of multiple caches will, however, require maintaining a more complicated data structure as query subscriptions are added and deleted.

In the case of interval tests, we need to modify the MAP to maintain information about the cache(s) to which each interval (in the sorted interval list) corresponds.

Complexity analysis. Assuming caches share templates, then the complexity can be further reduced from $O(N \times K \times log(S/K))$ to $O(K \times log(N\dot{S}/K))$.

6. Related work

Earlier work on database caching investigated predicate-based schemes and views to answer queries [18, 15, 8, 13, 9]. Recent papers have examined passive and active caching schemes for web applications and XML data [16, 6, 11]. Luo and Naughton described an active caching technique based on templates (forms) [16]. Consistency protocols for predicate caches have been investigated in [13], but the focus was on limited-size client-server database systems, and not on scalable deployments over the Web.

Related to the problem of update filtering is query containment— the problem of deciding whether a query is contained in the union of a set of views or cached queries. A wealth of previous work exists in the area of query containment and equivalence [17, 14, 5]. Previous work in the area of materialized view routing (i.e., answering queries by rewriting using materialized views) also describes techniques for matching and containment.

Most related to our work is the event matching approach of publish-subscribe systems [1]. Gryphon uses a parallel search tree (PST) to aggregate client subscriptions described as conjunctive predicate expressions. The difference in our approach is that instead of using a generalized PST we use specialized indexed data structures (MAPs) to aggregate the variant parts of subscription expressions based on the particular operator that appears in the variant part. This exploits the abundance of templates in Web applications, and enables compact representations of the data structures used to aggregate subscriptions and accelerate filtering.

7. Conclusions

Semantic database caching is a self-managing approach to dynamic materialization of "semantic" slices of back-end databases on edge servers. It can be used to enhance the performance of distributed Web servers, information integration applications, Web service platforms, and Web applications offloaded to edge-of-network servers. The scalability of semantic caching, however, relies heavily on protocols that maintain consistency between the back-end database and the distributed caches. In this paper, we focus the problem of "scaling" the update propagation protocols in such environments. In particular, we propose *template-based filtering* that efficiently aggregates cache subscriptions at the back-end by exploiting the similarity of "views" in edge query caches. We discuss the specifics of our approach, describe how we handle dynamic changes in cache subscription, and highlight the challenges and trade-offs that arise in this space.

References

[1] M. K. Aguilera, R. E. Strom, D. C. Sturman, M. Astley, and T. D. Chandra. Matching events in a content-based subscription system. In *Symposium on Principles of Distributed Computing*, pages 53–61, 1999.

[2] Akamai Technologies Inc. Akamai EdgeSuite. http://www.akamai.com/html/en/tc/core_tech.html.

[3] K. Amiri, S. Park, R. Tewari, and S. Padmanabhan. DBProxy: A Self-Managing Edge-of-Network Data Cache. Technical Report RC22419, IBM Research, April 2002.

[4] K. Amiri, S. Park, R. Tewari, and S. Padmanabhan. DBProxy: A Self-Managing Edge-of-Network Data Cache. In *19th IEEE International Conference on Data Engineering*, pages 821–831, March 2003.

[5] K. Amiri, S. Park, R. Tewari, and S. Padmanabhan. Scalable template-based query containment checking for web semantic caches. In *IEEE International Conference on Data Engineering*, pages 493–504, March 2003.

[6] L. Chen and E. Rundensteiner. XCache: XQuery-based Caching System. In *Proceedings of the Fifth International Workshop on Web and Databases*, June 2002.

[7] Colin C. Haley. IBM, Akamai Boost 'Virtual Capacity' . http://www.internetnews.com/infra/article.php/2200501.

[8] S. Dar, M. J. Franklin, B. T. Jónsson, D. Srivastava, and M. Tan. Semantic data caching and replacement. In *VLDB Conference*, pages 330–341, 1996.

[9] P. Deshpande, K. Ramasamy, A. Shukla, and J. F. Naughton. Caching multi-dimensional queries using chunks. In *SIGMOD Conference*, pages 259–270, 1998.

[10] A. Guttman. R-Trees: A Dynamic Index Structure for Spatial Searching. In *Proceedings of the ACM SIGMOD Conference on Management of Data*, June 1984.

[11] V. Hristidis and M. Petropoulos. Semantic Caching of XML Databases. In *Proceedings of the Fifth International Workshop on Web and Databases*, June 2002.

[12] IBM Corporation. Websphere Edge Server. http://www-4.ibm.com/software/webservers/edgeserver/.

[13] A. M. Keller and J. Basu. A predicate-based caching scheme for client-server database architectures. *VLDB Journal*, 5(1):35–47, 1996.

[14] P.-A. Larson and H. Z. Yang. Computing queries from derived relations: Theoretical foundations. Technical Report CS-87-35, Department of Computer Science, University of Waterloo, 1987.

[15] A. Levy, A. O. Mendelzon, Y. Sagiv, and D. Srivastava. Answering queries using views. In *PODS Conference*, pages 95–104, 1995.

[16] Q. Luo, J. F. Naughton, R. Krishnamurthy, P. Cao, and Y. Li. Active query caching for database web server. In *WebDB Conference (Informal Proceedings)*, pages 29–34, 2000.

[17] D. J. Rosenkrantz and H. B. Hunt. Processing conjunctive predicates and queries. In *VLDB Conference*, pages 64–72, 1980.

[18] T. K. Sellis. Intelligent caching and indexing techniques for relational database systems. *Information Systems*, 13(2):175–185, 1988.

PROXY+: SIMPLE PROXY AUGMENTATION FOR DYNAMIC CONTENT PROCESSING

Chun Yuan[1], Zhigang Hua[2], and Zheng Zhang[1]
[1]*Microsoft Research Asia,* [2]*Institute of Automation, Chinese Academy of Sciences*

Abstract Caching dynamic content can bring many benefits to the performance and scalability of Web application servers. However, such mechanisms are usually tightly coupled to individual application servers (or even applications) that prevent caching at more advantageous points. In this paper we propose an approach to enable dynamic content caching at enhanced Web proxies which requires only simple modifications to existing applications.

1. Introduction

Dynamic content will dominate the Web in the future. This necessitates architectural change in tandem. In particular, resources that are already deployed near the client such as the proxies that are otherwise underutilized for such content should be employed.

Legitimate strategies include offloading some of the processing to the proxy, or simply enhancing its cache abilities to cache fragments of the dynamic pages and perform page composition. While performance benefits including latency and server load reduction are important factors to consider, issues such as engineering complexity as well as security implication are of even higher priority. Our previous work [20] investigates what will be the best offloading and caching strategies and their design/deployment tradeoffs given the proxy resources at the edge of the network. We have shown that simply caching dynamic page fragments and composing the page at the proxy achieve close performance to other strategies. In fact, advanced offloading strategies can be overly complex and even counter-productive performance-wise if not done carefully.

This work continues the previous work to demonstrate the advantage and feasibility of proxy caching for dynamic content. We propose Proxy+ by adding a caching filter at Microsoft ISA proxy server [14] and offers fragment caching and page composition ability. Our core idea is to simply replicate the server-side caching functionality to the proxies, while carefully engineering the protocols so that consistency enforcement takes a free-ride from what the programmers have already

F. Douglis and B.D. Davison (eds.), Web Content Caching and Distribution, 91-108.

expressed when enabling server-side dynamic content caching. Our architecture requires only minor modifications to existing applications and is incrementally deployable. Finally, although our prototype is implemented under Microsoft ASP.NET [13] and ISA Server, we believe the method is equally applicable to other platforms.

The rest of the paper is organized as follows. Section 2 covers related work. Section 3 summarizes our previous results which motivates this work. Section 4 describes the architecture, components and protocol of Proxy+. Section 5 illustrates how existing ASP.NET applications can be modified to work with Proxy+. Section 6 reports our experimental results. Section 7 discusses some security issues. Finally we conclude in section 8.

2. Related Work

Optimizing dynamic content processing has been widely studied. Most work focused on supporting dynamic content caching on server side [9][10][19]. Currently there are already mature application server products offering this feature, e.g. Microsoft ASP.NET, BEA WebLogic [1], IBM WebSphere [7], Oracle9iAS [16]. Server-side caching can reduce server load and therefore improve response time when client stress is high. Moving the caching tier to proxies at the network edge that is closer to clients would bring more benefits as reported in [11][20].

Fragments caching is key to dynamic content caching, since it can improves page cacheability and cache efficiency. Many Java application servers allow programmers to mark a part of a page as cacheable using JSP tags. In ASP.NET such fragment can be explicitly put into a user control which has its own cache parameters and can be included by pages or other user controls. The cache is usually associated with the application server, thus preventing caching at more advantageous points. [5] also uses tags to support fragment caching on a reverse proxy which is assumed to be near the server, but it does not consider caching on remote proxies. Active Cache [3] is a rather general scheme to push server-side processing to proxies but it does not address specific issues with fragment caching. IBM WebSphere's Trigger Monitor [8] also support fragment caching, but on predetermined edge servers only.

ESI [4] proposes to cache fragments at the CDN stations to reduce network traffic and response time. ESI introduces some directives for programmers to author cacheable Web pages which will be interpreted by ESI engine on edge servers. A page may include fragments which will be retrieved separately from the server (when cache miss). Therefore for existing applications to employ ESI, original single output of a page has to be divided into parts which can be separately generated and retrieved. The top-level page output also need to contain ESI include directive for edge servers to get inner fragments. However this could violate original request processing workflow and semantics, because during original page generation the top-level page and the fragments are in the same request context while after factoring they are independently requested. Hence turning a fragment into a page would require understanding of the original page semantics and may require considerable reengineering effort.

Proxy+ relies on output tagging to distinguish fragments so that they can be independently cached. It requires only trivial extension to existing applications, provided that they have already made use of fragment caching on server side. Furthermore it does not interfere with the application's original workflow and need not to understand the details of the application. We let proxies notify a list of cache keys that are deemed relevant to the request so that network traffic and server load can be saved as much as possible.

3. Summary of Previous Result

In our previous work [20] we studied the performance of four Web application configurations, with three of them taking advantage of edge proxy to offload server processing.

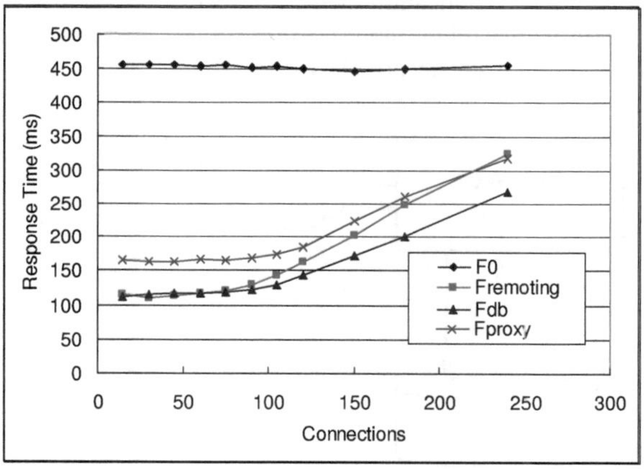

Figure 1. Response time versus number of connections. The roundtrip time is 400ms.

- F_0: the default setting, with all processing happening on the server
- $F_{remoting}$: push part of application logic (e.g. presentation tier) to the proxy
- F_{db}: push everything except database to the proxy
- F_{proxy}: push fragment caching and page composition functionality to the proxy

Our results (*Figure 1*) show that under typical user browsing patterns and network conditions, 2~3 folds of latency reduction can be achieved. Furthermore, over 70% server requests are filtered at the proxies, resulting significant server load reduction. Interestingly, this benefit can be achieved largely by F_{proxy} – simply caching dynamic page fragments and composing the page at the proxy. Taking implementation cost and security into account, our result can be summarized in *Table 1*.

Table 1. Comparison of different configurations

	Performance	Security requirements	Complexity
F_{db}	Best	Hight	Low
$F_{remoting}$	Second	Unsure → High	High
F_{proxy}	A closed 3rd	Low	Lowest

In other words, augmenting today's proxy caching capability has the best tradeoff. The F_{proxy} implementation in that paper took an ESI-like approach and was application-dependent, which motivates the work of this paper.

4. Proxy+ Architecture

Given that proxy augmentation offers the best tradeoff, we developed the prototype of Proxy+ on top of Microsoft ISA Server. The augmented proxy is able to do caching for ASP.NET applications which have made use of ASP.NET built-in output caching facility after minor modifications. In fact, these modifications will be trivial if necessary supports are absorbed by ASP.NET.

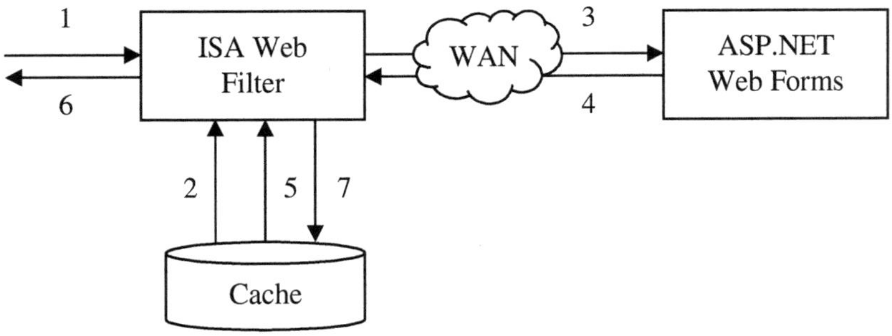

Figure 2. Architecture and workflow of Proxy+

The core idea is straightforward: replicate the output cache functionality at the proxy with an ISA Web filter. Therefore, the engineering focus is on how to get a "free ride" by mirroring caching semantics and its enforcement from server side's output cache. *Figure 2* shows the architecture and workflow of Proxy+.

The ISA Web Filter is responsible for directing the caching of multiple versions of pages and fragments as well as composing pages. Cache is the storage of previously requested pages and fragments. Both of them reside on the ISA Server. With ASP.NET, Web Forms constitute the presentation tier of the Web application running on the server side. In our framework, applications can cooperate with the ISA Server to take advantage of its enhanced caching ability after minor extensions to the classes that the Web Forms inherit from.

The system workflow is described as follows.

1. An HTTP request arrives at the proxy.

2. The filter computes the cache keys of the page and its fragments.
3. If all necessary items are valid in the cache, go to 5 and the result will be returned immediately. Otherwise, it attaches a list of keys identifying cached versions deemed relevant to the HTTP request header and forwards it to the server.
4. The application generates a (partial) response containing additional tags for delimiting cacheable fragments or behaving as placeholders to be substituted with cached content. In addition, necessary information, in the form of *Cache Variation Logic* (CVL) tags, which allows the proxy to compute the cache keys are sent over.
5. The filter parses the content (from the response or the cache) and fills the placeholder tags in the text with corresponding cached content, and installs any CVL tags. This way, CVLs are incrementally pulled over on-demand.
6. A complete response is sent back.
7. The fragments marked for caching are saved to the cache.

In the following text, we are going to illustrate the essential components in turn:

- We will start by reviewing ASP.NET's output cache and in particular its semantics of caching multiple versions.
- We will then present how cache keys are computed with the help from the application side. This entails two steps: modification at application to generate CVL tags, and proxy-side's use of such tags to produce cache keys.
- Next, we describe how the proxy assembles a page out of content in its cache and, if necessary, notify the server/application what are already available in its cache to avoid redundant transfer.

4.1 ASP.NET output caching

The presentation tier of an ASP.NET Web site consists of a set of Web Form pages (with .aspx extension) and user controls (with .ascx extension). They are responsible for generating Web pages (usually in HTML) to satisfy client requests. A user control represents a fragment of a page and can be included by many pages or other user controls.

ASP.NET allows many versions outputted by a page to be cached (at the server side). Programmers can use a high-level, declarative API or a low-level, programmatic API when manipulating the output cache. For example, the following @OutputCache directive (included in an .aspx or .ascx file) sets an expiration of 60 seconds for the cached output of the page or user control.

```
<%@ OutputCache Duration="60" VaryByParam="none" %>
```

Upon arrival of subsequent requests, the output cache will send the proper cached version as response directly. While this is the basic version, advanced features require programmers to specify how the output cache is varied by. @OutpputCache directive includes the following attributes referred to as Cache Variation Logic (CVL) that can be used to cache multiple versions of page output.

- **VaryByParam**: vary the cached output depending on GET query string or form POST parameters. Including the following example at the top of an .aspx file will cause different versions of output to be cached for each request that arrives with a different "city" parameter value. `<%@ OutputCache Duration="60" VaryByParam="city" %>`
- **VaryByHeader**: vary the cached output depending on the HTTP header associated with the request. The following example sets versions of a page to be cached, based on the value passed with the "Accept-Language" HTTP header. `<%@ OutputCache Duration="60" VaryByParam="none" VaryByHeader="Accept-Language" %>`
- **VaryByCustom**: vary the cached output by a custom string. This is the most advanced and powerful option. A special method called HttpApplication.GetVaryByCustomString() must be overridden which is used to map a string name to a value under some context. The following directive will cause the output to be cached for each request with a different value corresponding to "userstatus", which may be, for example, "login" or "logout". `<%@ OutputCache Duration="60" VaryByParam="None" VaryByCustom="userstatus" %>`

ASP.NET output cache computes a key based on CVL, and searches for the associated content in the cache. The key calculation is *internal* to the output cache. Consequently, proxy-side cache must be keyed with its own algorithm.

4.2 Cache key generation

In order to enable proxy-side output caching, the proxy must uses its own key generation algorithm. Our current implementation simply concatenates with semicolons the path name of the page (or user control) and the values that the output depends on in order to produce the cache key. For example suppose the page http://www.petshop.net/Category.aspx has the following CVL.

```
<%@ OutputCache Duration="60" VaryByParam =
"category_id" VaryByHeader= Accept-Language  %>
```

If it receives a request http://www.petshop.net/Category.aspx?category_id=cats, and the header field "Accept-Language" is "zh-cn", then the cache key is "/Category.aspx;cats;zh-cn".

And suppose the CVL of the user control named "header" at http://www.petshop.net is as follows.

```
<%@ OutputCache Duration="60"
VaryByCustom= userstatus  %>
```

It is included by "/Category.aspx" and is intended to show different interface for anonymous users and authenticated users. When requested, the programmer-defined method GetVaryByCustomString() will analyze the cookie and decide the user status. If the user has signed in, it will map "userstatus" to, for example, "login". Therefore, the final cache key of header's output would be "header;login". The way for the proxy to get the method is described in the next section.

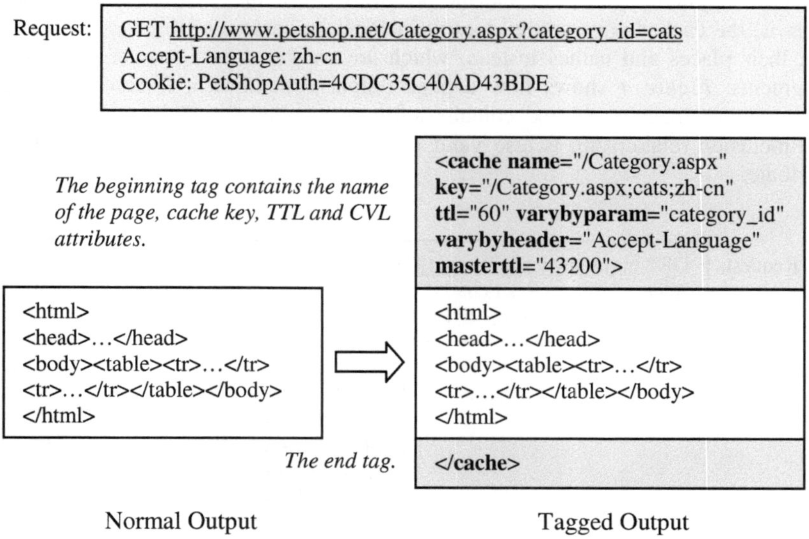

Figure 3. Tagging the output of a page

In Proxy+, to avoid redundant computation and transfer, whatever contents available at proxy will be notified to the server, using the cache keys. Server runs the same key generation algorithm to generate the cache keys again so as to skip the redundant content generation. This will be described in more detail in section 5.

Of course, in practice a proxy may interact with many servers, so the scope of a cache key is limited to the server which defines the CVL. In the above case, the cache key "/Category.aspx;cats;zh-cn" and "header;login" are only applicable to requests sent to the server www.petshop.net.

4.3 Tag generation and fragment caching

The CVL and TTL (time to live, i.e., Duration attribute in @OutputCache directives) of a page is communicated to the proxy through additional `<cache>` tags in the output. *Figure 3* illustrates how such tags are added to the normal output.

When receiving a tagged response, the proxy will cache the actual content with the key. It also recognizes the CVL attributes according to which cache keys for subsequent requests will be calculated. The CVL is installed if it sees it the first time, or updated if necessary. This way, CVL tags are pushed to proxy incrementally in an on-demand fashion. The "**masterttl**" attribute declares the lifetime of the master program that produces the output, while "**ttl**" defines the lifetime of the specific version of output only. Therefore in the above example, the CVL of the page named "/Category.aspx" is valid for 12 hours.

Fragments are treated in the same way, though a tagged fragment might be contained in another tagged page or fragment. In this case, the outer page needs to cache the content at its level and the positions and names of the inner fragments. That is, the cached page output doesn't include the content of the fragments inside but their places and names instead, which are used to insert new versions of the fragments. *Figure 4* shows how a page output containing a fragment is tagged (assume now "/Category.aspx" contains a fragment called "header" inside). Note that the inclusion relationship is also valid for 12 hours as limited by the "**masterttl**" attribute.

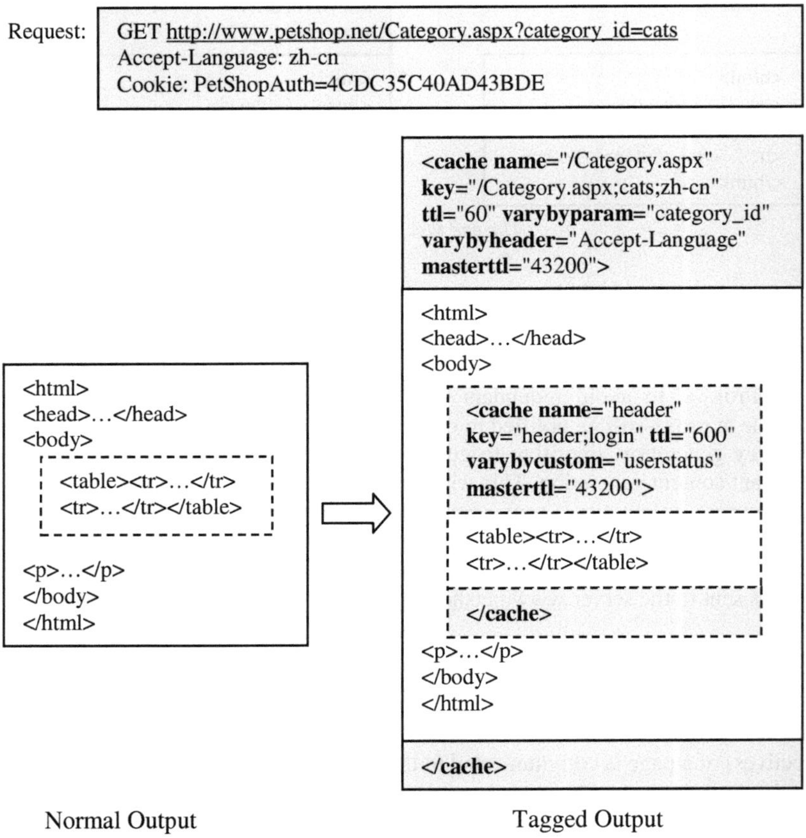

Figure 4. Tagging the output of a page as well as its inner fragment

The page and the fragment will be cached as shown in *Figure 5*. The `<include>` tag represents a placeholder that is to be replaced with the content of a fragment with the specified name. All `<cache>` tags will be removed from the output and the client will receive a response just like the normal output.

As for the fragment "header" whose CVL containing **VaryByCustom**, it is not enough to notify the attribute values to the proxy because it depends on the function

GetVaryByCustomString() to generate their cache keys. In Proxy+, the Web application specifies the location of the dynamic linking library by sending a new HTTP header, "X-GetVaryByCustom", with the response:

```
X-GetVaryByCustom: library-url
```

which must export the function:

```
string GetVaryByCustomString(HttpRequest req, string varyby);
```

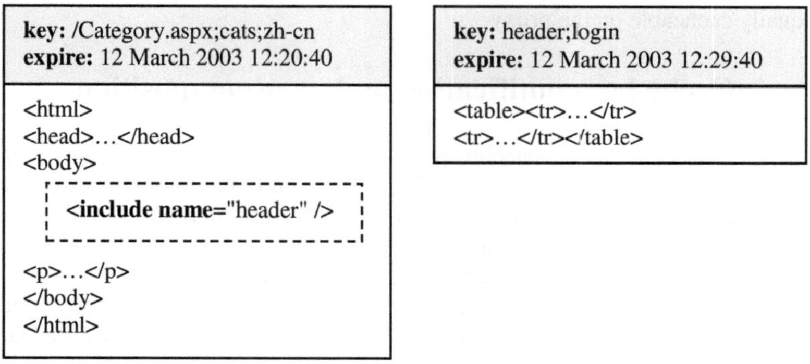

Figure 5. Cached output

Proxy is responsible for passing the complete request (the argument "req", including various header fields, cookies and body) and the value of **VaryByCustom** attribute associated with a fragment (the argument "varyby") to the function, which will return a unique string for identifying the version of the fragment. For example, besides generating the tagged output, the application would also add the following header to the response:

```
X-GetVaryByCustom:
http://www.petshop.net/varybycustom.dll
```

The proxy can download the DLL and import the function. Requesting pages/fragments with **VaryByCustom** attributes will cause cache miss when the DLL is not available (not downloaded or become obsolete).

When receiving a subsequent request, if any of the necessary keys is not found in the cache (i.e., cache miss), the proxy will forward the request to the server. Otherwise it will compose the items corresponding to the keys together and return a complete response. Continuing the example, when receiving the request http://www.petshop.net/Category.aspx?category_id=cats again (with "Accept-Language" header being "zh-cn") the proxy computes a key "/Category.aspx;cats;zh-cn" that would be found in the cache. Then it computes another key for the fragment the page needs to include. If the user hasn't signed out, GetVaryByCustomString(Request, "userstatus") will return a string "login" according to the authentication cookie in the request and thus the key would be "header;login", which means both items hit the cache. The proxy will insert the content of "header" into that of "/Category.aspx" and the full output is returned. If the user has signed out and the cache doesn't contain the key "header;logout", the request will be forwarded to the server.

GetVaryByCustomString() may not be able to accomplish the same function on the proxy as its counterpart on the server when it requires server-specific resources like database. For example on a personalized portal site, the method may read user settings and produce a news type id (such as "science") that the user is interested in for a news fragment, while user settings are not accessible from proxies. Therefore not all kinds of pages and user controls with **VaryByCustom** attributes on the server are equally cacheable on the proxy.

4.4 Cache keys notification and page composition

Unlike traditional HTTP caching proxies, proxy+ can cache some parts of the response even when the others miss. If the application could know what parts have been cached on the proxy and output only those that are missing, server resources would be saved. Therefore, in our architecture a proxy is allowed to notify the server a list of keys along with the request to indicate that the page/fragments with these keys have been cached so that the server doesn't need to generate the content again. The notification is done by appending a new HTTP header field, "X-CachedKeys", to the incoming request:

```
X-CachedKeys: cache-key1, cache-key2, ...
```

If the server-side application finds that the cache key of the page or fragment is listed in the header, it can skip the content generation process and instead put a placeholder tag (`<subst>`) along with the name and the cache key. The verification is processed by running exactly the same key generation algorithm as in proxy. Placeholder tags are intended to be substituted with the corresponding content from the proxy cache. For example, assume the proxy forwards the request http://www.petshop.net/Category.aspx?category_id=cats (with "Accept-Language" header equaling "zh-cn") with such a header added,

```
X-CachedKeys: header;login
```

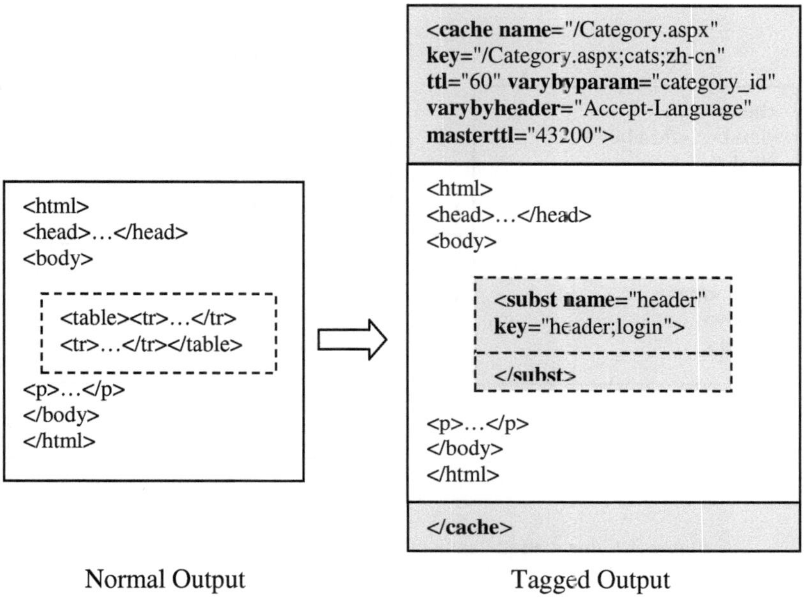

Figure 6. Using <subst> tag in place of cached inner fragment

If the user has signed in, the application will verify that the cache key of the inner fragment "header" is in the request header and can mark the output as in *Figure 6*. The pair of `<subst>` tag is to be replaced with the cached fragment having the key "header;login" on the proxy.

On the other hand, if the request has this header,

```
X-CachedKeys: /Category.aspx;cats;zh-cn
```

the output can be like *Figure 7*.

On the proxy the fragment "header" will be inserted into the cached output with key "/Category.aspx;cats;zh-cn" in the position of the placeholder and a complete page returned.

Note that the server is *not* required to skip the generation of the page or fragment even though its cache key is in the key list: it can optionally choose to include the actual output (with the `<cache>` tags). This flexibility allows a server to effectively remove the caching capability of an untrusted proxy, and to proactively update both the content and CVL when server deems necessary.

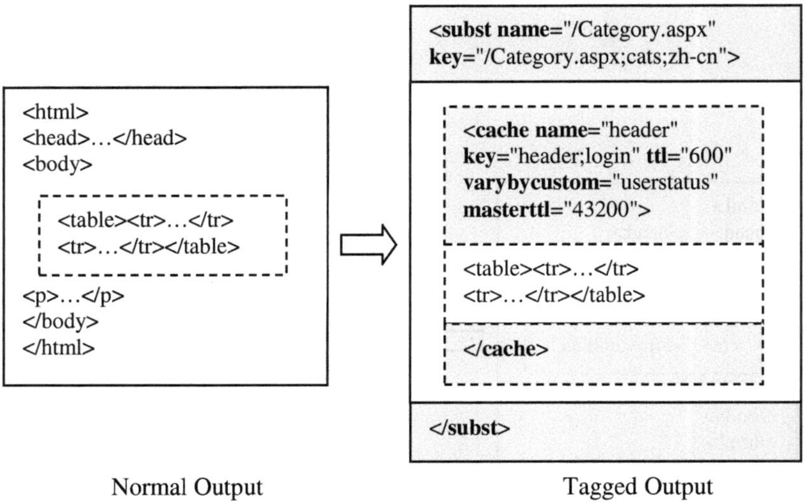

Figure 7. Using <subst> tag in place of cached outer page (fragment)

4.5 Summary of the protocol

The protocol can be summarized as follows:

- From server to proxy: uses `<cache>` tags to inform Proxy+ both the contents to be cached and the associated CVL tags. This allows on-demand and incremental installation of the CVL, and also affords server the opportunity to control the Proxy+ caching capability if necessary.
- From proxy to server: uses “X-CachedKeys” to inform already cached contents so that redundant computation and transfer maybe avoided.
- The advanced output cache feature **VaryByCustom** requires the server to specify an URL of a DLL that exports the function GetVaryByCustomString() which the proxy subsequently downloaded. This feature may not be possible if such function needs to access resources known only at the server side.
- Proxy+ is incrementally deployable: it behaves as any normal proxy with non-proxy+ aware applications (servers). The reverse is also true: no ill side-effect is caused when a proxy+ aware application interacts with a normal proxy.

The consistency control of Proxy+ cache is enforced exactly the same as the output cache on the server. Thus, our protocol accomplishes the goal of replicating output cache functionality on the proxy. It should be noted that using DLL to package custom cache key generation function is only for our convenience of implementation on Windows platform. A complete implementation may provide equal support for other platforms by, e.g., using a cross-platform language like Java or script language.

5. Application Modifications

We now turn our attention to the necessary modifications to applications. As it turns out, the changes are very minor, and even trivial if supports are built into ASP.NET.

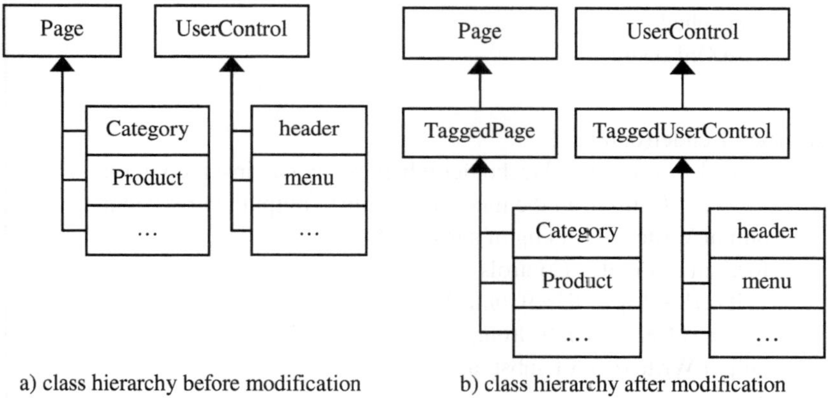

Figure 8. Web application presentation tier class hierarchy

Proxy+ architecture currently targets ASP.NET applications and assumes programmers have used ASP.NET Web Forms Page and UserControl class to implement dynamic content caching on server side. To enable output tagging for such applications, it is sufficient to simply extend Page and UserControl class to generate tags and avoid regeneration of cached content. The process doesn't interfere with the original workflow of the application *at all.*

In general a Web application builds its UI by using ASP.NET Web Forms. The components at the presentation tier are subclasses of the class System.Web.UI.Page and System.Web.UI.UserControl as illustrated in *Figure 8*a). The content is generated by handling appropriate events in the class. For example, "Load" is a typical event which is handled by many pages and user controls to populate content to UI.

The modification to the application must satisfy two requirements. The new application should be able to recognize the list of keys sent from the proxy and avoid regeneration of corresponding content. It also needs to insert additional tags as described to enable the proxy to do fragment caching and page assembly.

```
bool cached;

override OnInit(e) {
  cached = (my cache key) ∈ (the key list received with the request);
               // my cache key computed in the same way as on proxy
  base.OnInit(e);
}

override OnLoad(e) {
  if not cached
    base.OnLoad(e);
}

override Render(output) {
  if cached {  // make a placeholder where proxy can insert content;
               // still need the inner fragments to output their content
    output.Write (beginning of subst tag);
    foreach ctrl in this.Controls
      if ctrl is TaggedUserControl
        ctrl.RenderControl(output);
    output.Write (end of subst tag);
  } else {
    output.Write (beginnng of cache tag);
    base.Render(output);
    output.Write (end of cache tag);
  }
}
```

Figure 9. Pseudo code of the TaggedPage class

We modify an application as follows. Two new classes TaggedPage and TaggedUserControl are added; they are subclass of Page and UserControl respectively. All subclasses of Page and UserControl will inherit from them instead, as shown in *Figure 8*b). These two classes override the event dispatching and HTML outputting functions in their superclasses. According to the key list attached in the request, they will decide whether the specific event need to be dispatched to the original handler or not (to avoid regeneration of cached content) and what additional tags (`<cache>` or `<subst>`) and CVLs are to be inserted into the HTML output.

The pseudo-code of TaggedPage is listed in *Figure 9*. TaggedUserControl's code is essentially the same with minor differences. For applications built with other Web programming platforms (e.g. JSP), we believe the modifications would also be similar.

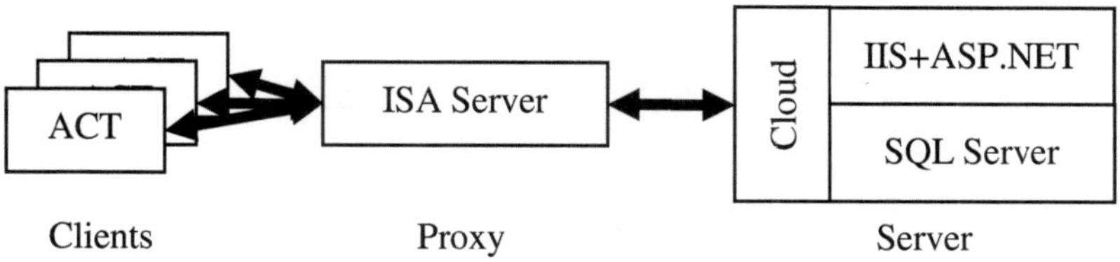

Figure 10. Experiment configuration

6. Experimental Results

We measure the performance of our Proxy+ prototype with a representative e-commerce benchmark called .NET Pet Shop [12]. The availability of the source code allows us to experiment Proxy+ *without* hack into ASP.NET itself. The experiment configuration is depicted in *Figure 10.*

Table 2. Distribution of the test workload

Activity	Percentage
Category Browsing	18%
Product Detail	16%
Search	18%
Home Page	18%
Shopping Cart	7%
Order	1%
Account/Authentication	22%

The clients run Microsoft Application Center Test (ACT) to simulate a number of concurrent Web browsers. ACT creates enough threads (specified by a connection number) to issue requests according to a test script that defines the test workload. The request distribution of the workload is shown in *Table 2*. The requests are sent to the proxy running Microsoft ISA Server with the output cache-enabled filter installed, and then forwarded to the backend Web server running Microsoft IIS, ASP.NET and SQL Server. The hardware settings of the proxy and the server are dual 1.7G Pentium 4 Xeon with 2GB RAM and dual 2.4G Pentium 4 Xeon with 1GB RAM respectively. The clients are also powerful enough not to become bottlenecks in our tests. All machines are connected in a switched 100Mbps Ethernet. A WAN emulator (Shunra\Cloud [18]) running on the backend server is used to set a constant latency between it and the proxy, while the client accesses the proxy over the LAN directly.

We compare the response time of the Web site accessed via the proxy with and without the filter enabled (Proxy+ and common proxy). *Figure 11* shows the average response time versus number of concurrent connections when the roundtrip network latency between the proxy and the server is set to 400ms.

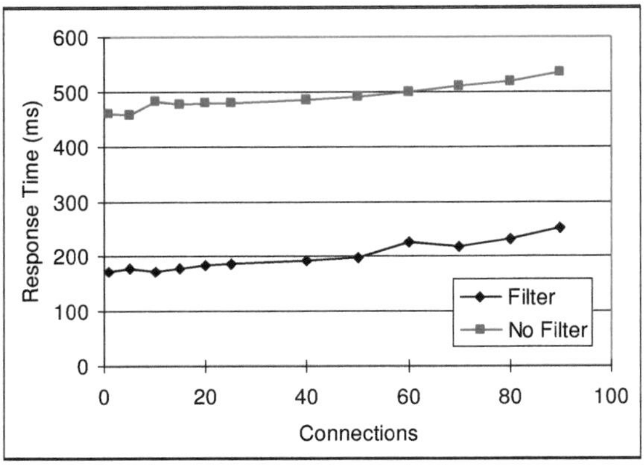

Figure 11. Response time with 400ms roundtrip latency

As can be seen, the response time of common proxy is always above the roundtrip latency because every request has to travel through the delayed link and so does its response. With Proxy+, about 60% of the response time can be saved, where the full page hit ratio (a page and all of its inner fragments hit the cache, thus avoiding server access) is about 70%.

Network traffic saving is another benefit besides response time improvement. In the above test, 87% of the traffic between the proxy and the server is reduced on average.

To measure the overhead of the filter, we repeat the test without setting any network latency and also compare them with the response time of accessing the Web site directly. The results are shown in *Figure 12*, where "No Proxy" means requests are issued directly to the Web server. Due to ISA Server's overhead, the response time of "No Filter" option is slightly more than that of "No Proxy". And because the filter's overhead happens to counteract the saved server side response time, "Filter" option has about the same response time as "No Filter". Therefore, the overhead can be roughly estimated as the server side response time (about 10ms as shown by "No Proxy" data since network delay is negligible) timing the hit ratio, that is, about 7ms, which is acceptable when network latency between proxy and server is considerable, for example, hundreds of milliseconds.

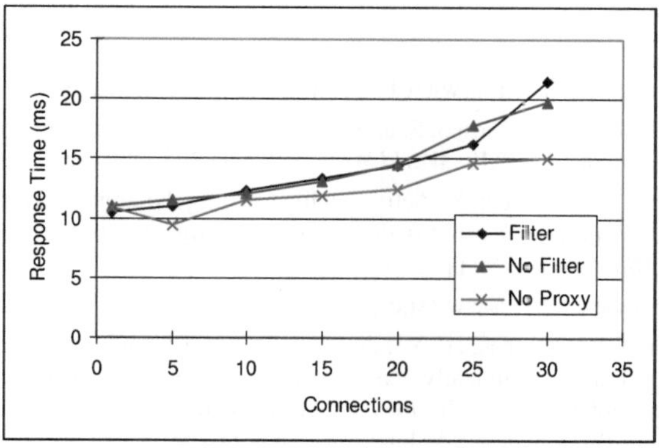

Figure 12. Response time without setting latency

7. Security aspect

Besides the security limitations that common HTTP caching proxies have, Proxy+ raises some different issues as well as interesting possibilities. Caching can be thought as filtering, consequently there is always a possibility of substituting contents. Proxy+ makes it possible for such actions – intentionally or otherwise, to be performed on a much finer granularity. We note that Proxy+ leaves the power and flexibility of control on the server. For a more systematic way to guard against proxy abuse we refer readers to approaches such as Gemini [15].

It poses a problem to proxies that they need to import DLLs provided by servers in order to cache pages/fragments with **VaryByCustom** attributes. Such DLLs must be signed by the server. If a proxy is unsure, it should run such DLL in a sandbox and deny its access to resources such as network and storage IO.

8. Conclusions

After establishing the argument that simple proxy extension will work just as well for dynamic content, we proposed Proxy+ architecture for dynamic content caching on augmented proxies near clients. The protocol uses simple extension to HTTP and can work coherently with common Web applications and proxies. Only minor modifications to existing applications are necessary to cooperate with Proxy+. Our experiment shows that a significant amount of response time and network traffic can be saved with Proxy+. Due to its low implementation cost and incremental deployability, we believe it is a competitive solution. We are currently extending Proxy+ fragment caching and page composition components from proxy servers to Web browsers so that clients with last-mile bottleneck can benefit from the bandwidth saving by this scheme [2][6][17].

References

[1] BEA WebLogic Server. http://www.bea.com/products/weblogic/server/

[2] Brabrand, C., Møller, A., Olesen, S. and Schwartzbach, M.I. Language-based caching of dynamically generated HTML. *World Wide Web* 5(4):305-323; Jan 2002

[3] Cao, P., Zhang, J. and Beach, K. Active cache: Caching dynamic contents on the Web. In: *Proc. of IFIP Intl. Conf. on Distributed Systems Platforms and Open Distributed Processing (Middleware'98)*, pp. 373-388.

[4] Edge Side Includes. http://www.esi.org/

[5] Datta, A., Dutta, K., Thomas, H., VanderMeer, D., Suresha and Ramamritham, K. Proxy-based acceleration of dynamically generated content on the World Wide Web: An approach and implementation. In: *Proc. of ACM SIGMOD Intl. Conf. on Management of Data*, Madison, Wisconsin, USA, June, 2002, pp. 97-108.

[6] Douglis, F., Haro, A. and Rabinovich, M. HPP: HTML macro-preprocessing to support dynamic document caching. *Proc. of USENIX Symposium on Internet Technologies and Systems*, December 1997, pp 83-94.

[7] IBM WebSphere Application Server. http://www-3.ibm.com/software/webservers/appserv/

[8] IBM WebSphere Edge Server. http://www-3.ibm.com/software/webservers/edgeserver/

[9] Iyengar, A. and Challenger, J. Improving Web server performance by caching dynamic data. In: *Proc. of the USENIX Symposium on Internet Technologies and Systems (USTIS'97)*, Monterey, CA, December 1997.

[10] Labrinidis, A. and Roussopoulos, N. WebView materialization. In: *Proc. of the ACM SIGMOD Intl. Conf. on Management of Data*, Dallas, Texas, USA, May 2000, pp. 367-378.

[11] Li, W.S., Hsuing, W.P., Kalashnikov, D.V., Sion, R., Po, O., Agrawal, D. and Candan, K.S. Issues and evaluations of caching solutions for Web application acceleration. In: *Proc. 28th Int. Conf. on Very Large Data Bases (VLDB)*, Hong Kong, China, Aug 2002.

[12] Microsoft .NET Pet Shop. http://www.gotdotnet.com/team/compare/petshop.aspx

[13] Microsoft ASP.NET. http://www.asp.net/

[14] Microsoft ISA Server. http://www.microsoft.com/ISAServer/

[15] Myers, A., Chuang, J., Hengartner, U., Xie, Y., Zhuang, W. and Zhang, H.. A secure, publisher-centric Web caching infrastructure. *Proceedings of IEEE Infocom* '01

[16] Oracle9iAS. http://www.oracle.com/appserver/

[17] Rabinovich, M., Xiao, Z., Douglis, F. and Kalmanek, C. Moving edge side includes to the real edge – the clients. *Proc. 4th USENIX Symposium on Internet Technologies and Systems*, March 2003

[18] Shunra\Cloud. http://www.shunra.com/cloud.htm

[19] Yagoub, K., Florescu, D., Valduriez, P. and Issarny, V. Caching strategies for data-intensive Web sites. In: *Proc. of the Int. Conf. on Very Large Data Bases (VLDB)*, Cairo, Egypt, 10-14 September, 2000.

[20] Yuan, C., Chen, Y. and Zhang, Z. Evaluation of edge caching/offloading for dynamic content delivery. In: *Proc. The 12th Int'l World Wide Web Conference (WWW 2003)*, Budapest, Hungary, 2003.

MULTICAST CLOUD WITH INTEGRATED MULTICAST AND UNICAST CONTENT DISTRIBUTION ROUTING

Dan Li, Arun Desai, Zheng Yang, Kenneth Mueller, Stephen Morris, and Dmitry Stavisky
Cisco Systems, Inc.

Abstract In this paper, we describe the concept and design of "application-layer multicast cloud," the first overlay network design that provides integrated content distribution routing between IP multicast and unicast via a user-configured group of multicast senders and receivers for content distribution across different layer-3 IP multicast groups, content distribution channels, and unicast distribution relay trees, that for the first time allows for application-layer (as opposed to layer-3 or layer-4) control of multicast security, failover, QoS, and bandwidth utilization. None of these were possible before and yet the need for such capabilities is great in Content Distribution Network (CDN) deployments.

Note that here "multicast cloud" refers to an "application-layer" user configured entity with (1) a number of properties that govern the IP multicast flow and the interaction between multicast and unicast content distribution, and (2) a set of hosts that can participate in multicast content replication that happens on multiple layer-3 IP multicast groups and across multiple layer-3 IP multicast islands.

1. Introduction

We expect that multimedia and data traffic will surpass the traditional voice traffic in mobile networks by the year 2005 [18]. High quality streaming multimedia content is likely to form a significant portion of this traffic. It is therefore important that large mobile networks find ways to manage client traffic and data efficiently.

The foundation of scalable content distribution is efficient one-to-many data transport, i.e., multicast[1]. Many content distribution network designs have centered

[1] "multicast" here refers to the 1-to-many service model, not necessarily the specific instance of IP multicast.

F. Douglis and B.D. Davison (eds.), Web Content Caching and Distribution, 109-118.

around "application-layer multicast" built on top of IP unicast, for its effective transport, incremental deployment, asynchronous delivery, application-aware routing, and versatility [1].

While some have embraced application-layer multicast and denounced the usefulness of IP Multicast, we think both have strengths and weaknesses. An optimal "multicast" system would take advantage of both of them. In particular, IP multicast helps application-layer multicast utilize physical broadcast media, build efficient distribution, and improve scalability [2].

In this paper, we present the architecture and design of a content distribution network (CDN) that effectively integrates IP unicast and IP multicast into a single application-multicast architecture, and offers the CDN operator superior application-layer control of the distribution traffic.

Here are some terminologies we use throughout this paper:

- The CDN consists of many devices deployed throughout the network, that we call the "Content Engines" or "CEs". Collectively, they perform the function of prepositioning content from "Origin Servers", often located at the corporate data center or scattered on the Internet, to the edge of the networks where the end-users reside, e.g., the retail stores or field offices.
- To manage the CDN devices and activities, the CDN administrator uses a central management station we call the "Content Distribution Manager" or "CDM". CDM communicates with every CE in the CDN, sending them configuration updates and collecting status information.
- To specify what content goes to which CEs, CDN administrator creates a "Channel" on the CDM. The channel defines a set of subscriber CEs as well as a head among them that we call the "Root CE" for the channel. Content enters into the CDN via the Root CE, which contacts the origin server to download the content.
- In channel routing, a CE that stores and forwards content for other CEs is called a "Forwarder CE". A CE that receives content from a forwarder CE is called a "Receiver CE". Note that a forwarder CE is often a receiver CE itself as well because the forwarder CE in turn needs to receive the content from its own forwarder to begin with.

This paper, then, is mostly concerned about how the content flows from the Root CE to all the other subscriber CEs in the most efficient manner, given the availability of unicast and multicast network connectivity among the subscriber CEs. Deciding the distribution flow for each channel is a process we call "Channel Routing".

In the rest of the paper, Section 2 describes a multi-tier overlay topology as a means for the CDN operator to specify the network topology in terms of unicast connectivity and adjacency. Section 3 presents the concept of "multicast cloud" as a means to specify the IP multicast topology. Section 4 details the channel routing process that synergistically combines IP unicast and IP multicast into coherent application-layer multicast. Section 5 describes the added benefit of our design in terms of superior application-layer traffic control. Section 6 is the related work and Section 7 concludes the paper.

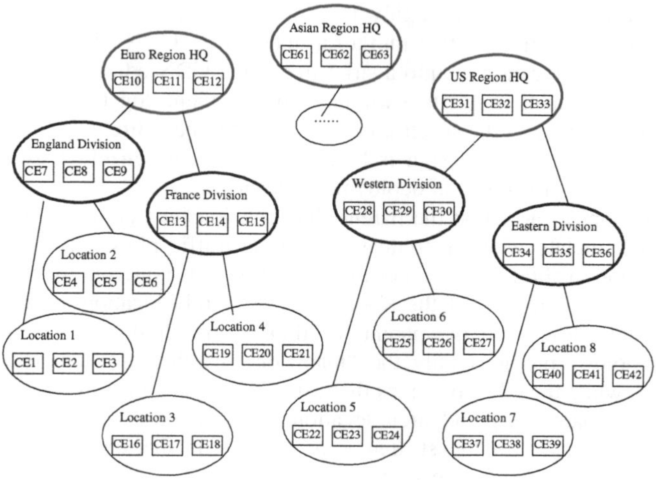

Figure 1: Sample overlay topology

2. Overlay Topology

The purpose of an "overlay topology" is to guide the construction of the application-layer multicast tree for content replication. The CDN administrator configures it on the CDM for the entire CDN, based on CEs' physical locations in the IP network. Then, for any particular channel, the CDN automatically forms a "channel distribution tree" among all CEs assigned to the channel, following the overlay topology as much as possible, --- a process called "channel routing".

In this architecture, the overlay topology has multiple tiers, with "tier-1" being the top tier (and normally closest to the IP backbone), and then "tier-2", "tier-3" and so on, toward the edge IP networks. On each tier, there are multiple locations, corresponding to data centers or POPs or topological vicinities in the IP network. Each location may have multiple CEs in it. Every location has a "parent" location on the higher tier, except tier-1 locations.

The CDN administrator configures the overlay topology at the CDM by first grouping CEs into locations based on their geographic adjacency and then specifying child-parent relationships between locations. A location can have a single parent. Locations with no parents are placed at tier-1. Locations whose parents are at tier 1 are placed at tier 2, and so on. Hence, this topology can be thought of as a forest of trees rooted at tier 1. Note that such an overlay topology will never have loops. See the sample graph below on a 3-tier overlay topology.

We chose to model the overlay network as a forest instead of a mesh for several reasons. First, a tiered overlay design matches well with the reality because it is our experience that most corporations and ISPs have a multi-tier IP network instead of a full mesh network. Second, the tiered design reduces the overall system complexity in terms of configuration and channel routing because the topology does not have loops. Third, this design is highly flexible in that it can degenerate into either extremes of the spectrum: either into a full mesh if the CDN administrator pools all CEs into a single location or into a strict tree if the CDN administrator makes each CE a separate location of its own and strings them together via parent-child relationships. Where in the spectrum the CDN is will be entirely up to the CDN user to configure. Hence this design is highly customizable, enabling us to build one architecture for a wide range of deployment scenarios.

It's important to note that the parent-child relationships between locations guide but does not dictate whether content replication traffic physically flows from the parent location to child location. For instance, if there is no CE in the parent location assigned to the channel or if the Root CE is in a child location to begin with, the content won't flow from the parent location to the child location. In short, the overlay topology provides only the "map" or "search path" for finding channel forwarders, while the CDN provides the rest of the intelligence. If there's no eligible forwarder in a parent or grandparent location, that location is completely bypassed in the application-layer multicast tree constructed for the channel. See also the "integrated channel routing" section.

3. Multicast Cloud

Separate from the overlay topology configuration, which governs how unicast content distribution flows, the CDN administrator also creates a "multicast cloud" on the CDM to govern how the multicast distribution flows and interacts with unicast distribution.

The multicast cloud specifies which CEs will be the multicast senders, which will be multicast receivers within the cloud, as well as properties of the multicast such as the multicast IP address for session advertisements, the range of multicast IP addresses for data transmissions, the IP TTL, whether the multicast medium is satellite or terrestrial, whether and how much FEC (forward error correction) to use, bandwidth and QoS settings, etc. We impose that a CE cannot be a sender in multiple clouds or a receiver in multiple clouds.

Then, the CDN administrator can assign multicast clouds to channels. One channel can have multiple clouds. Likewise, one cloud can be used in multiple channels. A channel can have both multicast receivers and unicast receivers. A receiver CE considers the channel a "multicast channel" if and only if the CE belongs to a multicast cloud as a receiver and the cloud is assigned to the channel. It's possible for such a channel to have some receiver CEs not belonging to any multicast cloud and receive the content via unicast.

On the sending side, a CE sends out content of a channel via multicast if the CE belongs to a multicast cloud as a sender and the multicast cloud is also assigned to the channel. On the receiving side, a CE tunes in to the multicast session advertisement address if the CE belongs to a multicast cloud. If the CE hears any session advertisement for content of a multicast channel it is subscribed to, the CE will tune into the channel multicast IP address to receive the content.

While the concept of multicast cloud and multicast channel is straightforward, the key challenge is in guaranteeing the scalability, reliability and eventual delivery of content. For example, if the receiver is down at the time of the multicast session, the multicast connectivity is broken, or the multicast sender crashed, the receiver CEs still must replicate the content timely. Furthermore, when a multicast cloud consists of thousands of CEs, the multicast loss repair becomes a scalability bottleneck, which often results in poor reliability and calls for a parallel unicast hierarchy for more efficient NAK aggregation and loss repair [3]. Addressing these challenges leads us to integrated channel routing for solutions.

4. Integrated Channel Routing

Next, we detail the flow of content distribution and the construction of distribution trees.

Store and Forward via a combination of unicast and multicast.[2] For CDN scalability and reliability, we use "store and forward" to distribute content from the Root CE to all the receiver CEs in the channel, meaning a receiver CE does not just go directly to the Root CE for content. Rather, it finds out who is its "forwarder CE" and goes to the forwarder either for unicast content replication in the case of unicast channels, or for multicast failover and repair service in the case of multicast channels. In turn, the forwarder CE downloads content from its own forwarder (which may ultimately be the Root CE), store the content on disk, and forward the content on to edge CEs that request content from it. If the forwarder CE is a multicast sender per the CDN configuration, it also proactively pushes out the content via multicast. In this case, any receiver in the multicast cloud will get the multicast content, regardless the sender is the receiver's direct forwarder or not.

With this design, content flows through a channel-specific distribution tree via a combination of unicast and multicast, enabling the CDN to reach across unicast IP networks to bridge isolated multicast IP islands. For example, in the sample graph in Section 2, all the country divisions may belong to a global satellite multicast network and receive content via multicast. Then the division receiver CEs turn around and act as unicast forwarders for the children locations within each country, or as multicast forwarders / senders if terrestrial multicast is available in that country. Similarly, a location may receive content via unicast from its forwarder and then internally replicate via multicast wherever IP multicast routing is turned on.

A Distributed and Dynamic Channel Routing Process. While the channel distribution tree is global, the channel routing process is actually distributed, as a *per-CE* function that answers the local question "who is my forwarder for this particular channel". On each receiver CE, the channel routing algorithm is run periodically to dynamically pick a forwarder for each channel.

The channel routing process considers both the overlay topology configuration and the dynamic availability of eligible forwarders. However, it incurs no "routing

---

[2] Our system does carry live video traffic over the same overlay network for delivery to the end-users, which does not require store-and-forward. Such live traffic is channel-routed in a similar fashion as described in this paper, but with additional care for live stream performance and reliability, that we won't detail in this paper.

probes", unlike SODA (self-organizing distribution architecture) [4], which is a major improvement (see also the "related work" section). Instead, it uses the byproduct from the content distribution process as the feedback to the channel forwarder selection. Every time a forwarder is chosen, the CE contacts the forwarder for content downloads. If the CE cannot reach the forwarder or experience poor service (e.g., frequent disconnections, indicative of a busy forwarder), the channel routing module remembers this fact about that particular forwarder and tries to pick another (better) forwarder next time around.

Location Leader. For any location where more than one CE have been subscribed to the channel, one and only one CE in the location will act as the "location leader" for the channel, i.e., to replicate content from outside of the location (via either unicast or multicast), while other CEs in the location will replicate content only from the location leader in the case of unicast. This ensures that only one copy of the content will cross the link connecting any two locations. In the case of multicast, i.e., some CEs in the location are part of a multicast cloud, these CEs will still receive multicast transfers even if the multicast sender is outside of the location. For them, such multicast replication will preempt the need of unicast replication from the location leader. However, if the multicast fails or has losses, the CEs recover using unicast within the location so there is never any extraneous traffic on the inter-location links.

For failover purposes, the other CEs in the location act as "backup location leader" in case the location leader is down. The channel routing algorithm deterministically (through consistent hashing), though distributedly, picks who is the location leader and generates an order list of the other CEs that are 1st backup, 2nd backup, and so on. In the Root location (i.e., the location where the Root CE belongs), the location leader is always the Root CE that the CDN administrator designated, while the "backup location leaders" are actually "backup Root CEs", which are allowed to go directly to the origin server to download content if the Root CE is down.

Note that a "location leader" is always a per-channel and per-location concept, a "forwarder" is always a per-channel and per-CE concept.

Forwarder Selection. A receiver CE finds its forwarder by examining the series of locations on the overlay topology "toward" the Root location, following the parent-child relationship.

First, find a forwarder within the CE's own location. The location leader should be the forward. If the location leader is down or too busy, use the backup location leader as the forwarder. If found, exit the algorithm.

If none found or the CE thinks it is the location leader itself, look for a forwarder in the next location "toward" the Root Location[3]. If still none found (e.g., because the CDN administrator assigned no CE of that location to the channel or because all the potential ones are unreachable or too busy), then look further in the yet next location "toward" the Root location, and so on. The recursion ends if a suitable forwarder is found or the algorithm reaches the Root CE's location.

Multicast Forwarder: for a multicast channel, the channel routing module will run

[3] The next location "toward" the Root Location may be below, instead of above, this location if the Root Location is below this location on the overlay topology graph.

the above search algorithm first trying to find a "multicast forwarder". Only when it failed to find any suitable multicast forwarder, will it run the algorithm again, this time, looking for "unicast forwarders". Note that any multicast forwarder is capable of unicast as well.

A multicast forwarder for a CE must satisfy all of the following rules, while a unicast forwarder satisfies only the first two rules.

1. The forwarder is on the topological location path toward the Root CE.
2. The forwarder belongs to the same channel.
3. The forwarder belongs to the same multicast cloud as a multicast receiver or sender.
4. The multicast cloud they belong to is assigned to the channel.

If the search reached the Root location still without success, as a last resort, the receiver CE may decide to contact the origin server directly for content, after proper retries and timeouts during the search.

The multicast forwarders form a parallel CE hierarchy next to the multicast cloud, providing scalability, reliability, and failover to the multicast data transfers. E.g., in the case of multicast loss, instead of all going to the multicast sender for repairs, the receiver CEs go to their respective multicast forwarder for any multicast loss repairs. In case the multicast sender is down or multicast connectivity is broken, the receiver CEs failover to unicast replication from the forwarder instead of from the multicast sender or Root CE.

5. Application-layer Traffic Control

Along with the overlay topology, multicast cloud, and channel configuration, the CDN administrator can set parameters that control many aspects of the distribution traffic, beyond the topologic directions of the traffic flow (which we have detailed in Sections 2 through 4). The many aspects of traffic control include distribution priority, bandwidth control, quality of service, and security.

Much of such traffic control is typically enforced on layer 3 and layer 4, and traditionally configured on layer 3 and 4 as well. That has been problematic in CDN deployment because it is too complex for the (layer 7) CDN administrator to understand and operate. Conversely, in our system, the configuration entities in the CDM become a natural vehicle for the CDN administrator to specify parameters for the overall CDN traffic. These parameters have direct semantics associated with the CDN applications and are configured along with the CDN applications, hence are easy to understand and use for the CDN administrators.

For example, the CDN administrator can configure four bandwidth limits on each

CE to control the content distribution traffic.[4]

- o Incoming content acquisition traffic from origin servers.
- o Incoming unicast content distribution traffic from forwarder CEs
- o Outgoing unicast content distribution traffic to receiver CEs.
- o Multicast content distribution traffic within a multicast cloud. Due to the nature of multicast, there is a common value for all CEs within the cloud. It's both the outgoing rate from the multicast sender and the incoming rate to the multicast receivers.

A CE may have all these limits defined because all four kinds of traffic can flow through the CE simultaneously. For each bandwidth limit, we also support "time of day", where the CDN administrator can configure different limits for different time segments that form a weeklong cycle. For example, the admin can say "incoming unicast traffic should not exceed 100kbps from 8am to 8pm Monday through Friday, but it can run as high as 10Mbps from 8pm throughout the night to 8am, Monday through Friday plus whole days Saturday and Sunday."

Other parameters such as multicast security and key management are also controllable from the CDN application layer. We will detail their designs in separate publications.

6. Related Work

SODA. Compared to our last-generation architecture: SODA (self organizing distribution architecture) [4], this design is a major improvement in that (1) the SODA routing probes generate extra traffic overload (sometimes quite heavy) on the network, (2) the routing results are often inaccurate or undesirable because the SODA heuristics may or may not reflect the true needs of every specific CDN deployment, and (3) the CDN administrator has no direct way of influencing the routing results.

In comparison, this design does not incur routing probes but rather monitor the regular distribution traffic to collect route information as a byproduct. This design also gives the CDN operator a great deal of flexibility in a large spectrum from the most automation to the most user control, simply based on the size and number of "locations" the CDN administrator defines. We find that most customers define a 4-tier overlay hierarchy with as many leaf locations as they have branch offices. Corporations that are more global may define up to 6 tiers.

FastForward. FastForward [5] is another commercial CDN design. It is primarily an application-layer broadcast network based on "publish and subscribe". In FastForward, the viewers tune in to receive the broadcast and the broadcast is delivered live to the network edge primarily via hierarchical unicast and marginally

[4] Other bandwidth limits, e.g., for media streaming and HTTP/FTP browsing, also exist in the system.

via multicast relay much like Mbone [6], via layer-3 and layer-4 tunneling.

Compared to our design, FastForward does not maintain any explicit notion of user-configured CDN multicast group for the purpose of content routing, central management, multicast security, and network monitoring. Secondly, FastForward concerns primarily the live media delivery to the end-user, while our system is a stored-and-forward network for preposition content ahead of time to facilitate video-on-demand (VOD) from the edge. The delivery to the end-user is separate and can be either live or VOD. In the case of live, our system can route the live stream through the overlay system the same way as described in Section 4, hence leveraging both IP multicast and IP unicast still. Lastly, FastForward pieces together IP multicast islands via layer-3 and layer-4 relays, while our system explicitly manages IP multicast from application layer instead of the network layer.

Overlay network. Overlay network research has been focused on providing a one-to-many content delivery service via the construction of an overlay network where each hop in the overlay is a unicast TCP link between two nodes, and leads to the network edge. See publications [7] [8].

Our architecture differs from the typical overlay networking research in that we are trying to make native IP multicast work in the overlay network while the typical overlay network research uses peer-to-peer or hierarchical unicast to provide a one-to-many delivery service, which is often also referred to as multicast or application-layer multicast.

In essence, our architecture improves upon existing overlay network to take advantage of both IP unicast and IP multicast in providing the one-to-many delivery service. The architecture also solves the integrated content routing problem between unicast and multicast, as well as application-layer traffic control.

7. Conclusion

In this paper we described an overlay network design that provides integrated content distribution routing between IP multicast and unicast via user-configured "multicast clouds". This paper embodies the principles in application-layer multicast as detailed in [2] and the major improvements based on our experience with SODA [4].

CDN designers are often faced with the decisions between fully automated and fully user controlled content routing, between IP unicast and IP multicast. While previous work takes advantage of one or the other approach and suffers the drawbacks of either approach, our design takes the best of both worlds and let them mitigate each other's drawbacks. Wherever available, IP multicast helps the CDN utilize physical broadcast media, build efficient distribution, and improve overall CDN scalability, while IP unicast bridges together IP multicast islands and provides failover for multicast.

In this design, the CDN administrator configures the overlay topology and multicast clouds, providing as much or as little topology guidance as the administrator desires. With the guidance from the user configuration, the channel routing algorithm dynamically and distributedly computes the best distribution traffic path through the network, taking into account the changing network conditions.

Hence, our system gives the CDN operator a lot of control over the form of the distribution tree, and yet still automates most of the route selection and failover process. Especially, the design has the flexibility to function in either extremes of the spectrum as well as in between the extremes, from fully user configured to fully automated, depending on the number and the size of the overlay locations the CDN operator specifies.

Furthermore, the CDN administrator not only influence the topologic directions of the traffic flow but also control many other aspects of the distribution traffic, including distribution priority, bandwidth limits, quality of service and multicast security. Our overlay design makes it possible to abstract the layer 3 and layer 4 traffic control parameters and translate into meaningful application-layer controls, making it much easier to use.

Our CDN builds a channel-specific distribution tree via a combination of IP unicast and IP multicast so as to reach across unicast IP networks to bridge isolated multicast IP islands. This is not a mere addition of unicast and multicast. More importantly, such integrated routing provided synergy between unicast and multicast content distribution. The multicast forwarders form a parallel unicast hierarchy next to the multicast cloud, providing scalability, reliability, and failover to the multicast data transfers, in the event of multicast loss repair and multicast failure.

This is a proven architecture, with efficiency, intelligence, and ease of use. We have successfully implemented this system and deployed in networks ranging from a few hundred to a couple thousand CEs. We are continuing to improve the system for even larger deployments.

8. References

[1] P. Francis, Yoid: Extending the Internet multicast architecture, http://www.aciri.org/yoid/docs/index.html

[2] D. Li and J. Jannotti, Application-layer multicast and enhancement with IP multicast, http://www-cs.stanford.edu/~dli/app-mcast-paper.pdf

[3] D. Li and D. R. Cheriton, OTERS (On-Tree Efficient Recovery using Subcasting): A reliable multicast protocol, in *Proc. 6th IEEE Int'l Conference on Network Protocols (ICNP'98)*, Oct. 1998.

[4] J. Jannotti, D. K. Gifford, K. L. Johnson, M. Frans Kaashoek, J. O'Toole Jr., Overcast: Reliable multicasting with an overlay network. *Proc. OSSDI*, 2000.

[5] FastForward, http://www.cs.ucsb.edu/ngc2000/ program/invited-francis.ppt

[6] Mbone, http://www-itg.lbl.gov/mbone/

[7] Overlay network designs: http://citeseer.nj.nec.com/jannotti00overcast.html, http://www.cs.berkeley.edu/~boonloo/classes/cs268/cs268.PDF, http://www.eurecom.fr/~btroup/RESEARCH/TOPICS/w_serkan.htm, http://nms.lcs.mit.edu/ron/, http://www.arl.wustl.edu/~sherlia/amcast.html, http://www.cs.virginia.edu/~hw6h/cs793/readlist.htm

[8] S. Shi and J. Turner, Routing in overlay multicast networks, *Proc. of IEEE INFOCOM,* 2002.

A LARGE ENTERPRISE CONTENT DISTRIBUTION NETWORK: DESIGN, IMPLEMENTATION AND OPERATION

Jacobus E. van der Merwe, Paul Gausman, Chuck D. Cranor, and Rustam Akhmarov
AT&T Labs – Research

Abstract Content distribution networks (CDNs) are becoming an important resource in enterprise networks. They are being used in applications ranging from broadcasted townhall-style meetings to the distribution of training and educational material. In this paper, we present the design, implementation and operation of a large enterprise CDN and describe some of the lessons we learned from our experiences building such a CDN.

1. Introduction

The use of streaming media is gaining popularity in many environments where it can be used for both live Webcasts and on-demand access to streaming content. This is especially true in corporate environments where streaming is used for employee townhall meetings and training courses. One of the main reasons for this gain in popularity is advances in encoding technology that enable reasonably good quality video to be obtained at fairly modest encoding rates. While its popularity is increasing, large scale distribution of streaming content remains a challenging problem in networks that includes a wide-area-network (WAN) component where link speeds can be a limiting factor.

In this paper we describe the design, implementation, and operation of a large scale corporate enterprise content distribution network (CDN). The CDN provides streaming services to more than 400 distributed sites connected by a private IP backbone in a large corporate environment with a total user population in excess of 50000 users. The CDN enables Webcasts throughout this environment, scaling from departmental meetings with tens of users to all-employee broadcasts involving potentially all users.

Creating a streaming CDN in this environment was technically challenging for the following reasons:

- The scale of the target environment is quite large, not only in terms of the number of sites to be covered, but also in terms of the potential number of simultaneous users that a Webcast such as a live all-employee broadcast can attract.

F. Douglis and B.D. Davison (eds.), Web Content Caching and Distribution, 119-128.

- Target sites are connected to the corporate backbone through WAN links that have widely varying link speeds. Additionally, WAN links are shared with other enterprise applications, so the streaming service bandwidth usage cannot be allowed to fully monopolize a link's bandwidth.

- Different parts of the target network have differing capabilities in terms of the transport options available for content distribution. The IP backbone is multicast enabled as are all of the sites connected to the backbone. However, approximately one third of the sites, while multicast capable internally, have no multicast connectivity to the backbone. Also, in our environment users who connect to the corporate intranet via VPN only have unicast transport available.

A key design goal in addressing these issues is to create a solution that minimizes and simplifies human operational involvement. The system should be fully automated with zero human intervention necessary during normal operation. Tools necessary to aid in fault management when problems occur should be available in a self-serve form on the Web so that capable users can quickly diagnose their own problems without having to wait for someone from a "help desk" to help them. We describe our approach and the system that was deployed in detail in this paper.

It is important to note that the challenges in running a successful streaming service in an enterprise environment are not all purely technical. As with any system deployed throughout a large organization, many players are involved. For a streaming service such as the one described here, that includes end-users, the network provider, a support organization (help desk), content owners and the content service provider. These roles will typically be present in any enterprise deploying streaming services. Some of the roles could be combined and performed by the same organization, or some of the roles might be performed by third parties in an outsourcing scenario. In this paper we address the problem mainly from the point of view of the CDN which might be operated by the network provider, the content service provider or a third party. However, to provide a holistic view we also describe the Content Service Provider requirements in some amount of detail and touch on interactions with other role players.

The outline of the paper is as follows. Section 2 describes the higher level services that the CDN accommodates and the underlying network infrastructure in which the CDN was realized. The design of the CDN and in particular the redirection system is described in Section 3. In Section 4 we describe operational issues and relate some of the lessons learned in "operationalizing" an enterprise CDN. We cover related work in Section 5 and end the paper with a conclusion and future work section.

2. Service Perspective and Network Infrastructure

The work presented in this paper describes the realization of a set of services that attempts to capture a common set of streaming requirements:

Presentation service: provides tools and functions for streaming presentations (by reservation), from pre-existing encoder rooms. Users can optionally select to use various rich-media components within the presentation browser window, like synchronized slides, Q&A applications and surveys. Presentations are typically live and can be archived for later On-Demand or scheduled Channel viewing (see below).

Channel service: provides a content channel for related programming. (Like a cable TV station, only over IP). For example, the public relations organization within a company might "own" a channel to distribute all of their content. Content is live or replays of pre-recorded content.

On-demand service: provides all functions needed for the publishing, management, storage and distribution of pre-recorded presentations that can be accessed in an on-demand fashion.

Venue service: provides high bandwidth streaming distribution to sites with large displays for group audiences. Typically used as a substitute for expensive satellite distribution mechanisms where practical.

To provide these streaming services, two functional components are required:

The publishing platform: provides content owner and end-user user interfaces and serves as a central publishing repository. The publishing platform also provides presentation tools, billing, and content management functions. The publishing platform thus serves as the main source of content for the CDN. It further interacts with the CDN by selecting distribution regions, setting content aging parameters, and turning encoder feeds on and off, automatically, as per content owner and service administrator configurations. As the focus of our work is on the distribution platform, the publishing platform will not be considered further in this paper.

The distribution platform (CDN): distributes Web and streaming content and is described in detail in the next section.

The corporate network in which the CDN is deployed consist of a private IP backbone distributed across the US. The backbone interconnects more than 400 distributed edge sites housing from tens to thousands of employees per site. Depending on its size and function, edge sites connect to the backbone network at line speeds ranging from 1.544 Mbps or less (T1 or fractional T1) to 45 Mbps or higher (full T3 or OC3). Within each site the LAN is typically a switched fast-Ethernet which also connects some legacy 10Mbps segments. Several data centers connect to the backbone network. The backbone runs a single multicast domain using Protocol Independent Multicast - Sparse Mode (PIM-SM) [4] routing which extends into most of the edge sites. Approximately 20% of the edge sites form Distance Vector Multicast Routing Protocol (DVMRP) [3] multicast islands isolated from the backbone multicast domain. Virtual-office users access the corporate network through a number of virtual private network (VPN) gateways that are also located in different data centers.

3. CDN Architecture

Figure 1 shows the architectural components of our CDN network. The CDN receives its content from encoders that are located in meeting rooms distributed throughout the organization. Media streams from each encoder are fed to a set of one or more

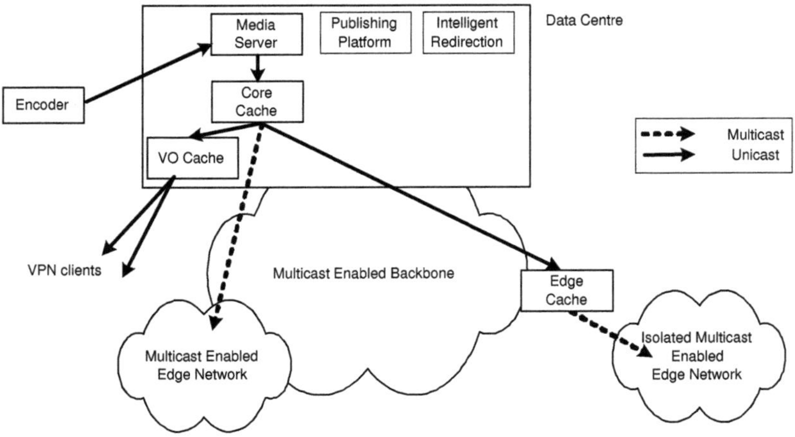

Figure 1. CDN architecture

media servers located in central data centers. As shown in Figure 1, the data center also houses the publishing platform and the redirection system. The media servers serve as the content origin for both live and on-demand content. As shown in Figure 1 the media server is front-ended with a core streaming cache. These caches are typically special purpose streaming appliances whose sole role is to stream data. In addition to being able to handle many more streams than the media server, the core cache is also the root of the distribution tree.

Once the media reaches the core cache, it can be streamed out in three ways depending on the location of the end-user. First, if the end-user is at an edge network that is part of the backbone multicast domain, then the core cache can use multicast through the backbone to stream it directly to the client. Second, if the end-user is located in an isolated multicast enabled edge network, then the core cache can unicast the media to an edge cache placed on the end-user's edge network. The client can then receive the media via multicast from the edge cache. Finally, for end-users residing in VPN-based virtual offices we have a special set of virtual office (VO) caches. The core cache streams the media to the VO caches via unicast, and then the VO caches stream the media onward to the VPN clients also using unicast

For our *Presentation Service* we are currently offering a fixed number of content *channels*. Based on a reservation mechanism in the publishing platform, at different times different encoders provide the actual content for each channel. Thus, content providers essentially "rent" time for their presentation on a fixed channel. To accommodate the varying amounts of bandwidth that is available to different sites, each channel is provided at multiple encoding rates. Our intelligent redirection system (described in detail below) is used to connect viewers to the appropriately encoded stream based on various parameters such as IP subnet and the bandwidth allocated to the site.

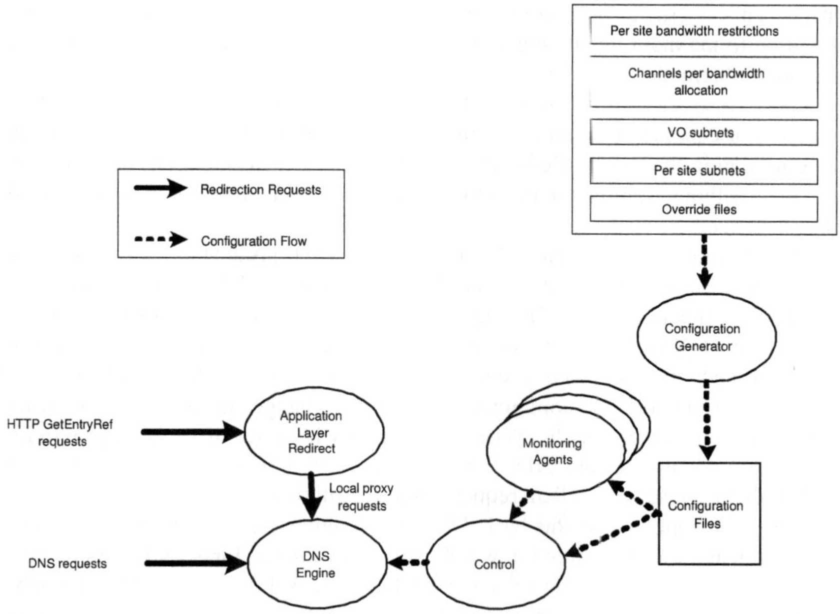

Figure 2. Redirection System

3.1 Redirection System

In order for our CDN architecture to function properly, it is essential that end-users are directed to the appropriate cache for their location, use the correct type of URL (i.e. a unicast URL or a multicast URL) and request the appropriate stream quality based on the bandwidth allocation for the site. Note that in the case of the isolated multicast islands, this is critical for correct operation as end users in those domains will not be able to receive multicast from the backbone multicast domain. We have developed a "content-aware," "location-aware" application-layer redirection system to seamlessly glue the system together while hiding all the technical details from users of the system. The architecture of our redirection system is shown in Figure 2.

To properly redirect clients, the redirection system needs to understand the topology of the intranet that it is operating in. The redirection system uses as input a set of configuration files obtained from the different role players in the company. These files contain a list of bandwidths allocated for each site, a specification of what streams a site with a specific bandwidth allocation should get, and the subnet information for each site as well as the VO subnet ranges. As far as the CDN is concerned this process is fully automated as these files are posted on a Web site from where it is pulled on a daily basis (or on-demand as needed). The bandwidth specification file is generated based on historic utilization trends in the WAN links. The process to create the per site subnets is fully automated and harvests subnet information directly from operational routers. This is crucial in a large network as subnets change on a daily basis and attempting to maintain this by means of a manual process would be error prone and therefore problematic.

Figure 2 also shows the "override files." These are a set of files that a system administrator can use to override settings in any of the regular feeds. For example, if an omission in the subnet file is detected, it can immediately be corrected through these files without waiting for the actual subnet discovery process to be corrected and/or debugged.

The configuration files are parsed and manipulated to translate the information they contain into a set of Intelligent Domain Name Services (IDNS) configuration files. IDNS [2] is a special purpose, intelligent DNS server. IDNS consists of three main components: a DNS engine that processes queries based on the IP address of the client and current network conditions, a control process that dynamically reconfigures the DNS engine as network and configuration conditions change, and a set of monitoring agents that track server health. When configured with a set of subnets, proximity information, and current status, IDNS will resolve DNS requests into an IP address or CNAME that is close to the client requesting the resolution.

For our CDN application this basic DNS-based redirection have been extended to enable application level redirection which works as follows. First, end-users request streaming content, typically by selecting a URL on a Web page hosted by the publishing platform. As is often the case with streaming, this Web object is a streaming meta-file (e.g. an ASX file) which contains the streaming URL. In our system the meta-file does not contain a static streaming URL, but rather contains an HTTP URL that points to one of our redirection server machines. The redirection server is then contacted to obtain the actual streaming URL (the HTTP GetEntryRef request in Figure 2).

Each redirection server runs a special purpose Web server that receives the Web request for the URL embedded in the meta-file. The Web server uses the IP address of the end-user and the channel being requested and encapsulates it into a DNS request to the locally running IDNS server. The IDNS DNS engine receives the request, extracts out the client's IP address and uses that to generate an answer to the query. In our case the answer is an encoded CNAME record. The CNAME record is sent back to the Web server, which decodes the record and uses it to dynamically produce a streaming meta-file (or a portion of a streaming meta-file) that contains the actual streaming URL. Note that as indicated in Figure 2, the redirection system also serves as a "normal" DNS server at the same time.

The CDN was designed to have no single point of failure. The CDN data center equipment is duplicated in two different data centers. Monitoring agents that are part of the redirection system (see Figure 2) monitor the health of the equipment in each data center, and feed this into the redirection system in real time. The media server/core cache combination in the active data center carries all of the load. Should either of these devices go down or connectivity to the data center as a whole be lost, the redirection system will move all traffic to the media server/core cache in the other data center. The redirection system itself is fully redundant with two separate servers in each data center. Redundancy between VO-caches is handled in similar fashion by the redirection system.

4. CDN Operation and Lessons Learned

The long term success of a service such as the one described here largely depends on its operational costs. As stated in the introduction, one of our main goals when designing the service was to reduce human involvement during normal operation of the system. The fully automated redirection system presented in the previous section is aligned with this goal. In this section we consider operational aspects of the system where human involvement can not totally be eliminated. None the less, we attempt to provide tools and systems to also reduce or simplify human involvement with these aspects.

Day-to-day operations involving the CDN fall in two categories:

Pro-active management and monitoring: involves monitoring all equipment via a network management tools, generating alarms when certain conditions occur on the CDN devices, and monitoring the redirection system for error conditions.

Reactive management: involves determining where the problem is when an end-user is unable to receive a Webcast. These type of problems (described in detail below) are somewhat unique in a streaming scenario.

In our environment, when an end-user selects a channel from the publishing platform and fails to receive the stream the cause can be any of a number of reasons. The user might be experiencing a **desktop** problem such as miss-configured software. Or there could be a **network** problem which could either be local to the user's location or somewhere else in the network (e.g. between the user's location and the backbone). The problem could be due to a **redirection** error. For example incorrect subnet information might have been entered into the configuration feeds resulting in users being redirected incorrectly. Finally there could be a problem with one of the **CDN** devices or the with the media **source** (the encoder).

Given this array of potential problems and our desire for a fully automated environment, we have developed a set of Web based "debugging" tools that allows users and support personal to quickly rule out and/or identify problems. In particular the tool for end-users tell them what their IP address is and whether the redirection system had any problems dealing with that address. The tool then tries to stream a small clip from the CDN infrastructure, which, if it fails, will point to some sort of a network or CDN problem, and, if successful, points to a probable multicast problem. Similarly, the tool for the support personnel allows them to see where the redirection system will redirect based on the user's IP address. Through the tool they can also verify that all CDN equipment involved in this distribution is functioning correctly and that network connectivity exists between the CDN equipment and the end-user.

Since the streaming service has been operational for only a few months, the webcasts hosted so far have been fairly modest. Audiences ranged from 50 to 300 persons for the Presentation Service and 100 to over 8000 persons for the Channel Service. Both services have distributed to audiences located in from three to over a hundred sites at one time, while simultaneously distributing to Virtual Office viewers.

In deploying the CDN infrastructure described in this paper we learned some lessons and indeed confirmed some of our suspicions. For the most part realizing the service happened without any major issues. That said, there were the normal difficulties involved with deploying a large distributed infrastructure. Configuration of all equipment was performed at a single location. This process itself was highly automated which significantly reduced the possibility of miss-configuration. The logistics of having to ship a large number of boxes first to the place where they are configured and then to where they are installed is not very efficient or cost effective though. If the ability existed to ship an un-configured device and separate media with per device configuration (e.g. a CD-rom or flash memory), thereby allowing simple remote configuration, that would have greatly simplified this part of the process. By necessity the physical installation of equipment at remote sites were performed by technicians who did not necessarily have training and/or experience with the specific equipment used. Having mostly appliances that could simply be connected and switched on allowed this to happen.

Once deployed the streaming caches are "centrally" configured by having the caches periodically poll a configuration server to check for new configuration files. Since most of the caches share the same basic configuration with only a small delta that is cache-specific, this approach proved very effective in updating the configuration of the complete CDN infrastructure in a timely fashion. (We have out-of-band access to all equipment to enable us to recover from any miss-configurations or to remotely administer network connectivity changes.)

The use of multicast enabled our streaming CDN service to be highly network efficient and is therefore highly desirable. However, most of the initial technical issues we encountered were due to the use of multicast. This included initial problems with multicast in the CDN vendor products. This is probably because those functions have not been used as extensively as their unicast counterparts. This was also true for the multicasting networking equipment (i.e. routers and switches). Multicast deployments of this scale are not that common in enterprise networks, with the resulting lack of real world operating experience for those products.

We mentioned in the introduction that many players are involved in realizing an enterprise streaming service. This means that when things go wrong it is not always clear who is responsible for fixing it. The debugging tools described earlier in this section not only help to identify the problem but also help to identify which group has responsibility for it, thus making the fault management process much more efficient.

Finally, in testing and operating the system we soon discovered that for technical or business reasons, some subnets and/or locations were not able to receive the streaming content. Not wanting to re-discover these problem areas every time and in keeping with our desire for automation, we have dealt with this problem by having a set of "known-problem-subnets." The IP address of an end-user experiencing problems is automatically checked against these subnets when she accesses the debugging tools, informing her whether the service is expected to work for her subnet.

5. Related work

While the use of CDN technology in enterprise environments are fairly common these days, we are not aware of published work that report on the operational aspects of an enterprise CDN. Our own work on the optimal placement of streaming caches that form an overlay distribution infrastructure in a VPN environment is clearly related [6]. However, that work was intentionally limited to a unicast-only environment and did not report on an actual CDN realization.

In terms of redirection, some related work is summarized in [1]. While it mentions streaming redirection, the focus is mainly on redirection for Web content. The fact that streaming normally involves the downloading of a meta-file, offers a natural redirection opportunity. While this is widely recognized, we are not aware of published related work that follows the same approach as our redirection system. Other redirection related work involves the products of streaming vendors (e.g. Network Appliance and Cisco). The lack of flexibility of these products, for example, in terms of dealing with large numbers of subnets and taking different constraints into account, prompted us to develop the solution presented in this paper. There are also more forward looking proposals for scalable redirection that would form an integral part of the Internet suite of protocols [5].

Finally, the fact that our system takes into account the bandwidth available at a specific site has some related work in [7]. They developed routing algorithms for overlay multicast networks that optimized the bandwidth usage between the nodes forming the overlay network. Our flat distribution "tree" directly from the core caches to the edge caches was realized mainly for pragmatic reasons. However, it appears to be sensible in an environment (such as ours) where the bandwidth constraints are mainly in the WAN links connecting the edge sites to the backbone network, rather than in the backbone itself.

6. Conclusion and Future work

In this paper we presented the design and operation of a streaming CDN in a large enterprise environment. This CDN is novel in its use of multiple transports and multiple explicit bitrates to accommodate varying bandwidth conditions in a large enterprise network. We showed the critical importance of an intelligent and flexible redirection system in making this work. We stressed the importance of automated tools in operations and debugging and described some of the tools we use.

From a service perspective in the future we plan to extend the service suite to include secure distribution of content into and out of the enterprise environment. The presentation service will be extended to allow support for ad-hoc presenter owned encoders, e.g. laptops or PCs anywhere in the enterprise network. We plan an **event service** to enable presentations with multiple origination sites and multiple audience sites.

For the content distribution platform we are currently investigating the use of a bandwidth probing tool to augment the per location bandwidth assignments. This is particularly relevant for virtual office users who might be accessing the network via a variety of access technologies (e.g. dial-up or cable modem). Running a bandwidth probe to determine the actual available bandwidth and taking that into account when doing redirection will ensure that each user receive the highest quality stream possible.

An issue related to our use of multicast is that, unlike with unicast streaming, there is no explicit quality feedback from the end-user players. This is due to the fact that when the media player is receiving multicast there is no control connection back to the streaming server/cache. This makes it difficult to determine the quality of the actual user experience unless users explicitly complain. We are investigating the use of tools that would provide explicit feedback about the quality of the stream even in a multicast environment.

References

[1] A. Barbir, B. Cain, R. Nair, and O. Spatscheck. Known CN request-routing mechanisms. RFC 3568, July 2003.

[2] A. Biliris, C. Cranor, F. Douglis, M. Rabinovich, S. Sibal, O. Spatscheck, and W. Sturm. CDN brokering. In *Proceedings of the Sixth International Workshop on Web Caching and Content Distribution*, June 2001.

[3] D. Waitzman and C. Partridge and S. Deering. Distance vector multicast routing protocol. RFC1075, Nov 1988.

[4] D. Estrin, D. Farinacci, A. Helmy, D. Thaler, S. Deering, M. Handley, V. Jacobson, C. gung Liu, P. Sharma, and L. Wei. Protocol independent multicast-sparse mode (PIM-SM): Protocol specification. RFC2362, June 1998.

[5] M. Gritter and D. R. Cheriton. An architecture for content routing support in the internet. In *Proceedings of the USENIX Symposium on Internet Technologies and S ystems (USITS)*, March 2001.

[6] Z. M. Mao, D. Johnson, O. Spatscheck, J. E. van der Merwe, and J. Wang. Efficient and robust streaming provisioning in VPNs. In *Proceedings of the 12th Int'l WWW Conference*, May 2003.

[7] S. Y. Shi and J. S. Turner. Multicast routing and bandwidth dimensioning in overlay networks. *Journal on Selected Areas of Communication*, 20(8), Oct 2002.

ARCHITECTURAL CHOICES FOR VIDEO-ON-DEMAND SYSTEMS

Anwar Al Hamra, Ernst W. Biersack, and Guillaume Urvoy-Keller
Institut Eurecom

Abstract Cost-effectiveness is of foremost importance for large scale VoD systems. We assume a VoD system where each video is split into two parts, the prefix and the suffix. We consider two new architectures: One architecture where the clients are equipped with set-top boxes that allow to store locally the prefix part of some/all popular videos and second architecture where the suffix is transmitted via satellite.

For each architecture, we develop a cost model to compute the delivery cost of videos. We show that these architectures are efficient and significantly reduce the system cost in many scenarios: (i) By more than 45% with set-top boxes at the client side, (ii) By more than 80% for satellite transmission of the suffix.

1. Introduction

Video-on-Demand (VoD) systems allow to support various applications such as distance learning, home entertainment, electronic commerce, to name but a few. However, the bandwidth-intensive nature of video calls for efficient and scalable architectures that serve many clients via a single multicast stream. Prior work on VoD can be classified into three categories:

- Open-loop systems [10, 2, 11]: In open-loop systems, the video is divided into many segments. Regardless the client requests, the server periodically and infinitely broadcasts segments each of which at its own rate.
- Closed-loop systems [7, 5, 4]: In closed-loop systems, clients contact the server to retrieve the video. Each request initiates a new unicast/multicast stream.
- Prefix caching assisted periodic broadcast [6, 3]: These systems combine open-loop and closed-loop approaches. This combination ensures a zero start-up delay and makes these systems suitable for both, popular and non popular videos.

In this paper, we present two new architectures to provide a scalable and efficient VoD service to a large client population. We assume that each video is split into two parts, the prefix and the suffix. In the first architecture, we provide clients with set-top boxes to store the prefix part of some/all popular videos. In the second architecture,

F. Douglis and B.D. Davison (eds.), Web Content Caching and Distribution, 129-138.

we transmit the suffix via satellite. For each architecture, we develop a cost model to compute the delivery cost of videos.

The rest of this paper is organized as follows: Section 2 presents related work. Section 3 describes the distribution network. In section 4, we derive the cost models for each of our architectures. Section 5 concludes the paper.

2. Contribution and Related Work

Providing a scalable and efficient VoD service to a wide client population has been extensively investigated in previous work. The basic idea to achieve scalability is to serve multiple clients via multicast.

Open-loop systems differ in the way they set the length and the transmission rate of each segment. Pyramid broadcasting [10] sets the same rate to all segments while the segment sizes follow a geometric series. Tailored transmission [2] sets the same length to all segments while the rate decreases as the segment number increases. You et al. [11] present an hybrid system that combines the two above methods.

While open-loop systems broadcast the video regardless of the request pattern of clients, closed-loop systems serve the video in response to clients' requests. With patching [7], the first client to arrive receives a dedicated stream from the server. A new client that arrives after the first one joins the initial unicast stream that is transformed into a multicast stream. At the same time, the new client receives a separate unicast stream for the part it missed from the initial stream. Gao et al. [5] extend this patching scheme with the inclusion of a threshold to reduce the cost of the unicast streams.

With the hierarchical merging system [4], when a new client arrives, the server initiates a unicast stream to that client. At the same time, the client listens to the closest (in time) stream (target) that is still active. When the client receives via unicast what it missed from the target stream, the initial unicast stream is terminated and the client listens only to the target stream, and the process repeats.

Guo et al. in [6], have developed a methodology to combine open-loop and closed-loop systems. They divide the video into two parts, the prefix and the suffix. The prefix is delivered via a closed-loop scheme while the suffix is multicast via an open-loop scheme.

Similarly, the PS model [3] combines both, open-loop and closed-loop systems. The PS model splits each video into a prefix and a suffix. The prefix is stored in prefix servers that can be placed at any level throughout the network other than the client side. The suffix is stored at the server, placed at the root of the network. The prefix servers send the prefix via *multicast controlled threshold* [5] while the server broadcasts the suffix via *tailored transmission* [2]. For more details on the PS model, please refer to [3].

In this paper we propose two new architectures for large scale VoD systems. We assume that each video is split into a prefix and a suffix. In the first architecture, we use the set-top box at the client side to store the prefix of some/all popular videos. In the second architecture, we transmit the suffix via satellite.

For each architecture, we derive from the PS model a cost model to compute the delivery cost of videos. In the cost models, we include not only the network transmission cost but also the *server* cost, which depends on both, the storage occupied and the number of input/output streams needed. We also account for the network transmission cost as a function of the number of clients that are simultaneously served by the multicast distribution (either from the prefix servers or the suffix server).

3. The distribution network

In our cost model, we assume that the topology of our distribution network is a m-ary tree with l levels (figure 1). A network tree model has many practical aspects. A tree model captures the hierarchical structure of a large-scale network, where large backbone routers service many smaller service providers which in turn service end-users. For example, a tree might include multiple levels, dividing a network into national, regional and local sub-networks.

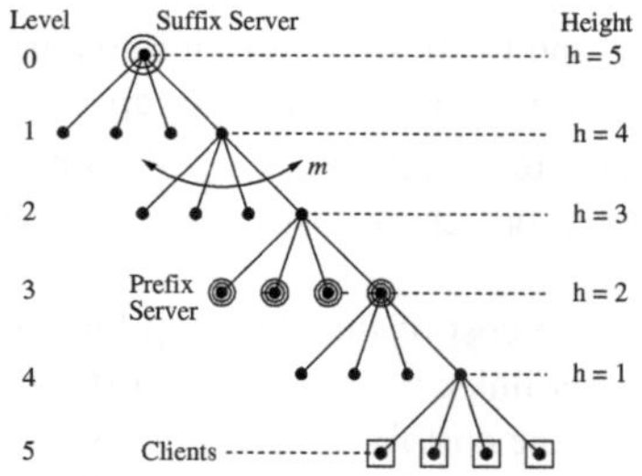

Figure 1. Video distribution network

The suffix server is assumed to be at the root of the tree. Prefix servers may be placed at any level of the distribution network other than the highest level (i.e. leaves). The clients are lumped together at the m^l leaf nodes. The number of clients watching simultaneously a video is not limited to m^l since a leaf node does not represent a single client but multiple clients that are for instance in the same building. In this paper, we assume homogeneous client populations.

4. Evaluation of Architectural Choices

4.1 Analytical model

In this section, we present the basic cost model and then derive the cost term for each architecture. We divide the cost of a VoD system into network and server cost. The network cost is proportional to the amount of network bandwidth needed for the transmission of the prefix and the suffix. The server cost depends on the disk storage used and the total number of input/output streams needed for both, the suffix server and the prefix servers. The total cost of the system can be computed as the sum of the total network and total server cost:

$$C^{system} = C^{system}_{netw} + \gamma C^{system}_{server} \quad (1)$$

To relate the network and the server cost, a normalization factor γ is introduced that allows us to explore various scenarios for the cost of the servers as compared to the cost of the transmission bandwidth.

The terms for the network and server cost are given by:

$$\begin{aligned} C_{netw}^{system} &= C_{netw}^{prefix} + C_{netw}^{suffix} \\ C_{server}^{system} &= C_{server}^{prefix} + C_{server}^{suffix} \end{aligned}$$

The server cost depends on both, the required amount of storage C_{sto} (in Megabit) and the amount of disk I/O bandwidth $C_{I/O}$ (in Megabit/sec).

$$\begin{aligned} C_{server}^{prefix} &= \max(C_{I/O}^{prefix}, \beta C_{sto}^{prefix}) \\ C_{server}^{suffix} &= \max(C_{I/O}^{suffix}, \beta C_{sto}^{suffix}) \end{aligned}$$

To relate the cost for storage and for I/O, we introduce the normalization factor β that is determined as follows: If our server has a storage capacity of d_{sto} [Megabit] and an I/O bandwidth of $d_{I/O}$ [Megabit/sec], then $\beta = \frac{d_{I/O}}{d_{sto}}$. Since the server will be either I/O limited or storage limited, the server cost is given as the *maximum* of $C_{I/O}$ and βC_{sto}.

To model the case where the cost of the "last-hop link" towards the clients is not the same as the cost of the other links, we can set the cost for the last link to the clients (lhc) to a value different from the cost for the other links.

This basic cost model has quite a few parameters. We present results only for a limited subset of parameter values that provide new insights. We will vary only the parameters γ and the last-hop cost *lhc*. We consider a distribution network with an out-degree $m = 4$ and a number of levels $l = 5$. We expect that such a topology is representative for a wide distribution system that covers a large geographical areas of the size of a country such as France or the UK. If one wants to model a densely populated metropolitan area such as NewYork, one would choose $l < 5$ (e.g. $l = 2, 3$) and $m > 4$ (e.g. $m = 10$). The other parameters are chosen as follows: For the disk I/O cost to disk storage cost ratio β, we choose $\beta = 0.001$ (a realistic value for the current disk technology such as the IBM Ultrastar 72ZX disk). The video length is $L = 90$ minutes.

4.2 Set-top box at the client side for prefix storage

Today, set-top boxes at the client side provide a large amount of storage capacity at a low cost. For instance, the digital video recorder developed by Tivo [9] allows to store between 20 and 60 hours of MPEG II coded video and can receive transmissions at high data-rates. In this subsection, we present a new distribution architecture (called **P-hybrid**) that uses the set-top boxes to store video prefixes.

We derive the P-hybrid model from the PS model [3]. Both models use the same protocols to distribute the prefix and the suffix. However, in contrast to the PS model, the P-hybrid model allows the prefix to be stored not only at the prefix servers but also at the set-top boxes. In the P-hybrid model, when the prefix is stored at the prefix servers, the prefix is delivered to clients via controlled multicast as with the PS model.

In this case, the cost of the prefix is the same for both models the P-hybrid and the PS ($C_{P-hybrid}^{prefix} = C_{PS}^{prefix}$).

The prefix can also be downloaded directly to the set-top boxes. In this case, the prefix cost with the P-hybrid model, as compared to the PS model, is limited to the download cost of the prefix to the set-top box.

Both, the PS and the P-hybrid models store the suffix at the central server and deliver it to clients via tailored transmission. Thus, both models have the same suffix cost ($C_{P-hybrid}^{suffix} = C_{PS}^{suffix}$).

Analytical model. To compute the cost of the P-hybrid model, we partition the set of videos into two disjoint subsets, namely S_1 and S_2, with $S_1 \bigcap S_2 = \emptyset$. S_1 represents the set of videos whose prefix is stored in the **prefix servers**. S_2 represents the set of videos whose prefix is stored in the **set-top boxes**. We calculate separately the cost for the videos in S_1 and S_2. The total P-hybrid system cost is the sum of the system cost over all videos in the two subsets:

$$C_{P-hybrid}^{system} = C_{P-hybrid}^{1}(S_1) + C_{P-hybrid}^{2}(S_2) .$$

For each video i in S_1, the P-hybrid model delivers the prefix and the suffix via the same protocols as the PS model does. Therefore, for each video i in S_1, the system cost can be computed using the PS model and the cost of S_1 is:

$$C_{P-hybrid}^{1}(S_1) = \sum_{i \in S_1} C_{\textit{P-hybrid}}^{system}(i) = \sum_{i \in S_1} C_{\textit{PS}}^{system}(i)$$

with

$$C_{PS}^{system}(i) = C_{PS}^{prefix}(i) + C_{PS}^{suffix}(i)$$

where $C_{PS}^{prefix}(i)$ and $C_{PS}^{suffix}(i)$ are given respectively in tables 1 and 2 in [3].

Concerning S_2, **the suffix cost** of each video $i \in S_2$ is computed as in the case of the PS model since in both architectures, the suffix is delivered to the clients via tailored transmission. In contrast, the prefix cost comprises only the download cost of the prefix.

$$\begin{aligned} C_{P-hybrid}^{2}(S_2) &= \sum_{i \in S_2} C_{P-hybrid}^{prefix}(i) + \sum_{i \in S_2} C_{\textit{P-hybrid}}^{suffix}(i) \\ &= \sum_{i \in S_2} C_{P-hybrid}^{prefix}(i) + \sum_{i \in S_2} C_{\textit{PS}}^{suffix}(i) , \end{aligned}$$

with

$$C_{P-hybrid}^{prefix}(i) = b\,\frac{D_i}{T_s} \sum_{j=1}^{l} m^j = b\,\frac{D_i}{T_s} \left(\frac{m^{l+1} - m}{m-1}\right) \quad (2)$$

In equation (2), b is the playback rate of the video, D_i is the length of the prefix of video i, and T_s is the **download interval** (time between two consecutive downloads) of the prefix to the set-top box. The term $\sum_{j=1}^{l} m^j$ accounts for the number of links traversed by the data at all levels during the download of the prefix. To minimize the cost of S_2, we must find the optimal set of prefix lengths $\{D_i\}_{i \in S_2}$ of the videos in S_2. For this purpose, we solve the following problem:

$$\min_{i \in S_2} f = C^2_{P-hybrid}(S_2)$$
$$s.t. \quad \sum_{i \in S_2} D_i \leq Cap$$

Where Cap is the storage capacity of the set-top box. The problem expressed above is a non-linear programming problem subject to linear inequality constraints. To obtain a solution, we apply the `fmincon` package of Matlab.

For a given partition of the videos between the two disjoint subsets S_1 and S_2, we apply the PS model to compute $C^1_{PS}(S_1)$ and we apply the `fmincon` package of Matlab to compute $C^2_{P-hybrid}(S_2)$. The P-hybrid system cost is the sum of S_1 and S_2. However, to find the optimal total system cost, we must find which video prefixes should be stored in the set-top box. To do so, we present the following heuristic algorithm to find the near optimal split of the set of K videos between S_1 and S_2. We sort the videos in decreasing order of popularity ($N_i > N_j$ if $i < j$) where N is the average number of simultaneous clients). We start with the case where all videos are in S_1 (all the prefixes are stored at the prefix servers). In this case, the P-hybrid model is equivallent to the PS model.

In S_1, the most popular video consumes the largest fraction of the system resources among all videos. Thereby, we move the videos from S_1 to S_2, one after the other, starting with the most popular one in the system. At each step, we compute the new cost of S_1 ($C^{1-new}_{PS}(S_1)$) and S_2 ($C^{2-new}_{P-hybrid}(S_2)$). We then compare the P-hybrid system cost at the current step $C^{system-new}_{P-hybrid}$ and at the previous step $C^{system-opt}_{P-hybrid}$. If $C^{system-new}_{P-hybrid} \geq C^{system-opt}_{P-hybrid}$ then we stop and the parameter values computed at the previous step are chosen.

Results. To obtain insights about the advantages of having a set-top box at the client side, we plot in figure 2(a) the cost ratio of the P-hybrid to the PS models for $\gamma = 1$ and homogeneous link cost (i.e. $lhc = 1$). Indeed, we evaluated the P-hybrid model for other scenarios such as $\gamma = \{0, 0.1\}$ and $lhc = 0.1$, and we found similar results. We set the download interval of the prefixes to $T_s = 10000$ minutes (about *one week*) and the number of videos to $K = 100$. We vary the storage capacity Cap of the set top-box from 100 to 1000 minutes of video. The popularities of the videos are Zipf distributed with $N_i = \frac{A}{i^\alpha}$ (N_i represents the average number of clients simultaneously viewing video i).

As we can observe from figure 2(a), as the storage capacity Cap increases, the P-hybrid system cost reduces as compared to the PS system cost. This reduction exceeds 45% provided a storage capacity $Cap = 1000$ minutes and a system with very few popular videos. Figure 2(b) shows the optimal partitioning of the set-top box amongst videos for different values of A, α and Cap. The larger the buffer space at the client, the larger the number of videos that share that buffer and the longer the length

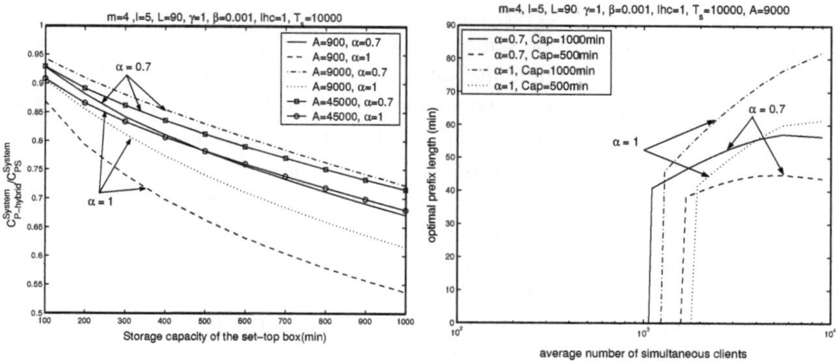

(a) $C^{system}_{P-hybrid}/C^{system}_{PS}$

(b) Optimal prefix length in the set-up box

Figure 2. $C^{system}_{P-hybrid}/C^{system}_{PS}$ and optimal prefix length in the set-up box for $\gamma = 1$ and $(lhc) = 1$.

of the prefixes stored locally. It might seem surprising that the length of the prefix does not necessarily increase monotonically with the popularity N of the video. In fact, the open-loop scheme (suffix transmission) performs well for very popular videos. Thus, it might be optimal to reduce the prefix length of the most popular video in order to free a place for the other popular ones.

Figure 2(b) also shows that, for given values of the parameter A and Cap, as α increases, the number of videos that have their prefix stored at the client side decreases while the length of the prefix becomes longer. Indeed, for a given Zipf distribution ($N_i = \frac{A}{i^\alpha}$), the popularity of video i decreases as α increases. As we mentioned before, the P-hybrid model reduces the system cost as compared to the PS model by storing locally the prefix of the most popular videos. In contrast to $\alpha = 0.7$, when $\alpha = 1$, there are fewer popular videos in the system that should have their prefix stored in the set-top box.

Figure 2(a) also shows that for a given value of A, increasing α increases the efficiency of the P-hybrid model. Actually, the P-hybrid model becomes more cost efficient as the cost of the most popular videos increases relative to the total system cost, which is the case when α increases.

4.3 Use of satellite for suffix transmission

Satellites are a very cost effective transmission medium for sending data to a large group of users. The cost of 1 Mbit/month satellite transmission bandwidth is about \$ 10,000 [8] whereas the cost for 1 Mbit/month terrestrial transmission bandwidth is \$1300 for 1 Mbit/sec during one month in case of a T1 line and \$350 for 1 Mbit/sec during one month in case of a OC-48 transmission link [1]. We now consider the case where the *suffix* is transmitted via satellite directly to the clients, while the prefix is transmitted from the prefix servers to the clients via the Internet. We refer to this distribution architecture as the **S-sat** model. In both the S-sat and the PS models, the prefix is stored at the prefix servers and delivered to clients via controlled multicast. As a result, the prefix cost is the same in both models ($C_{PS}^{prefix} = C_{S-sat}^{prefix}$). On the other hand, both models schedule the suffix via tailored transmission. However, in contrast to the PS model, the S-sat model transmits the suffix via satellite instead of a terrestrial network. As a consequence, the cost term for C_{net}^{suffix} for the S-sat model is $C_{net}^{suffix} = \sigma \cdot R_t^{min}$, where R_t^{min} is the total server bandwidth needed to schedule the suffix via tailored transmission and σ is a weight factor that allows to express the cost for the satellite transmission in *relative* terms with respect to the other cost elements such as terrestrial transmission or server storage and I/O. The I/O and storage cost for the suffix remain the same in both models.

In the following, we will use two different values for σ namely $\sigma = 100$ and $\sigma = 500$. In the light of the absolute prices given above, we consider both values as "conservative" in the sense that they are likely to overestimate the cost of a satellite transmission compared to a terrestrial transmission.

Results. We see in figure 3(b) that for the S-sat model, the prefix decreases more rapidly with increasing number of clients N since the transmission of the suffix via satellite is less expensive compared to a transmission over a terrestrial network. The smaller the value of σ, the cheaper the satellite transmission and the more cost effective the S-sat model. For $N > 10^2$, the S-sat model is very cost effective (figure 3(a)). For a very high number of simultaneous clients, the suffix becomes eventually as large as possible (89 minutes)[1] and satellite suffix transmission can reduce the cost by up to 80% (for $\gamma = 1, lhc = 1, \sigma = 100$). The cost reduction obviously depends on σ. For the case $\sigma = 500$, $\gamma = 1$, and $lhc = 0.1$, the satellite transmission is quite expensive compared to a terrestrial transmission and as a result, the suffix satellite transmission is not competitive. $\gamma = 0.1$ (figure 4(a)) makes the prefix servers cheaper, which allows to use more of them (equation (1), page 2). Such a cost reduction of the prefix servers benefits both the PS and the S-sat models. As a consequence, for $\gamma = 0.1$, the satellite transmission still remains, for large values of N, much more cost effective than the transmission of the suffix via terrestrial links.

If we completely ignore the server cost ($\gamma = 0$) and the last hop cost is reduced ($lhc = 0.1$), suffix transmission via satellite will remain more cost effective provided that satellite transmission is cheap ($\sigma = 100$, figure 4(b)).

We eventualy outline that our results showed (figure not shown) that using satellite suffix distribution as compared to terrestrial links has little impact on the placement of the prefix servers.

5. Conclusion and Future Work

We have presented two new architectures for large scale VoD systems. We assumed that each video is split into two parts, the prefix and the suffix. In the first architecture, we allow clients to store locally the prefix of some/all popular videos. In the second

[1] We limit the minimal length of the prefix to 1 minute in order to provide a zero start-up delay service.

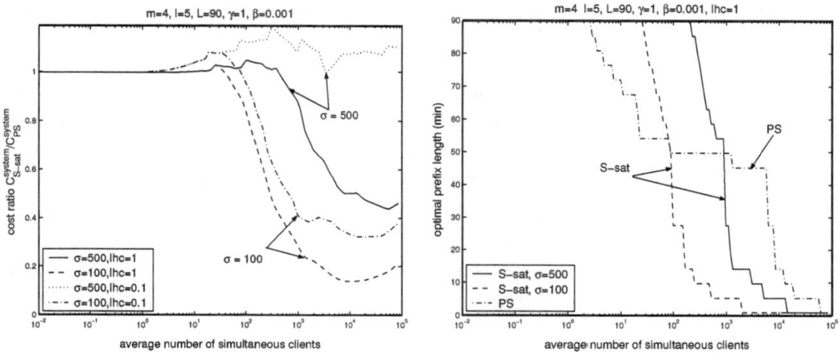

(a) $C_{S-sat}^{system}/C_{PS}^{system}$, $\gamma = 1$ and $lhc = \{1, 0.1\}$

(b) optimal prefix length, $\gamma = 1$ and $lhc = 1$

Figure 3. Cost ratio ($C_{S-sat}^{system}/C_{PS}^{system}$) for $\gamma = 1$, $lhc = \{1, 0.1\}$ and $\sigma = \{100, 500\}$ and optimal prefix length for both, PS and S-sat models for $\gamma = 1$, $(lhc) = 1$, and $\sigma = \{500, 100\}$ as a function of the number of clients N.

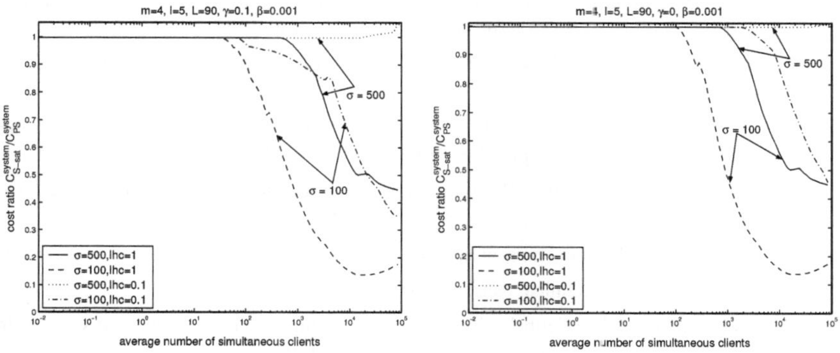

(a) $\gamma = 0.1$

(b) $\gamma = 0$

Figure 4. Cost ratio ($C_{S-sat}^{system}/C_{PS}^{system}$) for $\gamma = \{0.1, 0\}$, $(lhc) = \{1, 0.1\}$, and $\sigma = \{100, 500\}$.

one, we transmit the suffix via satellite. For each architecture, we developed a cost model to compute the delivery cost of videos. We applied these architectures to the PS model and we evaluated the overall reduction in the system cost. Our results showed that,

- Storing the prefixes of the most popular videos in the system at the client side can reduce efficiently the system cost by 30-45%.

- When the cost for the satellite transmission is low relative to the cost for the terrestrial transmission, using satellite to transmit the suffix can reduce the system cost by up to 80%.

In this work, we have assumed that (i) client requests for the same video are homogeneously distributed among all clients and (ii) the current video distribution network has a very regular structure with all clients being at the same distance from the root. A natural extension of this work would be to introduce heterogeneity in the video popularity and in the network contruction. As a future work, we intend to study the impact of these extensions on the two architectures that we introduced here. While these extensions will clearly change the absolute values that we presented, we do not expect that they will change the broad conclusions that we obtained for both architectures.

References

[1] S. Banerjee, J. Brassil, A. C. Dalal, S.-J. Lee, E. Perry, P. Sharma, and A. Thomas. Rich media from the masses. Technical Report HPL-2002-63R1, HP Lab, May 2002.

[2] Y. Birk and R. Mondri. Tailored transmissions for efficient near-video-on-demand service. In *Proceedings of ICMCS*, pages 226–231, June 1999.

[3] D. Choi, E. W. Biersack, and G. Urvoy-Keller. Cost-optimal dimensioning of a large scale video on demand system. In *Proc. of NGC*, October 2002.

[4] D. Eager, M. Vernon, and J. Zahorjan. Optimal and efficient merging schedules for video-on-demand servers. In *Proc. of 7th ACM Multimedia*, Nov. 1999.

[5] L. Gao and D. Towsley. Threshold-based multicast for continuous media delivery. *IEEE Transactions on Multimedia*, 3(4):405–414, Dec. 2001.

[6] Y. Guo, S. Sen, and D. Towsley. Prefix caching assisted periodic broadcast: Framework and techniques for streaming popular videos. In *Proc. of IEEE ICC 2002*, Apr. 2002.

[7] K. A. Hua, Y. Cai, and S. Sheu. Patching : A multicast technique for true video-on-demand services. In *ACM Multimedia*, pages 191–200, 1998.

[8] J. Nonnenmacher. Personal communication, Oct. 2002.

[9] TiVo. What is TiVo: Technical aspects, 2003.

[10] S. Viswanathan and T. Imielinski. Pyramid broadcasting for video on demand service. In *Proc. of Multimedia Conference*, San Jose, CA, Feb. 1995.

[11] P.-F. You and J.-F. Pâris. A better dynamic broadcasting protocol for video on demand. In *Proceedings of IPCCC*, pages 84–89, Phoenix, AZ, Apr. 2001.

DYNAMIC CACHE RECONFIGURATION STRATEGIES FOR A CLUSTER-BASED STREAMING PROXY*

Yang Guo, Zihui Ge, Bhuvan Urgaonkar, Prashant Shenoy, and Don Towsley
Department of Computer Science, University of Massachusetts at Amherst

Abstract The high bandwidth and the relatively long-lived characteristics of digital video are key limiting factors in the wide-spread usage of streaming content over the Internet. The problem is further complicated by the fact that video popularity changes over time. In this paper, we study caching issues for a cluster-based streaming proxy in the face of changing video popularity. We show that the cache placement problem for a given video popularity is NP-complete, and propose the *dynamic first fit* (DFF) algorithm that give the results close to the optimal cache placement. We then propose *minimum weight perfect matching*(MWPM) and swapping-based techniques that can dynamically reconfigure the cache placement to adapt to changing video popularity with minimum copying overhead. Our simulation results show that MWPM reconfiguration can reduce the copying overhead by a factor of more than two, and that swapping-based reconfiguration can further reduce the copying overhead compared to MWPM, and allow for the tradeoffs between the reconfiguration copying overhead and the proxy bandwidth utilization.

1. Introduction

The high bandwidth and the relatively long-lived characteristics of digital video are key limiting factors in the wide-spread usage of streaming content over the Internet. The problem is further complicated by the fact that video popularity changes over time. The use of a content distribution network (CDN) is one technique to alleviate these problems. CDNs cache partial or entire videos at proxies deployed close to clients, and thereby reduce network and server load and provide better quality of service to end-clients [1, 16, 8]. Due to the relatively large storage space and bandwidth needs of streaming media, a streaming CDN typically employs a cluster of proxies at

*This research was supported in part by the National Science Foundation under NSF grants EIA-0080119, ANI-0085848, CCR-9984030, ANI-9973092, ANI9977635, ANI-9977555, and CDA-9502639. Any opinions, findings, and conclusions or recomendations expressed in this material are those of the authors and do not necessarily reflect the views of the funding agencies.

F. Douglis and B.D. Davison (eds.), Web Content Caching and Distribution, 139-157.

each physical location. Each such cluster can collectively cache a larger number of objects and also serve clients with larger aggregate bandwidth needs. In this paper, we study caching issues for such a streaming proxy cluster in the face of changing video popularity.

Each proxy within the cluster contains two important resources: storage (cache) space and bandwidth. Each video file requires a certain amount of storage and bandwidth determined by its popularity. Assuming that video files are divided into objects, we study the *cache placement problem*, i.e., whether to cache an object, and if we do, which component proxy to place it on so that the aggregate bandwidth requirement posed on the servers and the network is minimized. Furthermore, since video popularities vary over time (e.g., many people wanted to watch the movie *The Matrix* again in preparation of the release of its sequel *The Matrix Reloaded* causing it to be very popular for a few weeks), the optimal cache placement also changes with time. *The proxy must be able to deal with dynamically varying popularities of videos and reconfigure the placement accordingly.*

In this paper, we first consider the offline version of the cache placement problem. We show that it is an NP-complete problem and draw parallels with a closely related packing problem, the 2-dimensional multiple knapsack problem (2-MKP) [11]. Taking inspiration from heuristics for 2-MKP, we propose two heuristics—static first-fit (SFF) and dynamic first-fit (DFF)—to map objects to proxies based on their storage and bandwidth needs. We then propose two techniques to dynamically adjust the placement to accommodate changing video popularities. Our techniques attempt to minimize the copying overheads incurred when adjusting the placement. The *minimum weight perfect match* (MWPM) reconfiguration method minimize the copying overhead associated with such a placement reconfiguration by solving a bipartite matching problem. In order to further reduce the copying overhead, we propose *swapping-based reconfiguration*, which mimics the hill climbing approach [6] used in solving optimization problems. The swapping-based reconfiguration also naturally allows us to trade off proxy bandwidth utilization against copying overhead.

We evaluate our techniques using simulation. We find that DFF gives a placement that is very close to the optimal cache placement, and that both DFF and SFF outperform a placement method that does not take bandwidth and storage requirements into account. We then examine the performance of MWPM reconfiguration, and show that it can reduce the copying overhead by a factor of more than two. Finally, we show that swapping-based reconfiguration can further reduce copying overhead relative to MWPM, and that further copying overhead reduction can be achieved by decreasing the proxy bandwidth utilization.

In summary, we study the dynamic cache reconfiguration problem for a cluster-based streaming proxy in the face of changing video popularities. Our contributions are twofold:

- We show that the cache placement problem is NP-complete, and propose DFF algorithm that is found to give the results close to the optimal cache placement.
- We propose MWPM and swapping-based techniques that can reconfigure the cache placement dynamically to adapt to changing video popularity with minimum copying overhead.

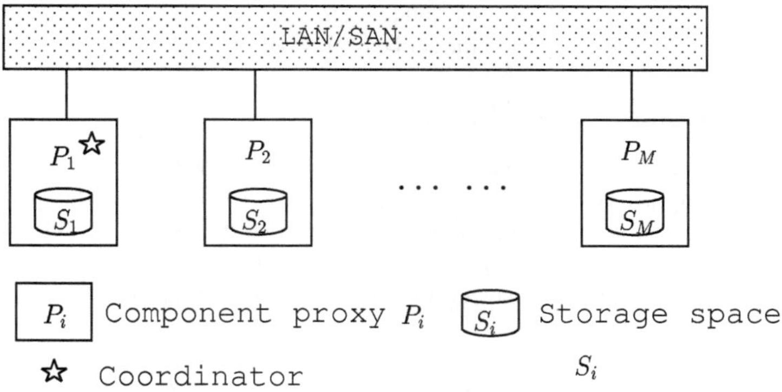

Figure 1. A cluster-based streaming proxy

The remainder of the paper is organized as follows. In Section 2, we describe the architecture of a cluster-based streaming proxy. We formulate the optimal cache placement problem and present the baseline strategies in Section 3. The techniques for dynamic reconfiguration of cache placement are presented in Section 4. Section 5 is dedicated to performance evaluation. Section 6 includes the related work, and Section 7 concludes the paper.

2. Architecture of cluster-based streaming proxy

A cluster-based streaming proxy consists of a set of component proxies as shown in Fig. 1. These individual proxies are connected through a LAN or SAN, and are controlled by a coordinator residing on one of the machines.

The coordinator provides an interface for clients and servers so that the cluster-based streaming proxy acts as a single machine proxy from the perspective of clients and servers. In addition, the coordinator provides the following functionalities to the component proxies inside a cluster-based proxy.

- Coordinate component proxies to serve client requests. A client request may require multiple cached objects from different component proxies with a certain timing relationship. The coordinator needs to orchestrate the component proxies to serve these requests.

- Monitor and estimate the popularity (access frequency) of multimedia objects.

- Compute the optimal cache placement based on the current content popularity, and conduct dynamic cache reconfiguration if the degree of the popularity change demands cache reconfiguration.

In this paper we focus on the dynamic cache reconfiguration issues for such a cluster-based streaming proxy. We assume perfect knowledge of client access information. Accurately monitoring the popularity of streaming objects and efficiently coordinating component proxies to provide requested service are significant problems in their own right, and lie beyond the scope of this paper.

3. Optimal Cache Placement

In this section, we investigate the optimal cache placement problem for a cluster-based streaming proxy for a given video popularity. We formulate this problem as an integer linear programming problem, and show that the problem is NP-complete. We then present two baseline heuristics for it. In Section 4, we will show how these baseline heuristics can be enhanced to realize dynamic cache reconfiguration with minimum copying overhead.

3.1 Optimal Cache Placement: Problem Formulation

Consider a cluster-based proxy with M component proxies. Let S_j and Φ_j denote the available storage space and bandwidth at the j-th component proxy. Suppose that the cluster-based proxy services a set of videos that are divided into K distinct objects. Let x_i and b_i denote the storage space and bandwidth requirement of object i. The proxy caches a subset of the K objects to reduce the network bandwidth consumption on the path from remote servers to the proxy cluster. We focus on the problem of which objects to cache and where.

Let c_{ij} be a selection parameter that denotes whether object i is cached at proxy j—c_{ij} equals 1 if proxy j holds a copy of object i and is zero otherwise. Further, let ρ_{ij} denote the amount of bandwidth reserved for object i at proxy j (ρ_{ij} indicates how much of the aggregate demand for the object is handled by component proxy j). The objective of caching objects at the proxy is to *minimize* the bandwidth consumption on the network path from the remote servers to the proxy, or equivalently, to *maximize* the share of the aggregate client requests that can be serviced using cached videos. The resulting *optimal cache placement (O.C.P.)* problem can be formulated as follows:

$$\max \quad B(\rho, c) = \sum_j \sum_i \rho_{ij} c_{ij} \tag{1}$$

$$\text{subject to:} \quad \sum_i x_i c_{ij} \leq S_j \tag{2}$$

$$\sum_i \rho_{ij} c_{ij} \leq \Phi_j \tag{3}$$

$$\sum_j c_{ij} \rho_{ij} \leq b_i \tag{4}$$

where $c_{ij} \in \{0, 1\}$, $i \in \{1, \ldots, M\}$, and $j \in \{1, \ldots, K\}$. $B(\cdot)$ denotes the allocated proxy bandwidth.

The solution to this problem yields values of c_{ij} and ρ_{ij} that completely describe the placement of objects and the bandwidth reserved for that object at a proxy. The solution may involve object replication across component proxies to meet bandwidth needs. Further, some objects may not be cached at any proxy, if it is not advantageous to do so.

PROPOSITION 3.1 *The optimal cache placement (O.C.P) problem is NP-complete. The problem is NP-complete even if all objects are of the same size.*

The proof is included in [13].

3.2 Cache Placement Heuristics

We note that O.C.P. is similar to the 2-dimensional multiple knapsack problem (2-MKP) [11]. 2-MKP has one or more *knapsacks* that have capacities along two dimensions, and a number of *items* that have requirements along two dimensions. Each item has a profit associated with it. The goal is to pack items into the knapsacks so as to maximize the profit yielded by the packed items, while the capacity constraints along both dimensions are maintained. The component proxies and video objects in O.C.P. may be viewed as akin to the knapsacks and the items in 2-MKP respectively; the profits associated with the video objects are their bandwidth requirements. However there is an important difference between the two problems—the requirements of an item along both directions in 2-MKP are indivisible meaning the item may be packed in exactly one knapsack; the bandwidth requirement of a video object in O.C.P. is divisible and may be met by replicating the object on multiple component proxies.

Heuristics based on *per-unit weight* are frequently used for knapsack problems. Consequently, we define the *bandwidth-space ratio* of object i to be b_i/x_i (the ratio of the required bandwidth and the object size), and the bandwidth-space ratio of proxy j to be Φ_j/S_j.

• **Static First-fit algorithm (SFF).** Static first-fit sorts proxies and objects in descending order of their bandwidth-space ratios. Each object (in descending order) is assigned to the first proxy (also in descending order) that has sufficient space to cache this object. If this proxy has sufficient bandwidth to service this object, the corresponding amount of bandwidth is reserved at the proxy, and the object is removed from the uncached object pool. On the other hand, if the proxy does not have sufficient bandwidth to service the object, the available bandwidth at the proxy is reserved for this object. The object is returned back into the un-cached object pool with the reserved bandwidth subtracted from its required bandwidth. The proxy is removed from the proxy pool since all of its bandwidth has been consumed. The algorithm is illustrated in Fig. 2.

• **Dynamic first-fit algorithm (DFF).** DFF is similar to SFF, except that the bandwidth-space ratio of a component proxy is recomputed after an object is placed onto that proxy and proxies are resorted by their new bandwidth-space ratios (in SFF, the ratio is computed only once, at the beginning). The intuition behind DFF is that the effective bandwidth-space ratio of a proxy changes after an object is cached, and recomputing this ratio may result in a better overall placement. In fact, as we will see in Section 5, DFF does perform better than SFF, and gives results close to the optimal cache placement.

So far we have focused on the optimal cache placement with fixed storage and bandwidth needs of a set of objects. In practice, bandwidth needs of objects vary over time due to changes in object popularities. The cache placement needs to be dynamically reconfigured in order to adapt to the changing popularities, e.g., newly popular objects

```
STATIC-FIRST-FIT (P, O)
sort P in descending order of bandwidth-space ratio
while ( O is not empty) {
  i = object with highest bandwidth-space ratio
  for (proxy j ∈ P in the sorted order) {
    if (x_i ≤ S_j) {
      c_ij = 1;      // cache object i
      S_j = S_j − x_i;
      if (b_i ≤ Φ_j)
        ρ_ij = b_i
        Φ_j = Φ_j − b_i
        remove object i from O
      else
        ρ_ij = Φ_j
        remove proxy j from P
        b_i = b_i − Φ_j;
        return modified obj. i into O
      break;
    } //end of if
  } //end of for loop
} //end of while loop
```

Figure 2. Static First-fit Placement Algorithm. Denote by P the collection of proxies that have bandwidth and space resources to provide caching service, and by O the collection of objects that have not been cached, or need additional bandwidth.

may need to brought in from the servers, and cold objects may need to be ejected. We denote the number of objects that need to be transmitted among component proxies and from servers to proxy as the *copying overhead* of cache reconfiguration. In the following section, we study how to realize the cache reconfiguration with the minimum copying overhead.

4. Dynamic Cache Reconfiguration

A straightforward technique for dynamic cache reconfiguration is to recompute the entire placement from scratch using DFF or SFF based on the newly measured popularities, and bring in the objects from neighboring component proxies or remote servers. We denote such cache reconfiguration approaches as *simple DFF reconfiguration* or *simple SFF reconfiguration*. These approaches may yield close-to-optimal bandwidth utilization. However, they may cause many objects to be moved across proxies, resulting in excessive copying overhead as indicated by the simulation experiments in Section 5.

In the following, we propose two cache reconfiguration techniques that can reduce the copying overhead by exploring the existing cache placement. We first present the *minimum weight perfect matching reconfiguration method* (MWPM reconfiguration) by formulating the minimum copying overhead reconfiguration problem as a minimum weight perfect matching on a bipartite graph. We then describe the *swapping-based reconfiguration* method that mimics the hill-climbing approach [6] used in solving optimization problems. The swapping-based reconfiguration naturally allows us to trade the proxy bandwidth utilization for the copying overhead.

4.1 MWPM cache reconfiguration

Let $P_1, P_2, \ldots P_M$ denote the M proxies in the cluster. Let Ψ denote the current placement of objects onto the proxy cluster and let Ψ_{new} denote the new placement that is desired. There can be as many as $M!$ ways to reconfigure the placement. Ideally, we would like to reconfigure the placement such that the cost of moving (copying) objects from one proxy to another or from the server to a proxy is minimized.

The above problem is identical to the problem of computing the *minimum weight perfect matching* on a bipartite graph. To see why, we model the reconfiguration problem using a bipartite graph with two sets of M vertices. The first set of of M vertices represents the current placement Ψ. The second set of vertices represent the new placement Ψ_{new}. We add an edge between vertex P_u of the first set and vertex P_v of the second set if P_u has enough space and bandwidth to accommodate all the objects placed on P_v in Ψ_{new}. The weight of an edge represents the cost of transforming the current placement on the proxy represented by P_u to the new placement represented by P_v (the cost is governed by the number of new objects that must be fetched from another proxy or a remote server to obtain the new placement).

To illustrate this process, consider the example in Figure 3 with two identical proxies and five objects. In the current placement Ψ, the first three objects are placed on P_1 and the remaining two objects on P_2. The new placement Ψ_{new} involves placing $\{1, 2, 3\}$ on one proxy and $\{1, 5\}$ on the other proxy. Since the two proxies are identical, we add edges between both pairs of vertices in the bipartite graph. This is because either proxy can accommodate all the objects placed on the other proxy by the new placement. The weights on the edges indicate the cost of transforming each proxy's cache to one of these sets (e.g., transforming P_2's cache from $\{4, 5\}$ to $\{1, 5\}$ involves one copy whereas transforming it to $\{1, 2, 3\}$ involves three copies; deletions are assumed to incur zero overhead). It can be shown that finding a *minimum weight perfect match (MWPM)* for this bipartite graph yields a transformation with minimum copying cost. A perfect match is one where there is a one-to-one mapping from vertices in the former set to the latter; a minimum weight perfect match minimizes the weights on the edges for this one-one mapping.[1] Thus, in Fig. 3, leaving P_1's cache as is, and transforming P_2's cache from $\{4, 5\}$ to $\{1, 5\}$ is the minimum weight perfect matching with a copying cost of 1.

4.2 Swapping-based cache reconfiguration

MWPM cache reconfiguration can significantly reduce the copying overhead, as indicated by the simulation experiments in Section 5. In this section, we describe a swapping-based reconfiguration technique that can further reduce the reconfiguration

[1]The bipartite graph can be constructed in $O(K + M)$ time and the minimum weight perfect matching is polynomial-time solvable with complexity $O(M^3)$.

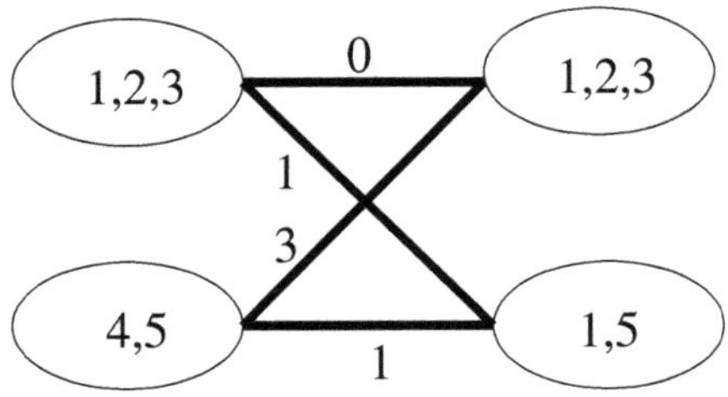

Figure 3. An example of using the minimum weight perfect matching to find the placement which minimizes object movement in a 2 node proxy.

overhead. The swapping-based reconfiguration takes the current placement as the initial point, and use the well-known hill-climbing method [6] to solve the optimal cache placement (O.C.P.) problem. The reconfiguration cost incurred in MWPM is used as a control parameter to limit the overhead incurred by the swapping-based technique.

The hill-climbing method is characterized by an iterative algorithm that makes a small modification to the current solution at each step to come closer to the optimal solution. To apply the hill-climbing method to the cache reconfiguration problem, two issues need to be properly addressed: (1) how to modify the existing cache placement at each step to improve the proxy bandwidth utilization, and (2) when to stop. In the following, we address the above two issues.

Object swapping. We propose to swap the position of two objects at each step to modify the current placement. Our approach is to select a pair of objects such that the utilized proxy bandwidth increases after swapping. In fact, we select the pair that maximally increases the bandwidth utilization so as to minimize the copying overhead of reconfiguration.

Proxies are classified into two categories: overloaded proxies and under-loaded proxies, as shown in Fig. 4. A proxy is said to be overloaded if the total bandwidth needs of objects currently stored at the proxy exceed capacity; under-loaded proxies have spare bandwidth. All objects not currently stored on any proxy are assumed to be stored on a virtual proxy with bandwidth capacity zero (thus, the virtual proxy is also overloaded). The abstraction of a virtual proxy enables us to treat cached and uncached objects in a uniform manner. Intuitively, a "cold" object from an under-loaded proxy is selected and swapped with a "hot" object on an overloaded proxy, so that the total bandwidth utilization increases. We apply the following rules in selecting object:

- *Cold object selection*: Randomly select an underloaded component proxy with probability proportional to the amount of spare bandwidth, and then choose the least frequently accessed object in this proxy.

- *Hot object selection*: Select the object with the highest bandwidth requirement cached in the overloaded proxies or the virtual proxy.

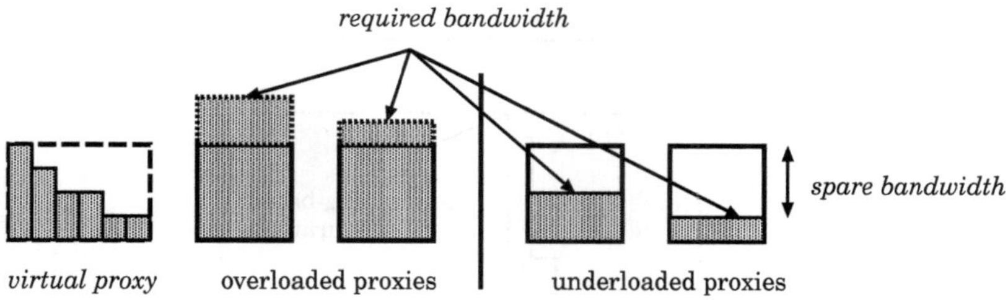

Figure 4. Selection of swapping objects in swapping-based reconfiguration

These two objects are then swapped, and the corresponding proxies are relabeled as under-loaded or overloaded based on the new cache contents.[2] Randomization is introduced to overcome thrashing observed in experiments where two proxies repeatedly swap the same pair of objects, causing the algorithm to be stuck in a local minimum with no further improvement.

Termination condition. We now examine the termination condition of the swapping-based reconfiguration algorithm. Let $util_{DFF}$ denote the bandwidth utilization achieved by DFF and $cost_{DFF}$ denote the copying overhead of achieving this placement as computed by the MWPM reconfiguration. The swapping-based reconfiguration uses $(1 - \delta) \times util_{DFF}$ as its bandwidth utilization target, where δ $(0 \leq \delta < 1)$ is a design parameter set by the proxy coordinator. Next, it runs the swapping-based heuristic to search for a placement that has a bandwidth utilization larger than $(1 - \delta) \times util_{DFF}$ but at a lower copying cost. The heuristic then chooses this placement, or reverts to the MWPM reconfiguration computed placement if the search yields no better placement. Thus, the swapping-based heuristic is run until one of the following occurs:

- The bandwidth utilization reaches the target $(1 - \delta) \times util_{DFF}$ at a cost lower than $cost_{DFF}$. Since a lower cost placement that closely approximates DFF is found, we pick this placement over MWPM DFF.

- The cost of the swapping-based heuristic reaches $cost_{DFF}$ but its bandwidth utilization is below $(1 - \delta) \times util_{DFF}$. No better placement is found, so we revert to the one computed using MWPM DFF.

By adjusting δ, we can trade the bandwidth utilization for the copying overhead. We will evaluate this in Section 5.3.

5. Performance Evaluation

In this section, we conduct simulation experiments to evaluate the performance of MWPM and swapping-based reconfiguration algorithms. We start by evaluating the

[2]No objects are physically moved at this time. Actual movement occurs after the algorithm is terminated.

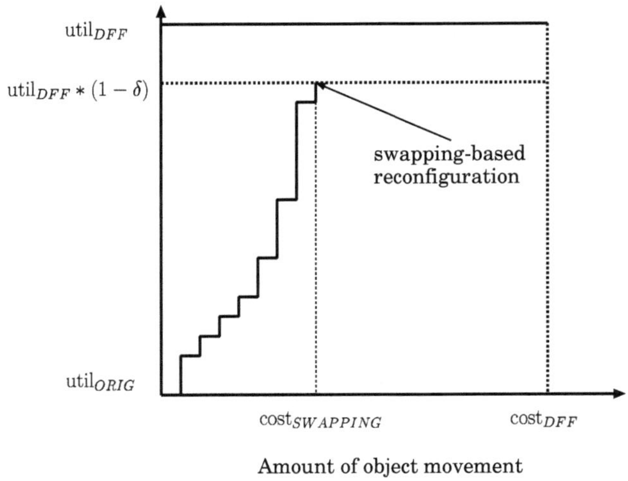

Figure 5. Swapping-based cache reconfiguration

cache placement algorithms, DFF and SFF. We find that DFF yields a placement very close to the optimal cache placement, and both DFF and SFF outperform a placement method that does not take the bandwidth and storage requirement of objects into account. We then examine the performance of MWPM reconfiguration, and show that it reduces copying overhead by a factor of more than two. Finally, we show that the swapping-based reconfiguration can further reduce copying overhead in comparison to MWPM, and further copying overhead reduction can be achieved by decreasing the proxy bandwidth utilization.

5.1 Simulation setting

Assume that clients access a collection of 100 videos whose lengths are uniformly distributed from 60 minutes to 120 minutes. The playback rate of these videos is 1.5 Mbps (CBR), and their popularity obeys the Zipf distribution, i.e., the i-th most popular video attracts a fraction of requests that is proportional to $1/i^{\alpha}$, where α is the Zipf skew parameter. Each video is divided into equal sized segments and each segment is an independent cachable object. We consider a streaming proxy that consists of 5 component proxies. The bandwidth available at component proxy j, Φ_j, is 100 Mbps for $1 \leq j \leq 5$. We denote by S_j the storage space at the j-th proxy. We set S_i/S_{i+1} to be a constant for $1 \leq i \leq 4$, and denote it as *the space skew parameter*, p_s. The bandwidth-space ratios of component proxies can be tuned by adjusting p_s.

5.2 Evaluation of cache placement algorithms

We use the lp_solve [15] linear programming package to solve the optimal cache placement problem exactly. We also use *space oriented placement*, SOP, as the baseline algorithm. Space oriented placement uses storage space as the only constraint. It places the objects in descending order of access frequency, into proxies that are sorted in descending order of storage space.

We choose the aggregate storage space of 5 proxies to be 40 Gbytes. The proxy configuration is chosen in such a way that the bottleneck varies between proxy storage space and bandwidth based on the the proxy space skew parameter and clients' access behavior. In practice, the number of videos are much greater than 100 videos, so we assume that the storage space much larger than 40 Gbytes is required. The object size is set to be that of a one minute video segment. We will investigate the impact of object size later. The aggregate client request rate is 4 request/min.

DFF outperforms SFF and SOP consistently and achieves a proxy bandwidth utilization comparable to that of the optimal cache placement. Note that the computation time for obtaining an optimal cache placement using lp_solve is more than two hours on a 1GHz CPU, 1GB memory Linux box, while it takes a couple of seconds for DFF and SFF. Fig. 6 depicts the allocated bandwidth and used storage space as a function of Zipf skew parameter. Here we set the component proxy space skew parameter, p_s, to one, i.e., the storage space is evenly distributed among the proxies. We observe that DFF, SFF, and SOC all achieve the optimal bandwidth utilization when the Zipf skew parameter is less than 0.5. Intuitively, when the Zipf skew parameter is small, the client requests are evenly distributed among different videos. The number of client requests for different videos are comparable, and the required bandwidth is therefore nearly equal for all objects. The storage space at the proxy is the bottleneck resource, and the maximum proxy bandwidth utilization is achievable as long as the proxy caches the "hot" objects and uses up the entire storage space.

DFF outperforms SFF and SOC as the Zipf skew parameter increases further. As the Zipf skew parameter increases, the discrepancy between the bandwidth required by "hot" and "cold" objects increases. As in the multi-knapsack problem, a right set of objects needs to be cached at each component proxy to fully utilize every component proxy's bandwidth. SOC uses the storage space as the only resource constraint, and caches the hot objects on the component proxy with the largest storage space. Since the first component proxy consumes the hottest objects, the aggregate required bandwidth of these object surpasses this component proxy's available bandwidth. Meanwhile, other component proxy's bandwidth is wasted since only "lukewarm" or "cold" objects are available. The aggregate utilized bandwidth decreases as the Zipf skew parameter increases further. SFF does a better job since it considers the bandwidth as another resource constraint and stops caching the objects into a component proxy once its bandwidth has been fully utilized. However, SFF fails to balance the bandwidth and storage space utilization at component proxies. For instance, when the Zipf skew parameter is 0.7, two of the component proxies use up the bandwidth while having free storage space in SFF. On the other hand, the storage space is used up on the other three component proxies while bandwidth remains free. This is shown in Fig. 6(b), where SFF doesn't fully utilize the storage space. In contrast, DFF distributes the hot objects among component proxies, and fully utilizes the storage space (which is the bottleneck resource compared to bandwidth for Zipf skew parameter equal to 0.7) and maximizes the utilization of proxy bandwidth.

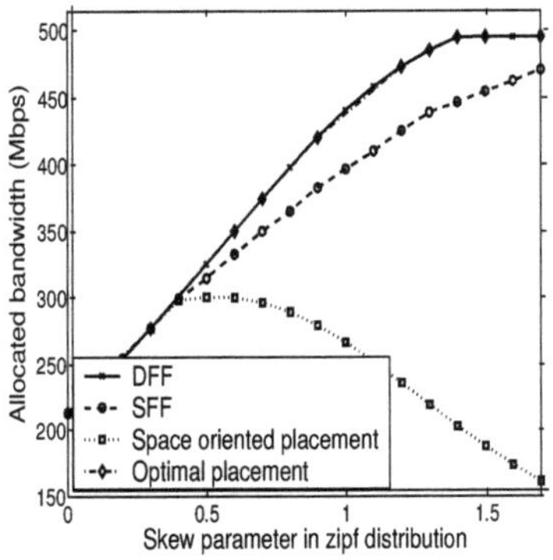

(a) Allocated bandwidth

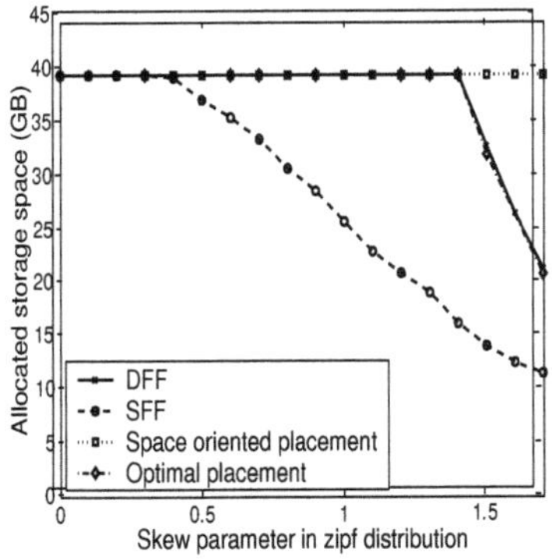

(b) Allocated storage space

Figure 6. Effect of Zipf distribution skew parameter

Effect of space skew parameter. We further evaluate the performance of DFF and SFF in the case of a large space skew parameter. The space skew parameter, p_s, changes the storage space distribution among machines. For instance, when $p_s = 2$, the smallest storage space is 1.29 Gbytes. Hence the bandwidth-space ratios of proxies are widely skewed.

Again, DFF outperforms SFF and SOC, and achieves a proxy bandwidth utilization close to the optimal cache placement. Fig. 7 depicts the allocated bandwidth and allocated space as a function of the space skew parameters when the Zipf skew parameter is 0.7. We observe from Fig. 7(a) that the bandwidth is not fully utilized by SFF when $p_s < 1.6$. Notice that the storage space is also not fully utilized by SFF, as shown in Fig. 7(b). The failure of balancing the bandwidth and storage space utilization causes this behavior. However, as the space skew parameter increases, the performance of SFF improves. This is because the component proxy with large bandwidth-space ratio should cache more hot objects in order to achieve maximum proxy bandwidth utilization, and SFF happens to do that.

Fig. 8 depicts the performance of DFF, SFF, and SOC with respect to the proxy bandwidth utilization with different space and Zipf skew parameter. We observe that the utilized proxy bandwidth of DFF is consistently larger than that of SFF and SOC. We will use DFF in the following subsections.

Effect of object size. The object size may affect the utilization of proxy storage space; we investigate its impact in this section. In the results reported in the previous experiments, the object size is that of a one minute video segment. When we increase the object size while fixing the aggregate proxy storage space at 40Gbytes, the proxy bandwidth utilization gradually decreases as shown in Fig. 9(a) because the component proxy storage space is much larger than the object size (the space skew parameter is selected to be one and Zipf skew parameter is 0.7).

However, when the proxy storage space is relatively small, the impact of object size on performance becomes significant. Fig. 9(b) plots the proxy bandwidth utilization

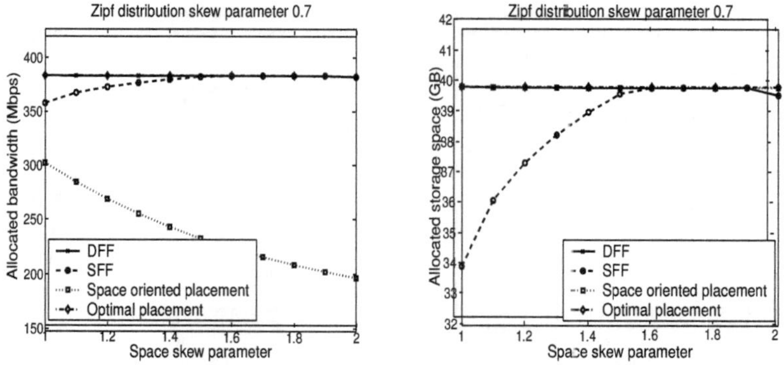

(a) Allocated bandwidth (b) Allocated storage space

Figure 7. Effect of proxy space skew parameter (zipf skew parameter = 0.7)

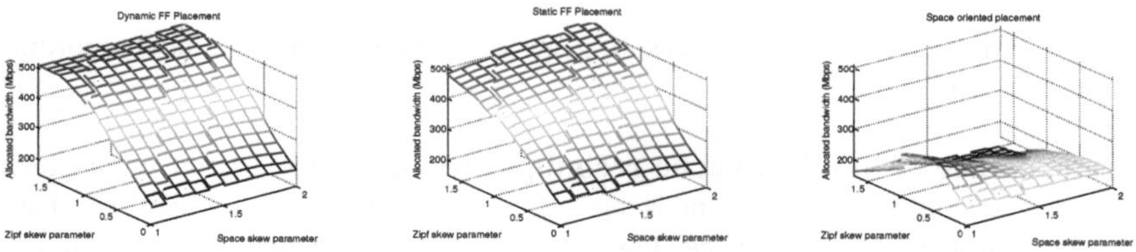

(a) Performance of DFF (b) Performance of SFF (c) Performance of SOC

Figure 8. Effect of zipf skew parameter and proxy space skew parameter on allocated bandwidth

vs. the object size when the aggregate storage space is 10 Gbytes. The bandwidth utilization degrades dramatically as the object size increases beyond that of a 10 minute segment. The smaller object size helps better utilize the storage space and thus better utilize the bandwidth. We suggest that the object size be chosen sufficiently small compared to the component proxy storage space. In the following experiments we use an object size of one minute.

5.3 Evaluation of cache reconfiguration algorithms

The optimal cache placement changes over time as the video popularity varies. Hence the cache placement needs to be dynamically reconfigured. During the cache reconfiguration process, the objects have to be moved between the component proxies, or be brought in from remote servers in order to maximize the proxy bandwidth utilization. The number of objects that need to be transmitted among the component proxies and from the servers is defined to be *copying overhead.* In the following, we evaluate the MWPM and swapping-based cache reconfiguration algorithm. We assume that the cache reconfiguration algorithm is executed periodically, and denote each period a round. We assume that 10% of the videos change their popularity ranks

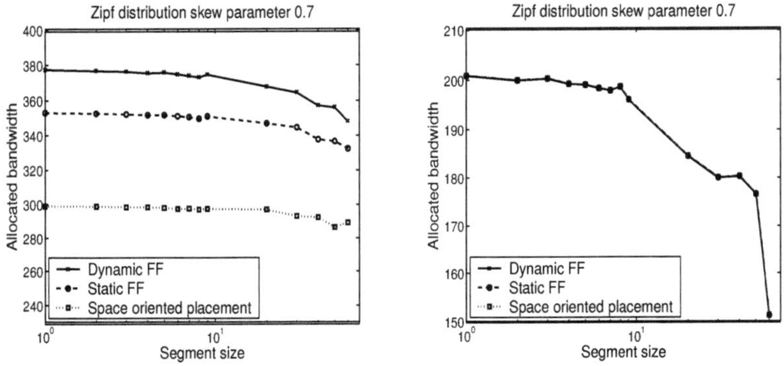

(a) Storage space = 40 Gbytes (b) Storage space = 10 Gbytes

Figure 9. Effect of object size (zipf skew parameter = 0.7)

at the end of each round. DSS is used to generate the new cache placement according to the new video popularity.

Performance of MWPM cache reconfiguration. MWPM can significantly reduce the reconfiguration cost and is most effective when all component proxies have the same amount of storage space. Fig. 10(a) depicts the copying overhead with and without MWPM. The Zipf skew parameter is selected to be 0.7 and the space skew parameter is 1. MWPM can reduce the copying overhead by a factor of 2.3. On average, about 520 objects (about 7 to 8 videos) need to be transmitted to adapt to the new video popularity.

We further investigate how MWPM performs as the space skew parameter changes. Intuitively, as the space skew parameter increases, the storage space at component proxies become increasingly more skewed. Hence fewer edges exist between the old and new placements in the bipartite graph. This reduces the effectiveness of MWPM algorithm. Define *copying overhead improvement ratio* to be the ratio of average copying overhead without using MWPM to that using MWPM. Fig. 10(b) depicts the copying overhead improvement ratio vs. the space skew parameter. The effectiveness of MWPM decreases quickly as space skew parameter increases, and MWPM can not improve the copying overhead when the space skew parameter is larger than 1.4.

Performance of Swapping-based reconfiguration. Swapping-based reconfiguration can further reduce the reconfiguration cost compared to MWPM. Fig. 11(a) depicts the copying overhead of swapping-based cache reconfiguration and MWPM with the Zipf skew parameter set to 0.7 and the space skew parameter set to 1. Here we choose δ to be 0.02, i.e., the swapping process stops once the allocated proxy bandwidth is within 98% of the allocated bandwidth computed by DFF. Swapping-based reconfiguration on average only needs to transfer 147 objects to adapt to the new video popularity. Fig. 11(b) depicts the copying overhead improvement ratio versus the space skew parameter. With the exception of space skew parameter of 1.1, the

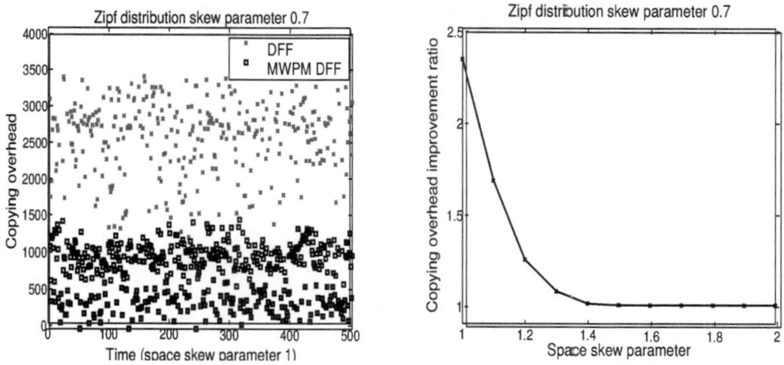

(a) Copying overhead (b) Improvement ratio

Figure 10. Performance of MWPM cache reconfiguration

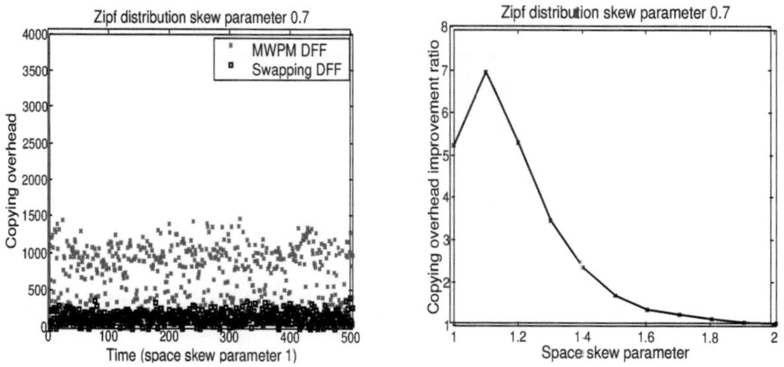

(a) Copying overhead (b) Improvement ratio

Figure 11. Performance of dynamic cache reconfiguration

improvement ratio decreases as the space skew parameter increases, which suggests that we should configure the component proxy as homogeneous as possible.

• **Tradeoffs between reconfiguration cost and target proxy bandwidth utilization.** Fig. 12 depicts the reconfiguration copying overhead versus the target bandwidth utilization with 95% confidence interval. This curve is concave, where the copying overhead decreases quickly as the target bandwidth utilization decreases. This suggests that the swapping-based cache reconfiguration algorithm offers good tradeoffs between reconfiguration cost and target bandwidth utilization.

• **Sensitivity to changes in video popularity.** Swapping-based reconfiguration is more robust to a change in video popularity than MWPM. To investigate the sensitivity of a change in video popularity on the performance of swapping-based reconfiguration and MWPM, we vary the percentage of videos whose ranks change at each round. Fig. 13 depicts the average copying overhead vs. the video popularity change in terms of the percentage of videos (with 95% confidence interval). We first notice that for

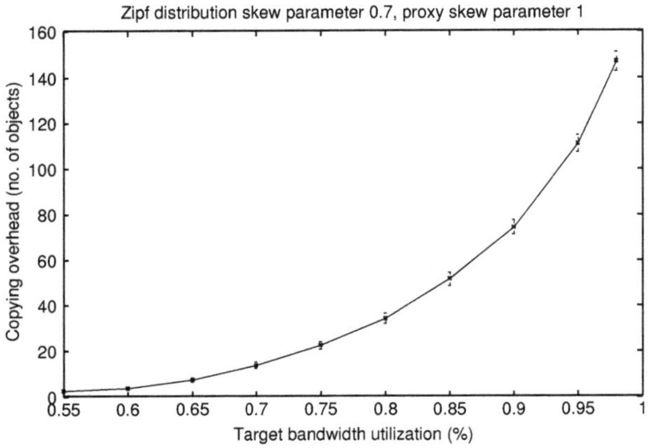

Figure 12. Tradeoffs of bandwidth utilization and copying overhead

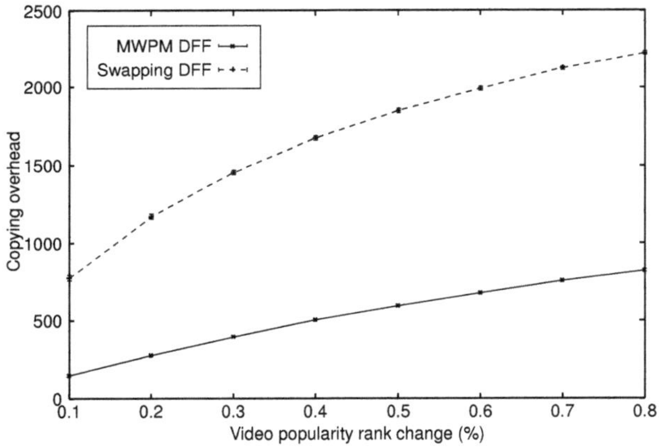

Figure 13. Sensitivity to the popularity change

both MWPM and swapping-based reconfiguration, the copying overhead increases as video popularity change intensifies. However, the copying overhead of MWPM is much larger than that of swapping-based reconfiguration, and the difference increases as the percentage of popularity change increases.

• **Time-of-Day Effect.** The client request rate also varies over time. In the following, we examine the performance of MWPM and swapping-based reconfiguration under changing request rate. Suppose that the request rate is 10 requests/min during the peak hours (from 10am to 4pm), and 1 request/min during the off-peak hours. From 7am to 10am and 4pm to 7pm are two transition periods, during which the request rate linearly increases (decreases) from off-peak rate to peak rate. We assume that 10% of the videos change their ranks every 5 minutes. We use 5 minutes as the

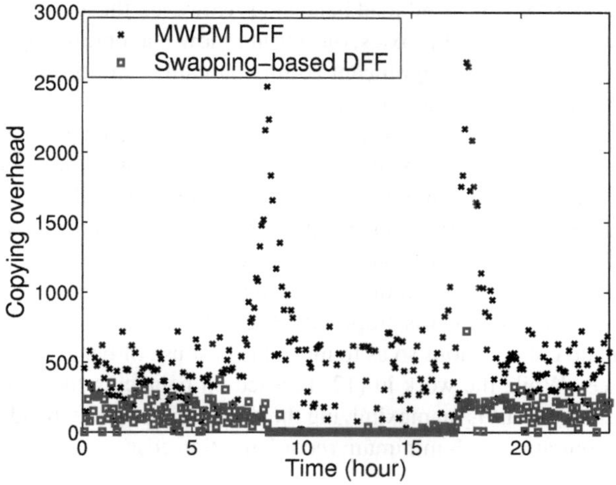

Figure 14. Impact of time-of-day effect and changing video popularity

time period of a round. Fig. 14 depicts the copying overhead incurred by MWPM and the swapping-based reconfiguration, respectively, during a 24-hours time period.

The copying overhead of MWPM reconfiguration increases dramatically during the transition periods. Both the client request rate and the video popularity can affect the optimal cache placement. During the transition periods, the optimal cache placement changes more drastically since both video popularity and access rate are changing, leading to the high MWPM reconfiguration cost. In contrast, swapping-based reconfiguration continues to perform well even during in the transition periods. In off-peak hours, the swapping-based reconfiguration incurs less copying overhead than MWPM reconfiguration. During the peak hour, the copying overhead is even lower since now the proxy bandwidth is the bottleneck resource. There are more "hot" objects hence a lot of swapping is not necessary.

6. Related Work

In general, two bodies of work are related to streaming cache design, namely web caching for regular objects and distributed video server design.

The web caching systems such as CERN httpd [3], Harvest [9], and Squid [21] are designed for classical web objects and do not offer any support for streaming media. Our work is closest to [1, 8] where placement and replacement of streaming media were studied for proxy clusters. The Middleman proxy architecture described in [1] consists of collection of proxies coordinated by a centralized coordinator. Middleman caches only one copy of each segment in the proxy cluster, and uses demand-driven data fetch and a modified version of LRU-K local cache replacement policy called *HistLRUpick*. The work presented in [8] considers a loosely coupled caching system without a centralized coordinator. The system caches multiple copies of segments for the purpose of fault tolerance and load balancing. However, the above efforts focused on optimizing storage space at proxies; our results show that optimizing storage space alone is sub-optimal and that significant additional gains can accrue by considering both bandwidth and storage space constraints.

The effectiveness of using bandwidth and storage space-based metrics was demonstrated for single proxy environments in [20, 2] where a central non-cooperating architecture is assumed. Our techniques extend these works and other studies of single proxy cache placement [17, 16], and are specifically designed for proxy clusters that are typical in today's content distribution networks.

There are other related work in the area of multimedia servers [7, 5, 4, 10, 18, 19, 14]. In [12], a dynamic policy was proposed that creates and deletes replicas of videos, and mixes hot and cold versions so as to make the best use of bandwidth and space of a storage device. While the work in [12] focused on balancing the load on multiple storage devices, our focus is on maximizing the utilization of proxy bandwidth and the dynamic reconfiguration with minimum reconfiguration cost.

7. Conclusions and future work

In this paper, we considered the problem of caching popular videos at a proxy cluster so as to minimize the bandwidth consumption on the proxy-server path. We proposed heuristics for mapping objects onto proxies based on bandwidth and storage space constraints, and showed how these heuristics give the comparable results to the optimal cache placement. We further propose MWPM and the swapping-based reconfiguration schemes that handle dynamically changing popularities with minimum copying overhead. Our simulation results demonstrated the benefits of our approach, and showed that the swapping-based reconfiguration would allow the tradeoffs between the reconfiguration copying overhead and the proxy bandwidth utilization.

Future research can proceed along several avenues. We would like to further refine the cache reconfiguration techniques. For instance, we currently do not take into consideration the distance that an object is moved in the cache reconfiguration. How to compute the copying overhead reflecting the actual overhead consumed over the network remains an interesting problem. We would also like to reduce the switching overhead when servicing a client request that requires multiple objects residing in different component proxies. The cache placement should take this into account and try to place the objects of the same video in the same component proxy if possible. Finally, accurate estimate of the client request rate is an important and interesting research problem.

References

[1] S. Acharya and B. Smith. Middle man: A video caching proxy server. In *Proc. NOSSDAV 2000*, June 2000.

[2] J. Almeida, D. Eager, and M. Vernon. A hybrid caching strategy for streaming media files. In *Proc. SPIE/ACM Conference on Multimedia Computing and Networking*, January 2001.

[3] T. Berners-Lee, A. Lutonen, and H. Nielsen. *Cern httpd.* http://www3.org/Daemon/Status.html, 1996.

[4] C. Bernhardt and E. Biersack. *The Server Array: A Scalable Video Server Architecture.* Kluwer Academic Press, 1996.

[5] W. Bolosky, J. Draves, R. Fitzgerald, G. Gibson, M. Jones, S. Levi, N. Myhrvold, and R. Rashid. The tiger video fileserver. *Proceedings of NOSSDAV 96*, 1996.

[6] A. Bryson and Y. Ho. *Applied Optimal Control.* Taylor & Francis, 1975.

[7] M. Buddhikot, G. Parulkar, and J. J. R. Cox. Design of a large scale multimedia storage server. *Journal of Computer Networks and ISDN Systems*, 27(3):503–517, 1994.

[8] Y. Chae, K. Guo, M. Buddhikot, S. Suri, and E. Zegura. Silo, rainbow, and caching token: Schemes for scalable fault tolerant stream caching. In *IEEE Journal on Selected Areas in Communications on Internet Proxy Services*, September 2002.

[9] A. Chankhunthod, P. Danzig, C. Neerdaels, M. Scwartz, and K. Worrell. A hierarchical Internet object cache. *Proceedings of the 1996 USENIX Technical Conference*, 1996.

[10] P. M. Chen, E. K. Lee, G. A. Gibson, R. H. Katz, and D. A. Patterson. Raid: High-performance, reliable secondary storage. *ACM Computing Surveys*, 26(2):145–185, 1994.

[11] P. Crescenzi and V. Kann, editors. *A compendium of NP optimization problems.* http://www.nada.kth.se/~viggo/problemlist/compendium.html.

[12] A. Dan and D. Sitaram. An online video placement policy based on bandwidth to space ratio (BSR). *Proceedings of ACM SIGMOD*, 1995.

[13] Y. Guo, Z. Ge, B. Urgaonkar, P. Shenoy, and D. Towsley. Dynamic cache reconfiguration strategies for a cluster-based streaming proxy. Technical Report 03-34, Department of Computer Science, University of Massachusetts Amherst, 2003.

[14] P. Lie, J. Lui, and L. Golubchik. Threshold-based dynamic replication in large-scale video-on-demand systems. *Multimedia Tools and Applications*, 2000.

[15] Lp_solve: Linear programming code. `ftp://ftp.es.ele.tue.nl/pub/lp_solve/`.

[16] R. Rejaie, H. Yu, M. Handley, and D. Estrin. Multimedia proxy caching mechanism for quality adaptive streaming applications in the internet. In *Proc. IEEE INFOCOM*, April 2000.

[17] S. Sen, J. Rexford, and D. Towsley. Proxy prefix caching for multimedia streams. In *Proc. IEEE INFOCOM*, April 1999.

[18] P. Shenoy. *Symphony: An Integrated Multimedia File System.* PhD thesis, Dept. of Computer Science, University of Texas at Austin, 1994.

[19] R. Tewari, R. Mukherjee, D. Dias, and M. Vin. Design and performance tradeoffs in clustered video servers. *Proc. IEEE International Conference on Multimedia Computing and Systems*, 1996.

[20] R. Tewari, H. M. Vin, A. Dan, and D. Sitaram. Resource-based caching for Web servers. In *Proc. SPIE/ACM Conference on Multimedia Computing and Networking*, January 1998.

[21] D. Wessels. *ICP and the Squid cache.* National Laboratory for Applied Network Research, 1999.

STREAM ENGINE: A NEW KERNEL INTERFACE FOR HIGH-PERFORMANCE INTERNET STREAMING SERVERS

Jonathan Lemon, Zhe Wang, Zheng Yang, and Pei Cao
Cisco Systems, Inc.

Abstract As high-speed Internet connections and Internet streaming media become widespread, the demand for high-performance, cheap Internet streaming servers increases. In this paper, we look into the performance limitations of streaming server applications running on PC servers with Linux, and propose a new kernel optimization called "stream engine" that combines both copy elimination and context switch avoidance to double the streaming server throughput. Our experiments with stream engine show that for Internet streaming, eliminating context switches is just as important as eliminating data copying. Using profile data, we also project the benefits of TCP offloading hardware implementing part or all of the stream engine optimization.

1. Introduction

As the adoption of broadband connections accelerates over the years, the demand and consumption of streaming media on the Internet also increase rapidly. The result is a constant increase in the throughput requirement of streaming servers.

Today's Internet streaming servers are mostly built with general-purpose servers. Unlike traditional VoD applications, Internet streaming protocols run over standard IP infrastructure and use TCP/UDP as the underlying data transport. The streaming protocols themselves undergo constant changes as vendors seek to perfect end-user experiences. Because of frequent changes in the protocols, streaming server vendors so far have avoided building special-purpose hardware for Internet streaming. Rather, PC-based servers, because of their low price/performance and good development environment, have been the platform of choice.

Unfortunately, PC-servers running generic operating systems are not well-suited for streaming workloads, which are characterized by high data rates and high protocol processing overhead. This paper demonstrates the limitations of traditional OS kernels for streaming servers through performance profiling of a server implementation, and proposes a new kernel interface, "Stream Engine", to address these limitations.

F. Douglis and B.D. Davison (eds.), Web Content Caching and Distribution, 159-170.

The novelty of the "stream engine" interface lies in its combination of data-copy avoidance and context-switch avoidance for applications like Internet streaming. While reducing data copying is a well-known technique to improve server throughput, it was only through implementation experiments that we learnt the significant overhead of user-kernel context switches in Internet streaming applications. The user-level context switches cannot be avoided by any application architecture change (for example, changing from a one-stream-per-thread server implementation to an event-driven server implementation). Rather, a new kernel mechanism and interface are needed.

We have implemented "stream engine" in the Windows Media streaming server running on a PC. The result is a two-fold increase in streaming server throughput, up to 1.2Gbps on current-generation dual-processor 1.7Ghz Pentium servers. In contrast, eliminating data copies by itself would only improve throughput by 30-40% (based on our experiments and back-of-envelop calculations). As a result, the "stream engine" optimization is incorporated in the Content Engine appliances, a shipping product in Cisco's Content Networking line of products.

Our experience with streaming engine also provides inputs on the requirements and interfaces for TCP-offloading hardware for Internet streaming. Most of the TCP offloading hardware devices developed today are for iSCSI acceleration [1, 21, 19, 6], and the interfaces are oriented toward SCSI-based storage. We show that for Internet streaming, the benefits from conventional TCP offloading engines are limited. Rather, TCP offloading hardware needs to interface with stream engine to offload packet assemblies and data transmissions from file buffer cache, and TCP offloading hardware that implements stream engine entirely on board can potentially quadruple the server throughput.

2. Characteristics of Internet Streaming Protocols

There are three main streaming protocols used on the Internet today: Windows Media Technology (WMT) from Microsoft, RealNetwork Streaming from RealNetworks, and QuickTime streaming from Apple Computers. Certain major content providers such as AOL and Yahoo also have their own streaming protocols. Despite their differences, the streaming protocols share some common characteristics.

Virtually all of the protocols utilize both a control connection and a data connection. Over the control connection, the media player sets up a streaming session with the media server, and passes along streaming control messages such as pause, fast-forward and rewind. In protocols that support variable bit rate streaming (i.e. adjusting streaming quality based on available bandwidth), the control connection is also used by the client to inform the server of changes in available bandwidth in the network. The control connection is typically a TCP connection established by the client. In the case of RealNetwork streaming and QuickTime streaming, the control connection uses the IETF standard RTSP (Real-Time Streaming Protocol) protocol [17] with vendor-specific extensions.

The data connection is used to transmit streaming media data from the server to the client. It can be based on either UDP or TCP. In virtually all streaming protocols, the data transmissions are "paced", that is, the data are not sent to the client all at once, but rather sent in blocks with delays between the blocks. The delays are determined by either the bit rate of the stream or the properties of the encoding scheme. The "pacing" of data transmissions is necessary to avoid overflowing the client-side buffer, particularly in the case of high-speed Internet connections. Yet, the "pacing" causes many user-kernel context switches.

Almost all protocols share the following usage patterns:

- The control connection is mostly idle during the streaming session while the data connection has packets going from the server to the client in a periodic, repetitive fashion;
- Most of the processing logic is devoted to the messages sent over the control connection, while most of the traffic is caused by data blocks traveling on the data connection;
- A message over the control connection almost always changes the behavior of the streaming server on the data connection;
- The processing logic for the control messages typically does not scan the streaming data.

For example, a typical streaming protocol behaves as the following. When a client wants to view a streaming video, it initiates a TCP connection to the streaming server. The TCP connection is the control connection; the client sends the player version info, the URL of the streaming media, and client credentials to the server over this connection. There maybe challenges/responses between the client and the server until the server authorizes the client. Afterward, the server extracts the properties of the streaming media such as length and bit rates from the media file and sends the properties to the client. The client and the server then negotiate a suitable bit rate to serve the media to the client.

The server then initiates a UDP data stream (the data "connection") to the client, and starts sending streaming data in blocks to the client. A fixed-duration delay exists between transmissions of data blocks to "pace" the transmission. If network conditions change and the client wants to reduce the bit rate of the stream, the client sends a message over the control connection, and the server changes the composition of the data blocks as a result, sending frames encoded in a lower bit-rate.

3. Implementation Experience

Our project is to implement a high-performance streaming server on PC-server architectures. We started with a quick initial prototype and then tried to identify and eliminate the performance bottlenecks.

3.1 Initial prototype

We implemented the above-described streaming protocol on PC servers running Linux 2.4.16. In the initial prototype, for "time-to-market" reasons, we adopted a one-stream-per-process programming model. (In Linux 2.4.16, threads have the same kernel overhead as processes, so a one-stream-per-thread model would have similar performance.) This programming model is much easier to code and debug than the event-driven model in which one process serves many clients and processes messages from all clients.

In the prototype, a parent process listens on the designated port and waits for incoming control connections. For each new client it spawns a child process to handle the client. The child process first performs the normal session exchanges with the client, then initiates the data connection to the client. Afterwards, the child process sends the first block of streaming data to the client, sleeps for a fixed duration, then sends the second block, and the process repeats. The sleep time is calculated using the block size and the bit rate of the streaming media.

The child process also monitors incoming messages over the control connection from the client. If a "seek", "fast-forward", or "rewind" message arrives, the process adjusts the starting location of the data block in the streaming file. If a "bit rate upgrade" or "bit rate downgrade" message arrives, the process changes the frame-selection procedure and selects frames encoded in different bit rates from the streaming media file.

While the implementation was straightforward and robust, its performance is less than desirable. On our initial experimental platform, which is a dual-processor 866MHz Pentium III PC server with 2GB of RAM and 20 disks, the throughput is only around 150Mb/s.

3.2 Identifying performance bottlenecks

We then embarked on the investigation to improve server performance. The first observation was that the system was limited by disk reads. It was suggested to us that perhaps a specially designed storage system is needed, instead of the regular UNIX file system that we have been using. Upon closer inspection, however, we realized that streaming files are accessed sequentially most of the time, which means that prefetching and large "disk allocation block size" can be very effective. Furthermore, servers today have a large amount of RAM and can easily afford a sizable file buffer cache for the video files. Since streaming workload tends to be heavily skewed [3], a large main memory cache can be quite effective at reducing the number of disk reads. Hence, we hypothesized that, combining the sequential access pattern and the large memory cache, a properly tuned UNIX file system implementation can work reasonably well for video files.

We went ahead and tuned the file system parameters, increasing the continuous-allocation size in the file system (i.e. the "extent-size") to 128KB and the pre-fetch size to 128KB. The tuning improved the server throughput to around 400Mb/s, under the workload where 100 video files are streamed concurrently, 10 on each disk. Each disk spindle can in fact support 120Mb/s of read throughput on those video files, rivaling the per-spindle throughput of some of the specially-designed streaming storage systems.

With the improvement in disk throughput, the system became CPU bound. We considered changing the programming model to event-driven, as doing so would reduce the number of processes and the associated kernel process-scheduling overhead. How-

Routine	Sampling Count	Percentage	Cumulative Percentage
file_read_actor	16508	18.6%	18.6%
csum_partial_copy_generic	12435	14.0%	32.6%
default_idle	11271	-	-
schedule	8141	9.1%	41.7%
USER	6592	7.4%	-
tcp_sendmsg	1868	2.1%	43.8%
__kfree_skb	1758	1.9%	45.7%
ace_interrupt	1721	1.9%	47.6%
skb_release_data	1386	1.5%	49.1%
ip_fw_check	1347	1.5%	51.6%

Table 1. Results from kernprof on the top 10 routines. The total sample count is 100000. The first column lists the routine name; USER means user-level code, and default_idle means CPU idle time. In this particular test, since we did not have enough client machines to drive the server to maximum load, the server's CPU is about 11% idle, as reflected in the sampling count for the routine "default_idle". We have found from other full-load tests that the system is indeed CPU bound. The second column is the total sample count of the routine as reported by kernprof. The third column is the percentage of the CPU usage that the routine incurs. The last column counts kernel routines only and is the cumulative sum of the CPU usage incurred by the kernel routines.

ever, changing the programming model is a significant effort, and before we plunge into that, we need to determine the potential payoff of the effort.

After noticing that most of the CPU is consumed by the kernel, we focused on performance profiling of the kernel using "kernprof" [18], the Linux kernel profiling tool. Kernprof uses the PC-sampling technique; for each kernel routine, it lists the number of times that a PC sample falls in that routine.

Table 1 shows the kernprof results of the top 10 routines from a typical streaming load test. During the test the system is serving 720 streams of a 500Kbps media file. Eight of the routines listed in the table are kernel code, and they account for 52% of the CPU usage.

From Table 1, we can see that data copying, which involves reading the file data from kernel space to user space (i.e. file_read_actor) and sending the data from the user space to kernel space (i.e. csum_partial_copy_generic), accounts for 33% of the CPU. Thus, if the data copies are eliminated completely, the server throughput can be improved by 33%.

This "back-of-envelope" calculation correlates with another experiment that we run. We changed the streaming server to use Linux's "sendfile" interface to send the streaming data blocks. Since Linux's "sendfile" implementation still incurs a copy from the file buffer cache to the network socket buffer (in other words, it eliminates just one instead of two copies), we expect to see half of the benefit of zero data copying. Indeed, the throughput improvement that we observed is between 15% and 20%.

The kernel's process scheduling routine, "schedule", which is the one most affected by the large number of processes, only accounted for 9% of the CPU. This significantly dampened our enthusaim for changing the programming model to event-driven. Even if the application is changed to an event-driven programming model and uses only one process, the kernel's CPU overhead would only be reduced by 9%. Of course, this is not a rigorous analysis; depending on how the event-driven application is constructed, it might have other benefits such as improved process cache hit ratios, etc. Nevertheless, given the significant coding effort involved in changing the application to event-driven model, it's clear that we should focus on eliminating data copying and other system bottlenecks instead of changing the programming model.

One aspect that the profiling data do not fully capture is the impact of user-kernel context switches on the system throughput. Even though researchers have long pointed out that context switches can be made cheap [9], expensive context switches are a fact of life for PC architectures running Linux. During the system's run-time we noticed that the number of context switches are high, over 10000 per second. This is due to the fact that the streaming server's main operation mode is to send a data block, sleep for some duration, wake up and send another one. Each data block transmission, sleep and wakeup are context switches between user-level and kernel.

Are there ways to eliminate the context switches? We notice that in the streaming server implementation, the actions of each child process are quite monotonic and repetitive, with few sophisticated processing logic. Hence, one possibility is to construct a state-machine for each child process to capture the monotonic operations, and "drop" it into the kernel. Combining this with data copy elimination could provide a significant performance boost.

Furthermore, such optimization is not limited to a particular streaming protocol. As discussed before, all streaming protocols share the common characteristics of repetitive data transmissions. Hence, the optimization applies to all streaming protocols. We call this optimization the "stream engine".

4. The Stream Engine Interface

We propose an optimization technique called "Stream Engine" for Internet streaming applications. This technique has been implemented as a kernel-level mechanism in our streaming server, and is suitable as an interface for TCP-offloading hardware for Internet streaming.

4.1 Definitions of a stream engine

A stream engine encapsulates the operations that the kernel (or potentially a TCP-acceleration hardware device) does for a streaming media stream. It is defined by a "context", which contains all the necessary information for the engine to perform its tasks. The user-level application passes the following pieces of information to the kernel, which stores them in the context of the stream engine:

- Data connection description: the transport mode (UDP or TCP) of the connection, and the file descriptor of the connection;

- Streaming source description: there are two types of streaming source:

  - type FILE, which contains the file descriptor of the streaming media file, the starting and stopping offsets (in bytes) of the data to be sent to the client;
  - type BUFFER, which contains the memory address of the buffer, length of the buffer, a flag indicating if the buffer is a circular buffer, and in the case of circular buffer, a user-level variable that is the end position of useful data in the buffer.

  Type FILE sources are used when streaming from an encoded streaming media file. Type BUFFER sources are used when streaming from a live source, in which case the encoded stream data are fed to the streaming server and stored in a user-level buffer by the receiving process.

- Transmission interval in milliseconds. The stream engine will always transmit one data block per interval.

- Data block assembly instructions, which tell the engine how to assemble data packets from the file data or buffer data. Clearly, the instructions depend on the streaming protocol and the encoding format. There can be two kinds of instructions:
  - "simple stepping", which means that the stream engine should send a fixed-size block of data at every interval with a certain stride between each block. The description includes the block size and the stride size.
  - a data transformation routine: the stream engine will read a block of data from the stream source into a buffer, then apply the data transformation routine to obtain a new buffer to be sent to the client.

  Simple stepping is used with encoding schemes that encode the streaming media file into equal-sized blocks. Some popular streaming media formats follow this scheme.

  The data transformation routine is versatile and can support any encoding scheme. In our implementation, we use it primarily for streaming quality adaption, where the routine extracts the correct frames from the data blocks and assembles them to be sent to the client. Due to limited resources, we did not look into mechanisms to support arbitrary data transformation routines in a safe manner (such as those proposed in SPIN [2]), but rather coded a few common ones for the streaming application to choose from.

- File descriptor of the control connection. Any incoming message on the connection terminates the stream engine's operation immediately, and the kernel returns to the application for further instructions.

We implemented stream engine as a blocking system call. The system call takes as an argument a data structure "stream_eng_ctx" which encloses the above pieces of information. The call returns when the streaming operation completes or when a message arrives at the control connection. The return parameters include the reason that the system call returned, and the total number of data blocks sent to the client.

In other words, the stream engine runs in the kernel until some action needs to be taken on the stream (e.g. a control message arrives from the client to change the stream rate or position), at which point it terminates and the call returns to the user application. The user application can then initiate a new stream engine call. Hence, changes in a video stream session, for example, changing the transmission rate, is accomplished by having a message sent to the control file descriptor to terminate the current stream engine, and then initiating a new stream engine.

In our streaming server implementation, the child process initiates a stream engine by calling the system call, when it is ready to start streaming data to the client. The child process then waits for the call to return. After the call returns, depending on the reason, the child process either processes the control messages and then starts a new stream engine, or terminates the stream and cleans it up.

In our current design, the admission control of the streams is handled by the user level application, and the kernel does not perform further resource checking. The reason is that in the streaming server, the admission control is not purely resource based, but rather depends on licensing terms and network configuration. However, the design does have the drawback that if the user application is not careful and over-commits the resource, the stream engine would not be able to pump data out at the specified data rate. In future versions we plan to investigate adding resource checks in the kernel for stream engines.

4.2 Performance improvements from stream engine

We applied the stream engine optimization to the streaming server described in Section 1.3. The end result is that the throughput is almost doubled. While the old implementation can service 720 streams of 500Kbps media at 90% of the system capacity, the new implementation can service 1360 streams of 500Kbps media at 90% of the system capacity. The improvement showed that eliminating context switches leads to about 60% increase in system throughput.

Table 2 lists the top routines from the kernel profile result for the new implementation. The top nine kenel routines account for about 50% of total CPU used. Comparing Table 1 and Table 2, one can see that stream engine eliminates the data copies and the context switch and process scheduling overhead. Instead, the overheads of checksum calculation (it has to be done somehow), timer handling (due to sleep/wakeup of the stream engines) and the NIC processing routines become more prominent. Some of these overheads can be alleviated by the use of TCP offloading hardware.

On a new state-of-art PC server with dual-processor 1.7Ghz Pentium system with 2GB of RAM and 20 disks, the streaming server with stream engine optimization can achieve 1.2Gbps throughput on streaming 300Kbps media files.

5. Implications for TCP Offloading Hardware

Recently there have been a number of commercial TCP Offloading Engine (TOE) products [22–1, 21, 19, 6, 12]. Their primary target is the IP storage market (e.g. iSCSI), and the goal is to offload TCP/IP processing from the host CPU. Some of the TOE products are implemented using network processors, while others are im-

Routine	Sampling Count	Percentage	Cumulative Percentage
stream_engine_op	16896	18.6%	18.6%
csum_partial	12075	13.3%	31.9%
default_idle	9343	-	-
timer_bh	2659	2.9%	34.8%
ace_interrupt	2657	2.9%	37.7%
__kfree_skb	2509	2.7%	40.4%
skb_clone	2297	2.5%	42.9%
kfree	2277	2.5%	45.4%
ace_start_xmit	2212	2.4%	47.8%
add_timer	1909	2.1%	49.9%

Table 2. Kernprof results from streaming server using stream engine. The total CPU sample size is 100000. The columns are similar to those in Table 1. Note that again the server is not driven to full load in this test.

plemented using ASICs. The TOE is usually embedded in a Network Interface Card (NIC) or a Host Bus Adapter (HBA).

We are investigating the use of TOE cards to improve the throughput of Internet streaming servers. Internet streaming, however, is different from IP storage and has quite different requirements.

Based on our experience, we believe that stream engine is an appropriate interface for TCP offloading engines. For Internet streaming applications, just offloading the processing of TCP/IP from the host computer brings limited benefits; it will not eliminate the data copying and the context switches. A full-fledged implementation of stream engine will bring the most performance benefits. Short of that, a TOE card must be able to work with the host system kernel's stream engine implementation. By this, we mean that the TOE's interface with the host system must support sending data from a linked-list of data buffers in the file buffer cache, with a starting offset and size of the data, and implementing the packet assembly routine on the file buffer data.

We illustrate our observations using back-of-envelope calculations based on full kernel profile results from our experiments:

- Using a TOE that does not work with stream engine. We calculate the benefit of a conventional TOE NIC that simply offloads the TCP/IP protocol processing from the host CPU. Assuming that the TOE offloads all checksum calculations, TCP stack operations, IP stack operations, NIC transmission and interrupt processing, the throughput can be improved by about 25%.

- Using a TOE that works with stream engine. If stream engine is implemented in the host system's kernel, and the TOE card can work with the implementation to offload the TCP/IP send/receive processing, checksum calculation,and NIC interrupts, the throughput can be improved by up to 60%. That is, in addition to the improvement from the stream engine implementation, the system's throughput can be improved further by 60%.

- Using a TOE that implements stream engine. If the TCP offloading hardware is powerful enough to implement the entire stream engine, we find that the load on the host CPU is reduced by 75%. In other words, if, in addition to TCP offloading, the TOE card also handles the stream engine logic including the timer processing, socket buffer management and disk read processing, then the host CPU only needs to handle a quarter of load. Assuming that the TOE card has enough parallel processing power to keep up with the host CPU, this implies that the system throughput can be potentially quadrupled.

In summary, any TOE for Internet streaming application needs to work with stream engine either as part or all of implementation. The above projections assume that the TOE card's CPU is powerful enough to keep up with the host CPU while offloading the work of the stream engine. In reality, the CPUs on the TOE cards are typically slower than the host CPU, and multiple TOE cards, each handling a subset of the streams, would be needed for maximum throughput.

6. Related Work

Research has long pointed out the need to eliminate data copying when building high-performance network servers [4]. Many kernel mechanisms and interfaces have been proposed to address data copy elimination, including fbufs [5], I/O Lite [16] and sendfile [10]. Our work confirmed the importance of copy elimination in high-performance streaming servers. The existing schemes however do not alleviate the extensive amount of context switches incurred by streaming applications. We showed that for streaming protocols, eliminating context switches is equally important.

Recently there have also been a number of projects looking at building high-performance streaming servers using either specialized operating systems or specialized hardware. For example, "Hi-Tactix" is a kernel built by Hitachi system development laboratory that is a streaming oriented real-time OS providing high I/O performance and transmission rate guarantees [11]. We are also aware of at least one small company building blade servers with specialized I/O backplane for Internet streaming purposes. Different from these efforts, we focus on improving streaming throughput on Linux by introducing a generic kernel optimization, and we are mainly interested in generic PC servers.

The debate between event-driven programming model and one-stream-per-process programming model is not limited to streaming servers. For example, it's well known that an event-driven Web server such as Flash [15] performs much better than a one-client-per-process Web server such as Apache [14]. However, event-driven programming models are typically harder to develop and harder to maintain. Our experience with the streaming server prototype and stream engine seems to suggest that, at least in the case of Internet streaming applications, changing the programming model to event-driven would provide limited benefits, while eliminating data copies and context switches provides much higher payoff and entails much shorter development time. We note that recently other researchers have arrived at similar conclusions [20].

Extensible operating systems [2, 7, 13, 8] would make it much easier to implement optimizations such as stream engine. For example, the SPIN operating system provides a ready mechanism for user applications to drop arbitrary data transformation routines into the kernel. Unfortunately, these operating systems are not yet suitable for commercial products. Hence, we designed stream engine for Linux to accommodate streaming server specific optimizations.

Finally, the design of better streaming protocols over the Internet is an active research area which this paper is not focused on. Rather, our study looks into what it takes for a server to perform well for existing Internet streaming protocols. We believe that the stream engine interface will benefit future variations of streaming protocols, as eliminating data copies and context switches is key to high throughput for many networking applications.

7. Conclusion and Future Work

In this paper, we introduce the "stream engine" kernel interface, and show that it is a suitable kernel optimization for Internet streaming applications. We have implemented the stream engine mechanism in the kernel and have converted one streaming server implementation to use it. We are currently in the process of converting other Internet streaming protocol implementation to use stream engine.

Our experience shows that in addition to data copying, user-kernel context switches can also reduce system throughput significantly, particular for applications that incur frequent context switches due to "paced" data transmissions. We are currently looking into whether stream engine can benefit other non-streaming protocols that incur "paced" data transmissions, and whether the interface needs to be further extended to support those protocols.

We are also working with TOE vendors to investigate having the TOE interface with a host system's stream engine implementation. In particular, we are looking into the partition of responsibilities between the host system and the TOE hardware and the kernel/hardware interfaces in this case.

References

[1] Alacritech Inc. The 1000x1 Internet protocol processor: Patented, third-generation TCP/IP Offload Engine (TOE) ASIC. In *http://www.alacritech.com/html/081902a.html*, Aug. 2002.

[2] B. N. Bershad, S. Savage, P. Pardyak, E. G. Sirer, M. Fiuczynski, D. Becker, S. Eggers, and C. Chambers. Extensibility, safety and performance in the SPIN operating system. In *15th Symposium on Operating Systems Principles*, pages 267–284, Copper Mountain, Colorado, 1995.

[3] J. V. der Merwe, S. Sen, and C. Kalmanek. Streaming video traffic: Characterization and network impact. In *Proceedings of 7th International Workshop on Web Content Caching and Distribution*, Aug. 2002.

[4] P. Druschel. Operating systems support for highspeed networking. Technical Report TR 94-24, Department of Computer Science, University of Arizona, Oct. 1994.

[5] P. Druschel and L. L. Peterson. Fbufs: A high-bandwidth cross-domain transfer facility. In *Symposium on Operating Systems Principles*, pages 189–202, 1993.

[6] Intel. Intel PRO/1000T IP storage adaptor. In *http://www.intel.com/network/connectivity/resources/doc_library/data_sheets/pro1000_T_IP_SA.pdf*, Feb. 2002.

[7] M. F. Kaashoek, D. R. Engler, G. R. Ganger, H. M. Briceno, R. Hunt, D. Mazieres, T. Pinckney, R. Grimm, J. Jannotti, and K. Mackenzie. Application performance and flexibility on exokernel systems. In *Proceedings of the 16th ACM Symposium on Operating Systems Principles (SOSP '97)*, pages 52–65, Saint-Malo, France, October 1997.

[8] I. M. Leslie, D. McAuley, R. Black, T. Roscoe, P. T. Barham, D. Evers, R. Fairbairns, and E. Hyden. The design and implementation of an operating system to support distributed multimedia applications. *IEEE Journal of Selected Areas in Communications*, 14(7):1280–1297, 1996.

[9] J. Liedtke. Improving IPC by kernel design. In *14th ACM Symposium on Operating System Principles (SOSP)*, Dec. 1993.

[10] Linux man page. Linux system call sendfile(2). In *http://www.die.net/doc/linux/man/man2/sendfile.2.html*, 2000.

[11] D. L. Moai. Cost-effective streaming server implementation using hi-tactix. In *ACM Multimedia 2002*, 2002.

[12] J. Mogul. TCP offload is a dumb idea whose time has come. In *Proceedings of the 9th Workshop on Hot Topics in Operating Systems (HotOS IX)*, 2003.

[13] A. B. Montz, D. Mosberger, S. W. O'Malley, L. L. Peterson, T. A. Proebsting, and J. H. Hartman. Scout: A communications-oriented operating system (abstract). In *Operating Systems Design and Implementation*, page 200, 1994.

[14] Open-Source Community. The Apache Software Foundation, 2003. http://www.apache.org/.

[15] V. S. Pai, P. Druschel, and W. Zwaenepoel. Flash: An efficient and portable web server. In *Proceedings of the 1999 USENIX Technical Conference*, Monterey, CA, June 1999.

[16] V. S. Pai, P. Druschel, and W. Zwaenepoel. IO-Lite: A unified I/O buffering and caching system. *ACM Transactions on Computer Systems*, 18(1):37–66, 2000.

[17] H. Schulzrinne, A. Rao, and R. Lanphier. Real time streaming protocol (RTSP). RFC 2326, IETF Network Working Group, 1998.

[18] SGI - Developer Central Open Source. Linux Kernprof (kernel profiling). In *http://oss.sgi.com/projects/kernprof/*, 2002.

[19] Trebia Networks, Inc. SNP-1000i dual port TCP offload engine. In *http://www.trebia.com/products/snp_1000i.shtml*, Oct. 2002.

[20] R. von Behren, J. Condit, and E. Brewer. Why events are a bad idea (for high-concurrency servers). In *Proceedings of the 9th Workshop on Hot Topics in Operating Systems (HotOS IX)*, 2003.

[21] WindRiver Systems, Inc. Complete TCP/IP offload for high-speed ethernet networks. In *http://www.windriver.com/products/tina/tcpip_offload.pdf*, Sept. 2001.

[22] E. Yeh, H. Chao, V. Mannem, J. Gervais, and B. Booth. Introduction to TCP/IP Offload Engine (TOE). In *http://www.10gea.org/SP0502IntroToTOE_F.pdf*, Apr. 2002.

STREAMING FLOW ANALYSES FOR PREFETCHING IN SEGMENT-BASED PROXY CACHING TO IMPROVE DELIVERY QUALITY

Songqing Chen[1], Bo Shen[2], Susie Wee[2] and Xiaodong Zhang[1]
[1]*College of William and Mary,* [2]*Hewlett-Packard Laboratories*

Abstract Segment-based proxy caching schemes have been effectively used to deliver streaming media objects. However, this approach does not always guarantee continuous delivery because the to-be-viewed segments may not be cached in the proxy in time. The potential consequence is the playback *jitter* at the client side due to the proxy delay in fetching these uncached segments, thus we call the problem *proxy jitter.* Aiming at improving the media delivery quality for segment-based caching schemes, in this paper we propose two simple and effective prefetching methods, namely, *look-ahead window based prefetching* and *active prefetching* to address the problem of proxy jitter. We focus on presenting streaming flow analyses on proxy and network resource utilizations and consumptions, performance potentials and limits of the two prefetching methods for different segment-based schemes under different network bandwidth conditions. Our study also provides some new insights into relationships between proxy caching performance and the quality of streaming. For example, we show that the objective of improving the byte hit ratio in a conventional proxy and the unique objective of minimizing the proxy jitter to deliver streaming media objects can have conflicting interests. Trace-driven simulations show the effectiveness of our prefetching methods, and further confirm our analyses.

1. Background and Motivation

Proxy caching technique has been widely and effectively used to cache the text-based objects in Internet so that subsequent requests to the same object can be repeatedly and directly served from the proxy that is close to the requesting client. Typical proxy systems include CERN httpd [7], Harvest [1], and Squid [11]. However, with the proliferation of the multimedia objects in Internet, the existing proxy infrastructure is challenged due to their typical large sizes and continuous streaming requirement. A segment-based approach developed for proxy caching is to cache media objects in segments instead of their entirety to reduce the user perceived startup latency and to reduce the network traffic to media servers and the disk bandwidth requirement on the media server. An example of such an approach is the prefix caching [10] that seg-

F. Douglis and B.D. Davison (eds.), Web Content Caching and Distribution, 171-186.

ments the media object with a prefix segment and a suffix segment and always caches the prefix segments. These segment-based caching schemes can be classified into two types in terms of segment sizes. The first one is to use uniformly sized segments. For example, authors in [9] consider the caching of fixed sized segments of layer-encoded video objects. More recently, the adaptive-lazy segmentation strategies have been proposed in [2], in which each object has itself segmented as late as possible by using an uniform segment length. We generally call them as the uniform segmentation strategy[1]. The second type is to use exponentially sized segments, where media objects are segmented in a way that the size of a segment increases exponentially from that of its preceding one [12]. We generally name them as the exponential segmentation strategy.

However, these segment-based proxy caching schemes do not always guarantee a continuous delivery because the to-be-viewed segments may not be loaded in the proxy in time when they are accessed. We call this problem *proxy jitter*. (On the client side, this problem is called "playback jitter"). A straightforward solution to eliminate the playback jitter is to store the full object by a viewing client. This special solution is also called "downloading" that requires a very large buffer size and introduces a prolonged startup delay. Unfortunately, this is not a cost-effective method. Many clients have limited buffer space and they rely on media proxies and/or servers to ensure the quality of service. Once a playback starts in a client, pausing in the middle caused by the proxy jitter is not only annoying but can also potentially drive the client away from accessing the content. The essential issue for a proxy using a segment-based caching scheme is to fetch and relay the demanded segments to the client in time.

The key to solve the proxy jitter problem is to develop prefetching schemes to preload the uncached segments in time. Some early work have studied the prefetching of multimedia objects [6, 5, 9, 8]. For layered-encoded objects [9, 8], the prefetching of uncached layered video is done by always maintaining a prefetching window of the cached stream, and identifying and prefetching all the missing data within the prefetching window a fixed time period before its playback time. In [6], the prefetching is used to preload a certain amount of data so as to take advantage the caching power. In [5], a proactive prefetching method is proposed to utilize any partially fetched data due to the connection abortion to improve the network bandwidth utilization. In [4], authors consider the playback restart in interactive streaming video applications by using an optimal smoothing technique to transmit data for regular playback.

To our best knowledge, prefetching methods have not been seriously studied in the context of segment-based proxy caching [3]. Specifically, neither have the previous prefetching methods considered the following unique conflicting interests in delivering streaming media objects. On one hand, *proxy jitter* is caused by the delayed prefetching of uncached segments, which clearly suggests that the proxy prefetch the uncached segments as early as possible. On the other hand, the effort of aggressively prefetching the uncached segments requires a significant increase of the buffer space for temporar-

[1]The term "uniform" in this strategy only applies to each object. However, different objects may have different segment lengths.

ily storing the prefetched data and the bandwidth to transfer the data. Even worse, the client session may terminate before such a large amount of prefetched segments are accessed. This fact suggests that the proxy prefetch the uncached segments as late as possible. Thus, an effective media streaming proxy should be able to decide when to prefetch which uncached segment aiming at minimizing the *proxy jitter* subject to minimizing the related overhead (wasted space and bandwidth).

In order to improve the media delivery quality for segment-based caching schemes, in this paper, we propose two simple and effective prefetching methods, namely, *look-ahead window based prefetching* and *active prefetching* to address the problem of proxy jitter. We focus on presenting streaming flow analyses on proxy and network resource utilizations and consumptions, performance potentials and limits of the two prefetching methods for different segment-based schemes under different network bandwidth conditions. Our study also provides some new insights into relationships between proxy caching performance and the quality of streaming. For example, we show that the objective of improving the byte hit ratio in a conventional proxy and the unique objective of minimizing the proxy jitter to deliver streaming media objects can have conflicting interests. Conducting trace-driven simulations, we show the effectiveness of our prefetching methods, and further confirm our analyses.

The rest of this paper is organized as follows. The look-ahead window based prefetching method and the active prefetching method are presented in Section 2 along with the related streaming flow analyses. We further discuss insights of media proxy streaming in Section 3. We evaluate the proposed methods in Section 4. We make concluding remarks in Section 5.

2. Prefetching Methods for Segment-based Proxy Caching

In this section, we present two prefetching methods, namely look-ahead window based prefetching and *active prefetching*, along with their streaming flow analyses. Both prefetching methods consider achieving the two objectives jointly: (1) to fully utilize the storage resource, and (2) to minimize *proxy jitter*.

We have made the following assumptions in our analyses.

- A media object has been segmented and is accessed sequentially;
- The bandwidth of the proxy-client link is large enough for the proxy to stream the content to a client smoothly; and
- Each segment of the object can be delivered by a media server in a unicast channel.

Considering that the prefetching is conducted segment by segment, we define the following related variables.

- B_s: the average streaming rate of object segments in bytes/second (can be extracted from the object's inherent encoding rate), which can be viewed as the average network bandwidth consumption of the proxy-client link;
- B_t: the average network bandwidth of the proxy-server link;

- k: the total number of segments of an object;
- L_i: the length of the ith segment of an object, and $L_1 = L_{base}$ is the size of the first segment, also known as the base segment size. For the uniform segmentation strategy, $L_i = L_1$; for the exponential segmentation strategy, $L_i = 2^{i-1}L_1$, where $k \geq i > 1$; (Note: in [12], $L_i = 2^{i-2}L_1$ $(i > 2)$ since the segment is numbered from 0, while we number it from 1 for the consistency with the uniform segmentation strategy.)
- W: the amount of the prefetched data in bytes that are not used by the client due to an early termination (a client may only want to view a portion of the media object). The prefetching requires resources of (1) the buffer to temporarily store them; and (2) the network bandwidth used to transfer them. Both the buffer size and the network transferring cost are proportional to W. Denoting the networking cost as $f_n(W)$, the buffer cost as $f_b(W)$ and the wasted resource as as $f_w(W)$, we have $f_w(W) = f_n(W) + f_b(W)$. As long as W is known, the wasted resource cost can be determined.

2.1 Look-ahead window based prefetching method

The major action of the look-ahead window based prefetching is to prefetch the succeeding segment if it is not cached when the client starts to access the current one. The window size is thus fixed for the uniform segmentation strategy and is exponentially increasing for the exponential segmentation strategy.

Considering the streaming rate B_s and the network bandwidth B_t, we propose the look-ahead window based prefetching method under the following three different conditions.

1 $B_s = B_t$.

Ideally, in this case, the streaming flow can be smooth from the server to proxy and then to a client without an effort of prefetching, thus, no buffer is needed. However, in practice, we do need a buffer denoted as $Buffer_{network}$ to smooth the network jitters caused by network traffic and congestion. The size of $Buffer_{network}$ can be experimentally determined in practice. For example, it can be large enough to buffer media segments of 5 seconds. In this case, the preloading of the next uncached segment always starts 5 seconds before the streaming of the current segment is complete.

2 $B_s < B_t$.

In this case, no prefetching is needed, but we need a buffer to hold overflowed streaming data in the proxy from a media content server, which is denoted as $Buffer_{overflow}$. The preloading of the next uncached segment does not start until the streaming of the current segment is complete. The size of $Buffer_{overflow}$ is thus at least

$$L_i - \frac{L_i}{B_t} \times B_s = L_i \times (1 - \frac{B_s}{B_t}). \quad (1)$$

For different segmentation strategies, the buffer size can be calculated as follows:

- For the uniform segmentation strategy, the size of $Buffer_{overflow}$ is at least $(1 - \frac{B_s}{B_t})L_1$.
- For the exponential segmentation strategy, the size of $Buffer_{overflow}$ is at least $(1 - \frac{B_s}{B_t})L_i$, which increases exponentially.

3 $B_s > B_t$.

In this case, both segment prefetching and buffering are needed. A buffer denoted as $Buffer_{starvation}$ is used in the proxy to prevent the streaming starvation of a client. Assume the prefetching of the next uncached segment S_{i+1} starts when the client starts to access the position x in the current segment S_i. Thus, x is the position that determines the starting time of prefetching, called the prefetching starting point. To denote y as $y = L_i - x$ and to guarantee the in-time prefetching of the next uncached segment, we have

$$\frac{y + L_{i+1}}{B_s} \geq \frac{L_{i+1}}{B_t}, \tag{2}$$

which means

$$y \geq \frac{L_{i+1} \times (S - T)}{T}. \tag{3}$$

Since $y = L_i - x$, thus

$$x \leq L_i - \frac{L_{i+1} \times (B_s - B_t)}{B_t}. \tag{4}$$

We can calculate the prefetching starting point as the percentage of the current segment by dividing x by L_i, which leads to

$$\frac{x}{L_i} \leq 1 - \frac{L_{i+1}}{L_i} \times (\frac{B_s}{B_t} - 1) \tag{5}$$

Equation (5) means to prefetch the next uncached segment when the client has accessed the $1 - \frac{L_{i+1}}{L_i} \times (\frac{B_s}{B_t} - 1)$ portion of current segment. Accordingly, the size of the minimum buffer size for $Buffer_{starvation}$ is $\frac{y}{B_s} \times B_t$, which is $L_{i+1} \times (1 - \frac{B_t}{B_s})$. Once we know the minimum buffer size, we know that in the worst case, the fully buffered prefetched data may not be used by the client, which means the maximum amount of wasted prefetched data, W, that has the same size as the buffer. Thus, we always give the minimum buffer size by the following analysis.

For different segmentation strategies, the situations are as follows:

- For the uniform segmentation strategy, by Equation (4), we have $\frac{x}{L_i} \leq 2 - \frac{B_s}{B_t}$. It implies that B_s could not be 2 times larger than B_t. The minimum size of $Buffer_{starvation}$ is $L_1 \times (1 - \frac{B_t}{B_s})$. The prefetching of the next uncached segment starts when the client has accessed to the $2 - \frac{B_s}{B_t}$ portion of the current segment.

- For the exponential segmentation strategy, by Equation (4), we have $\frac{x}{L_i} \leq 3 - 2 \times \frac{B_s}{B_t}$. It implies that B_s could not be 1.5 times larger than B_t. The minimum size of $Buffer_{starvation}$ is $L_{i+1} \times (1 - \frac{B_t}{B_s})$, which increases exponentially. The prefetching of the next uncached segment starts when the client has accessed the $3 - 2 \times \frac{B_s}{B_t}$ portion of the current segment.

Above analysis shows that this look-ahead window based prefetching method does not work when $B_s > 1.5B_t$ for the exponential segmentation strategy, and it does not work when when $B_s > 2B_t$ for the uniform segmentation strategy.

In addition, since $B_s > B_t$, we have

$$\begin{aligned} B_s > B_t \Rightarrow & \quad \frac{B_s}{B_t} && > 1 \\ \Rightarrow & \quad 2 \times \frac{B_s}{B_t} - \frac{B_s}{B_t} && > 1 \\ \Rightarrow & \quad -\frac{B_s}{B_t} && > 1 - 2 \times \frac{B_s}{B_t} \\ \Rightarrow & \quad 2 - \frac{B_s}{B_t} && > 3 - 2 \times \frac{B_s}{B_t}. \end{aligned} \tag{6}$$

The left side of Equation (6) represents the prefetching starting point for the uniform segmentation strategy, while the right side denotes that for the exponential segmentation strategy. Thus, Equation (6) states that the prefetching of the next uncached segment for the exponential segmentation strategy is always earlier than that for the uniform segmentation strategy, causing a higher possibility of wasted resources.

Since the condition of $B_s > B_t$ is quite common in practice, the look-ahead window based prefetching method has a limited prefetching capability in reducing the proxy jitter. Next, we will address its limit by an active prefetching method.

2.2 Active prefetching method

If the prefetching is conducted more aggressively, we are able to further reduce proxy jitter, and of course, which will also consume more resources. The basic idea of our second method, active prefetching, is to preload uncached segments as early as the time when the client starts to access a media object. We define the following additional notations for this prefetching method.

- n: the number of cached segments of a media object;
- m: when the in-time prefetching of $n + 1^{th}$ segment is not possible (which will be discussed soon), the proxy should start to prefetch a later segment m once the client starts to access an object, where $m > n + 1$.

We also re-define the prefetching starting point, x, as the position in the first n cached segments (instead of a position in the nth segment for the look-ahead window based prefetching method) that is accessed by a client. As soon as this prefetching starting point is accessed, the prefetching of $n+1^{th}$ segment must start in order to avoid the proxy jitter.

Based on Equations (2) to (5), we obtain the prefetching starting point x as

$$x \le \sum_{i=1}^{i=n} L_i - \frac{L_{n+1} \times (B_s - B_t)}{B_t}. \tag{7}$$

Next, we will discuss the active prefetching alternatives for different segmentation strategies.

Uniform segmentation strategy. As we know, when $B_s > 2B_t$, it is impossible for the uniform segmentation strategy using look-ahead window based prefetching to completely avoid proxy jitter. However, the active prefetching may work this situation. Based on Equation (7), we obtain

$$x \le nL_1 - \frac{\frac{B_s}{B_t} - 1}{n} \tag{8}$$

for the uniform segmentation strategy.

Equation (8) not only gives the prefetching starting point when $n+1 \ge \frac{B_s}{B_t}$, it also implies that if $n+1 < \frac{B_s}{B_t}$, the in-time prefetching of $n+1^{th}$ segment is not possible! So when $n+1 < \frac{B_s}{B_t}$ and the segments between $n+1^{th}$ and $\frac{B_s}{B_t}^{th}$ are demanded, the *proxy jitter* is inevitable. Thus, when $n+1 < \frac{B_s}{B_t}$, to minimize the *proxy jitter*, the proxy should start the prefetching of a later segment, denoted as m, rather than the $n+1^{th}$ segment since it can not be prefetched in time. The prefetching of the mth segment should be faster than the streaming of the first m segments, which leads to

$$\frac{L_m}{B_t} \le \frac{mL_1}{B_s}. \tag{9}$$

We get $m \ge \frac{B_s}{B_t}$, and the corresponding minimum buffer size is thus

$$\frac{(m-1)L_i}{B_s} \times B_t = (1 - \frac{B_t}{B_s})L_1. \tag{10}$$

For the uniform segmentation strategy, the active prefetching works as follows when $B_s > 2B_t$. (When $B_s \le 2B_t$, it works the same as window-based prefetching method.)

- $n = 0$: No segment is cached. The proxy starts to prefetch the $\frac{B_s}{B_t}^{th}$ segment. Before the client accesses to this segment, the *proxy jitter* is inevitable. The minimum buffer size is $(1 - \frac{B_t}{B_s})L_1$.

- $n > 0$ and $n+1 < \frac{B_s}{B_t}$: The proxy starts to prefetch the $\frac{B_s}{B_t}^{th}$ segment once the client starts to access the object. If the segments between $n+1^{th}$ and $\frac{B_s}{B_t} - 1^{th}$

are demanded, the *proxy jitter* is inevitable. The minimum buffer size is $(1 - \frac{B_t}{B_s})L_1$.

- $n > 0$ and $n + 1 \geq \frac{B_s}{B_t}$: The prefetching of $n + 1^{th}$ segment starts when the client accesses to the position of $(n+1-\frac{B_s}{B_t})L_1$ of the first n cached segments. The minimum buffer size is $(\frac{B_s}{B_t} - 1) \times \frac{L_1}{B_s} \times B_t$, i.e., $(1 - \frac{B_t}{B_s})L_1$. The *proxy jitter* can be totally avoided.

Exponential segmentation strategy. For the exponential segmentation strategy, when $B_s > 1.5B_t$, the calculation of the prefetching starting point based on Equation (7) leads to

$$x \leq \sum_{i=1}^{i=n} L_i \times (1 - \frac{2^n}{2^n - 1} \times (\frac{B_s}{B_t} - 1)). \tag{11}$$

The right side of Equation (11) must be greater than 0, so Equation (11) not only determines the prefetching starting point, it also means that when $n < log_2(\frac{1}{2-\frac{B_s}{B_t}})$, the in-time prefetching of $n + 1^{th}$ segment is not possible, and when $B_s \geq 2B_t$, the in-time prefetching of any uncached segment can never be possible!

When $n < log_2(\frac{1}{2-\frac{B_s}{B_t}})$, to minimize the *proxy jitter*, the proxy should start the prefetching of some later segment, denoted as m since the $(n+1)^{th}$ segment can not be prefetched in time. Thus, the transferring of the mth segment must be faster than the streaming of the first m segments, which leads to

$$\frac{L_m}{B_t} \leq \frac{\sum_{i=1}^{i=m} L_i}{B_s}. \tag{12}$$

Thus, we get $m \geq 1 + log_2(\frac{1}{2-\frac{B_s}{B_t}})$ and the corresponding minimum buffer size is

$$2 \times \frac{L_{m-1}}{B_s} \times B_t = L_1 \times \frac{{B_t}^2}{2B_sB_t - {B_s}^2}. \tag{13}$$

For the exponential segmentation strategy, the active prefetching works as follows when $B_s > 1.5B_t$. (When $B_s \leq 1.5B_t$, it works the same as window-based prefetching method.)

1 $B_s > 1.5B_t$ and $B_s < 2B_t$.

- $n = 0$: No segment is cached. The proxy starts to prefetch the $1 + log_2(\frac{1}{2-\frac{B_s}{B_t}})^{th}$ segment once the client starts to access the object. The *proxy jitter* is inevitable before the client accesses to this segment. The minimum buffer size $L_1 \times \frac{{B_t}^2}{2 \times B_s \times B_t - {B_s}^2}$.
- $n > 0$ and $n \leq log_2(\frac{1}{2-\frac{B_s}{B_t}})$: The proxy starts to prefetch the $1 + log_2(\frac{1}{2-\frac{B_s}{B_t}})^{th}$ segment once the client starts to access this object. The

minimum buffer size is $L_i \times \frac{B_t}{B_s}$, where $i = 1 + log_2(\frac{1}{2-\frac{B_s}{B_t}})$. The *proxy jitter* is inevitable when the client accesses segments between the $n+1^{th}$ segment and $1 + log_2(\frac{1}{2-\frac{B_s}{B_t}})^{th}$ segment.

- $n > 0$ and $n > log_2(\frac{1}{2-\frac{B_s}{B_t}})$: The prefetching of the $n+1^{th}$ segment starts when the client accesses to the $1 - \frac{2^n}{2^n-1} \times (\frac{B_s}{B_t} - 1)$ portion of the first n cached segment. The minimum buffer size is $L_{n+1} \times \frac{B_t}{B_s}$, which will increase exponentially.

2 $B_s \geq 2B_t$.

No prefetching of the uncached segments can be in time. Whenever a client accesses to an uncached segment, the *proxy jitter* can not be reduced.

3. Segment-based Proxy Caching Strategies with Least Proxy Jitter

We have shown that even the active prefetching can not always guarantee the continuous media delivery. However, the analysis of the active prefetching also indicates that for any strategy, with enough segments being always cached in the proxy, the prefetching of the uncached segments can always be in time whenever necessary. The number of the segments being always cached in the proxy for this purpose, which we denote as q, is dependent on the segmentation strategy, B_s and B_t.

In this section, we determine q for different segmentation strategies to guarantee the continuous delivery without considering a cache space limit when $B_s > B_t$. Then we further look into its insights with a limited cache size.

3.1 Minimum number of segments cached for proxy jitter free

We use the notations as before, the prefetching of the $q+1^{th}$ segment to the kth segment must be faster than the streaming of the whole object, which means

$$\frac{\sum_{i=1}^{i=k} L_i}{B_s} \geq \frac{\sum_{i=1}^{i=k} L_i - \sum_{i=1}^{i=q} L_i}{B_t}, \tag{14}$$

where $\sum_{i=1}^{i=q} L_i$ is the amount of data should be cached. Thus,

$$\sum_{i=1}^{i=q} L_i \geq (1 - \frac{B_t}{B_s}) \sum_{i=1}^{i=k} L_i. \tag{15}$$

Applying Equation (15) to different segmentation strategies, q can be determined as follows.

- Uniform segmentation strategy

Since the L_{base} is the base segment length, the number of cached segment must be

$$q = \lceil \frac{(1 - \frac{B_t}{B_s}) \sum_{i=1}^{i=k} L_i}{L_{base}} \rceil. \tag{16}$$

- Exponential segmentation strategy

  Since $L_i = 2 \times L_{i-1}$, the number of cached segments must be

$$q = \lceil log_2 (\frac{(1 - \frac{B_t}{B_s}) \sum_{i=1}^{i=k} L_i}{L_{base}}) \rceil + 1. \tag{17}$$

3.2 Trade-off between low proxy jitter and high byte hit ratio

In a practical segment-based streaming environment with a limited cache size, it is common that for some objects, they have a large number of segments cached (e.g. very popular objects), which is very likely to be larger than their q values, while for some objects (e.g. not popular objects), they may have a lower number of segments cached in the system, which is possibly to be less than their q values. If we evict some rear segments of these popular objects whose cached number of segments is larger than their q values, the byte hit ratio of the system decreases. However, the released space can be used to cache the objects whose cached number segments is less than their q values, which reduces the proxy jitter. This scenario implies that there is a trade-off between the high byte hit ratio and the low proxy jitter with a limited cache size.

With the trade-off consideration, we further change the existing uniform and exponential segmentation strategies as follows with two additional notations:

- K_{min}: the minimum number of segments that if cached, no startup latency will be perceived by the clients as in [12];
- K_{jitter}: the minimum number of segments that if cached, no proxy jitter or startup latency will be caused. It is the maximum of K_{min} and q.

In the modified segmentation strategies, instead of caching the first K_{min} segments for reducing the startup latency when the object is initially accessed, the larger value of K_{min} and the minimum number of segments for guaranteed in-time prefetching, q, is calculated, as K_{jitter}, and the first K_{jitter} segments of the object is cached. However, when the cache space is demanded, these K_{jitter} segments other than the first K_{min} need to compete for the cache space according to their caching utility value. Thus, the decrease of the byte hit ratio can be used to trade for the reduction of the proxy jitter to some extent.

4. Performance Evaluation

To evaluate the performance of these prefetching methods under different segmentation strategies, an event-driven simulator has been built. We simulate the exponential segmentation strategy as in [12]. The uniform segmentation strategy differs from the

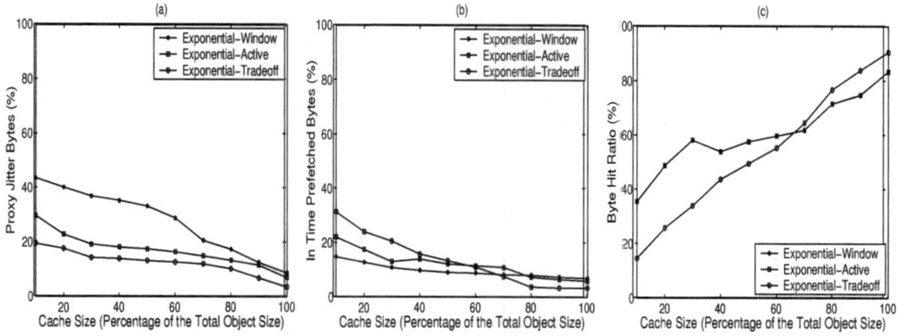

Figure 1. WEB: (a) Proxy Jitter Bytes, (b) Bytes Prefetched In Time and (c) Byte Hit Ratio

exponential segmentation strategy in that all objects are segmented in a uniform length of 1KB according to [2].

To evaluate the effectiveness of the prefetching methods, the major metric we use is *proxy jitter bytes.* It is the amount of data that are not served to the client in time by the proxy, thus causing the potential playback jitter at the client side, normalized by the total bytes the clients demand. It denotes the quality of the streaming service to the client. The *in-time prefetched* byte is used to represent how many bytes could be prefetched in time, normalized by the total bytes the clients demand. The *byte hit ratio* is defined as the amount of data delivered to the client from the proxy and normalized by the total bytes the clients demand. It is used to show the impact on caching performance when different prefetching methods are used.

Based on the previous analysis, we know that once the bandwidth of proxy-server link (or connection rate will be used in the following context) is less than half of the streaming rate, it may be impossible for the exponential segmentation strategy to prefetch any uncached segments in time. Thus, in this evaluation, the streaming rate of each object is set in the range of half to two times of the bandwidth of proxy-serve link. (Note that other than the results presented in the following context, we have also evaluated the performance with larger ranges of streaming rate and proxy-server bandwidth. The results show the similar trends with larger gaps between the uniform and exponential segmentation strategies, which we omit due to the page limit.)

4.1 Workload summary

To evaluate the performance, we conduct simulations based on two synthetic workloads. The first, named WEB, simulates accesses to media object in the Web environment in which the length of the video varies from short ones to longer ones. Since clients' accesses to videos may be incomplete, that is, a session started may terminate before the full media object is delivered, we use the second workload, named PART, to simulate this behavior. In this workload, 80% of the session terminates before 20% of the object is delivered. Both workloads assume a Zipf-like distribution ($p_i = f_i / \sum_{i=1}^{N} f_i, f_i = 1/i^{\alpha}$) for the popularity of the media objects, where α is

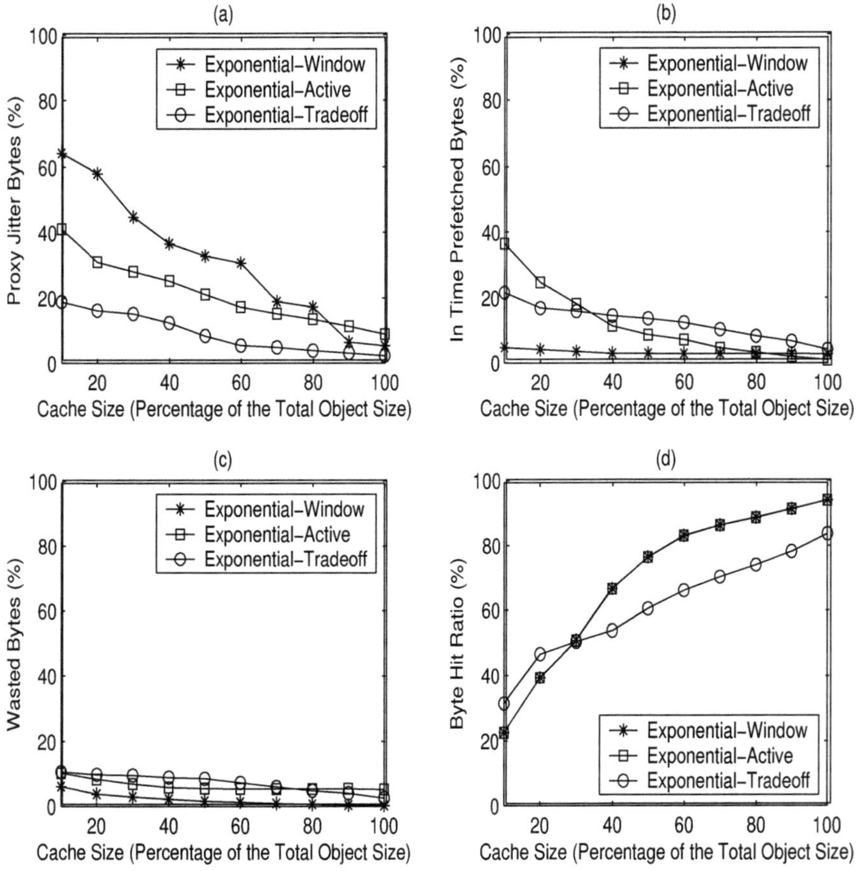

Figure 2. PART: (a) Proxy Jitter Bytes, (b) Bytes Prefetched In Time, (c) Bytes Wasted and (d) Byte Hit Ratio

set to 0.47 for both workloads. They also assume request inter arrival to follow the Poisson distribution ($p(x, \lambda) = e^{-\lambda} \times (\lambda)^x/(x!), x = 0, 1, 2...$) with a λ of 4.

Both WEB and PART have a total number of 15188 requests in a day for 400 different media objects. The total size of all media objects is 51 GB. The viewing length of objects ranges from 2 to 120 minutes.

4.2 Exponential segmentation strategy

In all the following figures, notation *Window* represents the look-ahead window based prefetching method, and notation *Active* represents the active prefetching method. The *Tradeoff* represents the prefetching method with the consideration of the trade-off between the byte hit ratio and the proxy jitter.

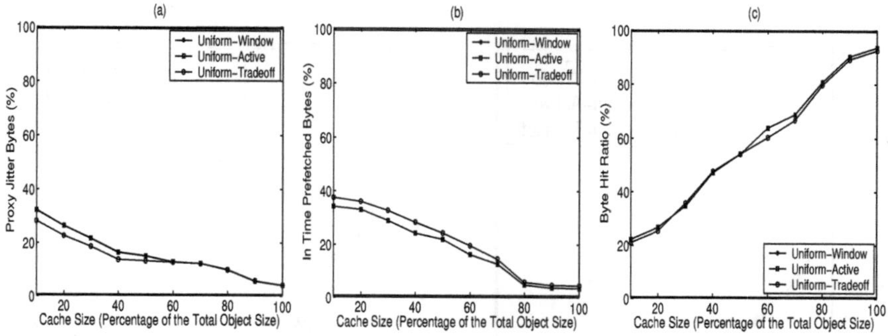

Figure 3. WEB: (a) Proxy Jitter Bytes, (b) Bytes Prefetched In Time and (c) Byte Hit Ratio

Figure 1 shows the performance results by using workload WEB for the exponential segmentation strategy. Figure 1(a) shows that *Exponential-Tradeoff* has the least proxy jitter bytes served to the clients and much less than that of *Exponential-Window* and *Exponential-Active*. As shown in Figure 1(c), although the byte hit ratio achieved by *Exponential-Tradeoff* decreases when the cache size is less than 60% of the total object size, the overall streaming service quality provided to the client is improved.

Figure 1(b) shows that *Exponential-Tradeoff* achieves the largest amount of data prefetched in time on average and decreases when the cache size increases. This is because more data has been cached in the proxy with more available cache space.

Figure 1(c) shows that the byte hit ratios for *Exponential-Window* and *Exponential-Active* are the same. This is expected as our previous analysis indicates that if the connection rate is larger than half of the streaming rate, only the amount of prefetched data for those two prefetching methods are different.

Figure 2 shows the results by using the workload PART for the exponential segmentation strategy. Figure 2(a) shows that *Exponential-Tradeoff* still achieves the least percentage of proxy jitter bytes served to the clients. Figure 2(b) shows that on average *Exponential-Tradeoff* achieves the largest amount of data prefetched in time.

For the partial viewing cases, a new metric, the *wasted* byte, is used to evaluate how much of the in-time prefetched data that is not actually used since the client terminates earlier. Figure 2(c) shows that all three prefetching methods produce some wasted bytes due to a large percentage of prematurely terminated sessions. *Exponential-Tradeoff* results in more wasted bytes than others since it gets more in-time prefetched data.

4.3 Uniform segmentation strategy

As aforementioned, in the experiments, the streaming rate of each object is set in the range of half to two times of the connection rate, thus the performance improvement of the uniform segmentation strategy does not look as significant as the exponential segmentation strategy. Also, for the uniform segmentation strategy, if the connection rate is larger than half of the streaming rate, the active prefetching and the look-ahead window based prefetching methods perform exactly same.

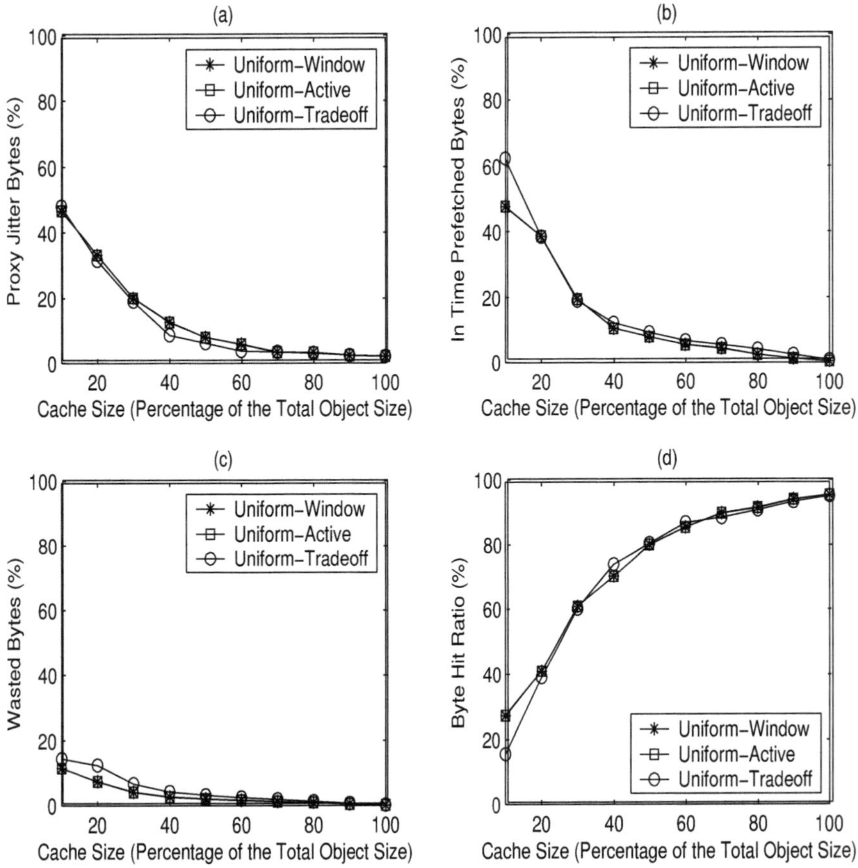

Figure 4. PART:(a) Proxy Jitter Bytes, (b) Bytes Prefetched In Time, (c) Bytes Wasted and (d) Byte Hit Ratio

Figure 3 shows the performance results by using workload WEB. Figure 3(a) shows that *Uniform-Tradeoff* has the least *proxy jitter* bytes on average. Although Figure 3(c) shows it achieves a lower byte hit ratio, the quality of the streaming service it provides to the client is the best. Figure 3(c) shows that the byte hit ratios achieved by *Uniform-Window* and *Uniform-Active* are the same. This is consistent to our previous analysis. On Figure 3(a) and Figure 3(b), they also achieve the same performance results. Figure 3(b) shows that *Uniform-Tradeoff* gets the largest amount of in-time prefetched data.

Figure 4 shows the results by using the partial viewing workload PART. Similar trends have been observed as in Figure 3, but with a smaller performance gap among different methods. Figure 4(a) shows that *Uniform-Tradeoff* still achieves the least amount of proxy jitter bytes, giving the best quality of streaming delivery to the clients. Figure 4(a) shows that *Uniform-Tradeoff* also gets the largest amount of data prefetched in time, while Figure 4(c) shows that *Uniform-Tradeoff* has more wasted data due to more in-time prefetched data.

5. Conclusion

In this study, focusing on the segment-based caching strategies with a uniform segment length or an exponentially increasing segment length, we have examined the prefetching issues in the segment-based caching strategies. We have presented and evaluated the look-ahead window based prefetching method, the active prefetching method and the prefetching method with the consideration of the trade-off between improving the byte hit ratio and minimizing the proxy jitter analytically and experimentally. The prefetching method with the trade-off consideration is the most effective in reducing the proxy jitter and can provide the best quality of streaming delivery to the clients.

We are implementing the prefetching in our media surrogate testbed and performing real tests.

References

[1] C. Bowman, P. Danzig, D. Hardy, U. Manber, M. Schwartz, and D. Wessels. Harvest: A scalable, customizable discovery and access system. Technical Report CU-CS-732-94, University of Colorado, Boulder, 1994.

[2] S. Chen, B. Shen, S. Wee, and X. Zhang. Adaptive and lazy segmentation based proxy caching for streaming media delivery. In *Proceedings of ACM NOSSDAV*, Monterey, CA, June 2003.

[3] S. Chen, B. Shen, S. Wee, and X. Zhang. Analysis and design of segment-based proxy caching strategies for streaming media objects. In *Proceeding of ACM/SPIE Multimedia Computing and Networking*, Santa Clara, CA, January 2004.

[4] J. K. Dey, S. Sen, J. F. Kurose, D. Towsley, and J. D. Salehi. Playback restart in interactive streaming video applications. In *IEEE Conference on Multimedia Computing and Systems*, Ottawa, Canada, June 1997.

[5] J. Jung, D. Lee, and K. Chon. Proactive web caching with cumulative prefetching for large multimedia data. In *Proceedings of WWW*, Amsterdam, Netherlands, May 2000.

[6] J. I. Khan and Q. Tao. Partial prefetch for faster surfing in composite hypermedia. In *Proceedings of the 3rd USENIX Symposium on Internet Technologies and Systems*, San Francisco, CA, March 2001.

[7] A. Luotonen, H. F. Nielsen, and T. Berners-Lee. Cern httpd. http://www.w3.org/Daemon/Status.html.

[8] R. Rejaie, M. H. H. Yu, and D. Estrin. Multimedia proxy caching mechanism for quality adaptive streaming applications in the internet. In *Proceedings of IEEE INFOCOM*, Tel-Aviv, Israel, March 2000.

[9] R. Rejaie, M. Handley, H. Yu, and D. Estrin. Proxy caching mechanism for multimedia playback streams in the internet. In *Proceedings of International Web Caching Workshop*, San Diego, CA, March 1999.

[10] S. Sen, J. Rexford, and D. Towsley. Proxy prefix caching for multimedia streams. In *Proceedings of IEEE INFOCOM*, New York City, NY, March 1999.

[11] Squid proxy cache. http://www.squid-cache.org/.

[12] K. Wu, P. S. Yu, and J. Wolf. Segment-based proxy caching of multimedia streams. In *Proceedings of WWW*, Hong Kong, China, September 2001.

SUBSCRIPTION-ENHANCED CONTENT DELIVERY

Mao Chen, Jaswinder Pal Singh, and Andrea LaPaugh
Department of Computer Science, Princeton University

Abstract In existing content delivery systems user accesses are popularly used for predicting the request pattern of contents. In novel web applications such as publish/subscribe services, users explicitly provide statements of interest in the form of subscriptions. These subscriptions provide another source of user information in addition to access patterns. This paper addresses the content delivery problem when user-stated interest is available. Each request by a user is either based on a notification about the availability of content that matches the user's subscriptions, or general browsing that is not based on the publish/subscribe service. We propose two approaches to content delivery that exploit both proactive push-time placement and passive access-time replacement based on the subscription information, the access pattern of subscribers, and that of non-subscribers. In our simulation-based evaluation, the two approaches are compared to an access-based caching only algorithm and to three approaches that were proposed for pure notification-driven accesses in our earlier study [5]. The results demonstrate that incorporating subscription information judiciously can substantially reduce the response time, even when only a small portion of accesses is driven by notifications and the subscription information does not reflect subscribers' accesses perfectly. To our knowledge, this work is the first effort to investigate general content delivery and caching enhanced by using subscription information.

1. Introduction

The information needs of content consumers form the key to intelligent caching and content delivery over the Internet. Existing content delivery systems typically determine these information needs based on access patterns. However, many content-intensive web services are based on users' subscriptions, which are statements of interest. This explicit user interest can therefore also be used as a basis for content delivery. Little exploration of subscription-enhanced content delivery has been done.

F. Douglis and B.D. Davison (eds.), Web Content Caching and Distribution, 187-204.

This paper considers an application scenario in which users may request contents either based on notifications that match their subscriptions or as part of general browsing that is not based on the publish/subscribe service. An example is news delivery. Many news sites such as CNN provide notification service that is based on matching the contents to users' subscriptions. Users may request news pages upon receiving the notifications or browse news sites independent of the notification service. The two types of accesses are called *notification-driven accesses* and *non-notification-driven accesses* or *general browsing* in this paper. In addition, a user may not read all the pages that match the user's subscriptions (and of which they therefore receive notifications).

Our goal is reducing the response time to end-users regardless of the type of access they make, by placing the contents in edge servers that are deployed close to end-users. Each server connects to a group of users, aggregates users' subscriptions, monitors users' accesses, and serves as the cache that is consulted when the requested contents are not in the client caches.

"Pushing to the edge" is a philosophy taken by many content delivery systems [1]. Our system has three important features. First, each content server performs autonomous placement based only on local information at the server. Therefore, our methodology is scalable when content production and consumption is highly dynamic and globally distributed. Second, our approach seamlessly combines proactive publisher-initiated placement and passive client-side caching. Third, this work addresses the storage management at edge servers based on two complementary information sources: the subscriptions from subscribers and the access patterns of all users.

Subscriptions form the basis for estimating future requests from subscribers. This estimation can be done at content generation time by matching the content to subscriptions from all users, which creates an opportunity for early content distribution (the matching of content to subscriptions may be done in an intermediary publish-subscribe service that sits between the user and the content source, and may be based on the entire content or subsets of the content or metadata). The proactive placement benefits all readers of the content, whether they explicitly provided their interest in advance or not.

On the other hand, subscription information by itself is insufficient for designing efficient content delivery algorithms. This is because subscriptions just imply the static distribution of subscribers' requests. Moreover, it is unrealistic to assume a subscriber will read every page that matches the user's subscriptions, so subscriptions may not even reflect the request distribution of subscribers perfectly. Therefore, the access patterns of users provide knowledge that is complementary to subscriptions for estimating the actual temporal and frequency request patterns.

We developed two approaches that exploit the type of access, access information, and subscription information. One approach combines all the information sources into a single replacement algorithm for managing the caches, while another one divides a server cache into two portions and processes two types of accesses using different algorithms. Our earlier study [5] presented a set of content distribution approaches for the case where all accesses are notification-driven (i.e. general browsing accesses were not considered in that study). This study determines which of the new approaches and adapted versions of the old approaches is superior for the new scenario.

As no real-world workload is available to us for this publish/subscribe-based application model, our evaluation study is simulation-based. We extend the simulator and workloads that are built in our earlier study based on the measured behaviour of a busy news site [22]. The new simulator and the workload model the relative composition of two types of accesses and the probability for a subscriber to read the pages that match their subscriptions. Compared to a caching approach based only on observed access patterns, all of our approaches are able to reduce the miss ratio substantially. Interestingly, this is true even when only a small portion of accesses is notification-driven and the probability for subscribers to follow notifications is low.

The contributions of this paper are as follow:

- Presenting the first study for enhancing general content delivery by taking advantage of publish-subscribe information when it is available;
- Proposing several content delivery mechanisms based on access type, subscription distribution, and access pattern, and comparing them under different scenarios to determine the best one;
- Experimentally demonstrating the importance of subscription information in reducing the miss ratio in local cache servers, even when such information is quite limited;
- Modeling the hybrid request sequence and the subscription quality, and incorporating them into a publish/subscribe workload generator.

The next section presents our content delivery strategies, including the new mechanisms proposed specifically for the scenario of this study as well as methods from our earlier study originally for pure notification-driven accesses but also show promise in this scenario. Section 3 introduces a news delivery simulator and the workload that models the hybrid user access and subscription quality. The simulation results are presented in section 4. Section 5 discusses related work to Internet content delivery. Section 6 draws conclusions and indicates future directions for this work.

2. Subscription-based content delivery

There are two obvious opportunities to deliver a page to a proxy-server. When a page is generated, a matching engine can determine that the page matches the subscriptions of some users of a proxy, and the page can be proactively placed in the server's cache. Alternatively, as in conventional caching systems, a page can be placed on the cache misses based on the fact of rather than the prediction or likelihood of users' accesses to a page. Our content placement is value-based. A page is given a value determined from the access pattern, the subscription information, and/or the type of access. A content distribution engine can use one or both placement opportunities, using a combination of the three information sources.

2.1 Access-based caching: the baseline approach

Caching algorithms passively place pages in a local server only at cache misses. The value of a page is purely based on the access pattern of the page. We use a replacement algorithm called Greedy-Dual* (GD*) [14] as the baseline algorithm. GD* determines the value of a page, *V*(*p*), from the access frequency, the access recency, the cost to fetch a page, and the size of the page, as in equation 1:

$$V(p) = L + \left(\frac{f(p) \cdot c(p)}{s(p)} \right)^{1/\beta} \qquad (1)$$

Where
L: inflation value to capture the access recency
$f(p)$: number of accesses on the page
$c(p)$: cost to fetch a page from the publisher
$s(p)$: page size
β: balance factor of popularity and temporal correlation

In our implementation, the reference count of a page is discarded when the page is evicted. The fetch cost is the network distance from the cache server to the origin publisher. The constant parameter β is set manually according to preliminary experiments.

On a cache hit of page *p*, GD* increases the reference count *f*(*p*) and recalculates *V*(*p*) based on the current inflation value *L*. A cache miss results in a placement of the requested page. If there is no room for the page, the old pages in the cache are evicted in the order of the pages' values; the inflation value *L* is set to be the value of the last evicted page.

By viewing any type of access equally, GD* can be used in either pure notification-driven content delivery or in content delivery that allows hybrid notification-driven accesses and general browsing.

2.2 Single cache and single replacement method

One category of our content delivery strategies combines subscription information and access information into a single evaluation function that is used in both push-time and access-time placement.

Two approaches in our earlier work

Our earlier study for pure notification-driven access identified two good approaches that combine access and subscription information. The first method replaces the frequency factor in equation 1 by equation 2, assuming a page is valuable if many users' subscriptions match the page and many users have requested the page. This approach is referred to as Subscription-GD*-1 (SG1).

$$f(p) = s + a \qquad (2)$$

Where

s : the number of subscriptions matching page p

a : the number of accesses of page p

Another approach, called Subscription-GD*-2 (SG2), takes into account the relation between the subscriptions and the accesses of a page. If every user requests a page of that user's interest exactly once, the difference between the number of subscriptions matching a page and that of the accesses to the page should be the amount of requests of the page in future. SG2 replaces the frequency factor in equation 1 by equation 3.

$$f(p) = s - a \qquad (3)$$

Where

s : the number of subscriptions matching page p

a : the number of accesses of page p

Both SG1 and SG2 base their decision on whether to store a page *P* at a proxy server *S* purely on page value. When there is not enough room at *S*, all the pages whose value is smaller than that of *P* are candidates for eviction. If the total size of all the candidates is smaller than the size of *P*, the placement algorithm aborts storing *P* at *S*. On a cache miss, failing to store a requested page means that the proxy server just fetches the page from the origin site and forwards it to the user without storing the page in the server's cache.

SG1 is directly applicable to the application scenario of this paper, while SG2 needs adaptation. Equation 3 assumes the number of subscriptions that match a page is always larger than the number of accesses of the page up to any point in time. The assumption holds when all requests are driven by notifications. However, if only some accesses are notification-driven as assumed in this study, the result of equation 3 may be negative, which is invalid in the evaluation function 1.

A simple way to deal with the invalid case is to set all negative results to 0 when using equation 3. In this way, when the number of accesses exceeds that of the subscriptions, SG2 ignores the frequency information and evaluates a page purely based on access recency. To distinguish from SG2, the above adapted version of SG2 is referred to as Restrictive-Subscription-GD*-2 or RSG2 in this paper.

First new approach: HUG

Our first new approach in this paper applies the same placement algorithm at both push-time and access-time, as do SG1 and RSG2. Different from SG1 and RSG2, the evaluation function in the new approach distinguishes the information associated with two types of accesses. The new approach is called Hybrid-User-GD* (HUG) in this paper, since it incorporates the type of access into the GD* framework using equation 1.

HUG analyzes the frequency information of two types of accesses separately. For notification-driven accesses, the difference between the number of subscriptions matching a page and that of notification-based accesses up to the current time implies the amount of notification-driven accesses in the future. In contrast, the number of non-notification-based accesses in the future is derived from that in the past, according to the traditional LFU algorithms. The following equation estimates the number of future requests of a page using the sum of two factors each of which is associated with one type of access. The evaluation function of HUG is equation 1 after substituting $f(p)$ with equation 4.

$$f(p) = a_{NS} + (s - a_S) \qquad (4)$$

Where

a_{NS} : number of accesses of page p from non - subscribers

s : number of subscriptions matching page p

a_S : number of accesses of page p from subscribers

Equation 4 interprets the access frequency of two types of accesses in opposite ways: a high volume of past notification-driven accesses indicates low leftover value of a page for subscribers, while heavy general browsing implies a strong interest of web surfers in the page.

2.3 Dual-caches approaches

Previous dual-caches approach

Our earlier study presents a Dual-Caches approach that divides a server cache into two portions and uses one portion at push-time and another portion at access-time placement independently. The approach is referred to as *DC* in this paper.

The push-time module of DC uses equation 5 to evaluate pages and decides whether to store a newly published page in the local server using the same replacement policy as SG1. The access-time portion is managed using GD*.

$$V(p) = \frac{f_S(p) \cdot c(p)}{s(p)} \qquad (5)$$

Where

$f_S(p)$: the number of subscriptions matching the content of page p

$c(p)$ and $s(p)$ have the same meaning as in the equation 1

DC adaptively re-partitions a server's cache according to the publishing pattern and the access pattern. When a page in the push portion (*PP*) is requested, the storage of the page is assigned to the access portion (*AP*). When the push module fails to store a new page and some pages in *AP* have not been requested since the last replacement in *AP*, the storage of the pages is assigned to *PP*. To avoid the imbalance in the partition, the storage assigned to each portion is bounded.

In an environment with hybrid user accesses, DC still works if it does not distinguish between notification-driven accesses and general browsing.

Second new approach: DC-TA

The key idea of the dual-caches approaches is exploiting different information sources in separate cache portions. DC analyzes the subscription information only at push-time and the access pattern only at access-time, hence DC divides a cache based on the type of operations, pushing and caching.

The access type is also a dimension to partition a cache space. Our second new approach divides a server's cache into two equal portions that serve notification-driven and non-notification-driven accesses respectively. In our presentation this approach is called Dual-Caches partitioned by Type of Access (DC-TA).

Using DC-TA, the cache portion that is consulted for notification-driven accesses is called the notification-based portion (*NP*). The other portion that serves general browsing is referred to as the browsing portion (*BP*) in this paper. NP is managed using SG2 since it is one of the best approaches for pure notification-driven accesses according to our earlier study. BP works as a traditional cache using GD*. The page evaluation in NP is based on the subscription information and the accesses from subscribers, while the evaluation in BP relies only on the access pattern of general browsing.

NP and BP cooperate in locating requests of end users. If BP holds a page that is requested in a notification-driven access, the locating algorithm returns the page from BP. Similarly, a non-notification-driven request can be satisfied in NP. The cooperative locating between NP and BP is carried out in background and thus is transparent to end-users. From the point of view of users, the two portions seamlessly form a cache as a whole.

A push-time placement for page *p* in NP is always before any request of *p*, hence *p* cannot be in BP yet. The access-time fetching in either portion occurs only when the page is not found in both BP and NP. Therefore, the cooperative locating in the two portions guarantees that no page resides in both portions at the same time.

However, NP and BP do not have to exchange knowledge about the two types of accesses beyond the existence of pages in each portion. In other words, each portion is read-only for the other portion. This design is suitable for an integrated caching system shared by multiple independent service providers. For example, one portion of a server's cache is assigned to a content delivery service provider such as Akamai [1], and another portion is licensed to a publish/subscribe service provider. The locating engine in DC-TA aggregates only the location information of pages in all the portions, but leaves other privacy-sensitive information to each portion.

3. Simulator and workload

3.1 Simulator and workload for news delivery

Because of the difficulty of acquiring real-world data about the applications considered in this paper, our evaluation study is simulation-based. Our simulator mimics news delivery based on analyses and observations about the real-world data in the literature. The simulator assumes a single publisher as a news site and a content delivery system consisting of 100 globally distributed proxy-servers. The network topology of the proxy servers and the publisher is a random graph built using GT-ITM [12]. Each proxy-server is assumed to be deployed close to a set of end-users. The workload includes three traces of a 7-day simulation.

The first trace is a publishing sequence. The publisher generates 30,147 distinct pages in total within 7 days. The second trace is an access trace that includes around 195,000 requests. This trace records the requested page, the issuing time and the closest proxy-server of each request. The popularity distribution of requests follows Zipf-law[1] with α as 1.5 according to the observation about news requests in [22]. More details about the methods for building the publishing and the access traces are discussed in [5]. For the validation purpose of this study, the access sequence is assumed to consist of two types of accesses.

The third trace is the subscription sequence of end-users. The subscriptions are assumed to be static information that is known in advance. This assumption is reasonable since the update rate of user subscription is usually much lower than the rate of user access. Given the above assumption, the only subscription information of interest is the number of subscriptions matching each page. This information can be built from the access sequence.

3.2 Generating subscription information

Equation 6 determines the number of subscriptions that match a page i at a server j based on the total number of requests to i at j and the correlation between the subscriptions and the accesses. The correlation is determined by two factors: the ratio of the number of notification-driven accesses to that of all accesses for the page at the server, and the ratio of the number of the notification-driven accesses to that of the subscriptions for a page. The first parameter models the composition of two types of accesses, while the second one quantifies the accuracy of subscriptions or subscription quality.

$$SF_{i,j} = \frac{P_{i,j} \cdot F_{i,j}}{SQ_{i,j}} \qquad (6)$$

Where

$SF_{i,j}$: number of subscriptions matching page i at server j

$P_{i,j}$: number of requests to page i at server j

$F_{i,j}$: fraction of notification - driven requests to page i at server j

$SQ_{i,j}$: probability for subscribers to read page i at server j

The global parameter F is the estimate of the fraction of notification-driven accesses for all page/server pairs. The global parameter SQ stands for global subscription quality, and it is the estimate of the probability for a subscriber to access a page that matches the user's stated interest. Given a distribution and a global

[1] $R_i = 1/i^{\alpha}$, the request rate on the page with rank i

parameter F or SQ, one can generate $\{ F_{i,j} \}$ and $\{ SQ_{i,j} \}$ for all pages and servers, and hence build the subscription information using equation 6.

Given the lack of evidence in literature about the distributions of $\{ F_{i,j} \}$ and $\{ SQ_{i,j} \}$ as defined, we explore three different distributions to model $\{ F_{i,j} \}$ and $\{ SQ_{i,j} \}$. The first distribution is a step-wise function with the disjoint point at the estimate value of the sequence. The second distribution is a uniform distribution. The last one is a Gaussian-like distribution. All three distributions have a domain of [0, 1]. Since the experiments using the three distributions yield similar results, we focus our discussion on the results using a Gaussian-like distribution to model $\{ F_{i,j} \}$ and $\{ SQ_{i,j} \}$.

3.3 Tagging accesses as notification-driven or not

Since HUG and DC-TA require knowledge about the type of access, each access in our request streams should be identified as either notification-driven or non-notification-driven.

Given $P_{i,j}$ and $F_{i,j}$ as defined in equation 6, $P_{i,j} \cdot F_{i,j}$ accesses are randomly chosen from the access sequence to page i at server j using a uniform random number generator. The chosen accesses are tagged as notification-driven, while others as general browsing.

4. Experimental results

4.1 Metric and experimental setup

The publishing site usually resides in a different backbone network; then the communication latency between users and a local proxy-server is usually much smaller than that between users and the publisher. Consequently, a low miss ratio in local servers can be translated into a small response time. We use *global miss ratio* (*M*) on the 100 servers to evaluate the performance of an approach using the following definition:

$$M = \frac{\sum_{i=1}^{100} M_i}{\sum_{i=1}^{100} R_i} \quad (7)$$

Where

M_i : number of misses on server i

R_i : number of requests on server i

To investigate the importance of subscriptions in content delivery systems, the performance of any of our subscription-based approaches is expressed as the relative reduction in M as compared to the baseline approach GD*, as defined in equation 8. A positive value of I indicates an improvement of a method over the caching-only approach GD*.

$$I_i = \frac{M_{GD*} - M_i}{M_{GD*}} \cdot 100\% \qquad (8)$$

Where

I_i : the relative improvement of method i

M_i : global miss ratio of method i

M_{GD*} : global miss ratio of GD*

In our simulation, the storage capacity of a cache is set to be 5% of the total number of unique bytes requested at the server in the whole simulation. The performances of the approaches are tested under different user models that are characterized by the two global parameters, F and SQ.

4.2 Comparing RSG2 and HUG

Our earlier study experimentally identifies SG2 as one of the best content delivery and caching approaches we proposed for pure notification-driven accesses. RSG2 and HUG, the two variants of SG2, differ in whether the two types of accesses are distinguished in the access frequency analysis. Therefore, the two approaches are same when F is 1, in which case all accesses are notification-driven.

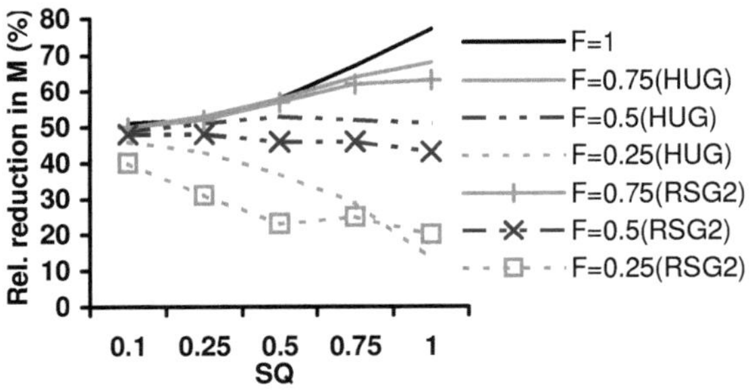

Figure 1. Performances of RSG2 and HUG

As shown in Fig. 1, RSG2 and HUG reduce the miss ratio over GD* in all tested cases. The superiority of HUG to RSG2 in most cases demonstrates the importance of distinguishing access type in evaluating pages.

A larger F indicates that most accesses are driven by notifications and thus implies more importance of the subscription analysis. The positive correlation between F and the performances of RSG2 and HUG in Figure 1 exhibits that the both approaches are able to use the subscription information adequately to enhance content delivery.

Interestingly, higher subscription quality itself does not always benefits the performances of the approaches. When F is 0.25, the performances of RSG2 and HUG degrade with SQ. A combination of small F and large SQ is likely to result in a small number of subscriptions using equation 6, which reduces the importance of subscription information.

4.3 Behavior of DC-TA

Comparing DC and DC-TA

Figure 2 shows the performances of the two dual-caches approaches. Both DC and DC-TA yield better results in terms of miss ratio than GD* in all cases. As for RSG2 and HUG, the performances of DC and DC-TA become worse with SQ when F is 0.25. As explained before, this is because a combination of small F and large SQ results in little subscription information available in the workload.

Under any setting of F and SQ, DC yields high gains than DC-TA. Recall that the major distinction between DC and DC-TA is the criterion to partition pages into two cache portions. DC groups all pages that have not been referenced in one portion, and other pages in another portion. In contrast, when a placement is triggered on a cache miss, DC-TA may put the new page into either NP or BP according to the access type. Consequently, DC-TA may evaluate a page using different methods at different reference times of the page. We conjecture that the partition rule based on type of accesses may be one reason for the worse performance of DC-TA.

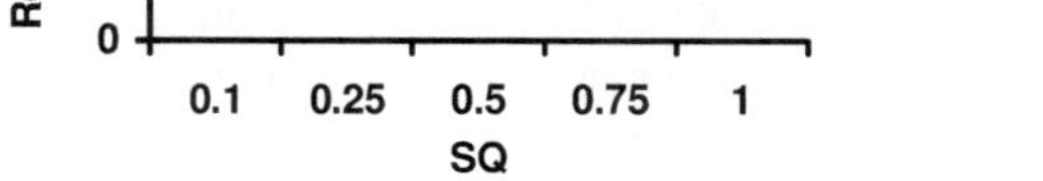

(a) Performances of DC

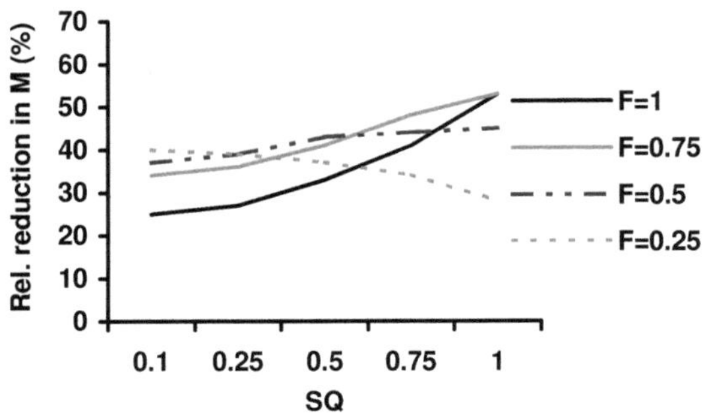

(b) Performance of DC-TA

Figure 2. Comparison of performances of two Dual-Caches approaches

4.3.2 Influence of cache partition to DC-TA

Different from RSG2, HUG, and DC, the performance of DC-TA does not increase with F. Given a SQ, DC-TA achieves the best or the second best result when F is 0.5. Since DC-TA uses a 50%-50% partition on a server's cache, we conjecture that the good performances of DC-TA when F is 0.5 are resulted from the suitable cache partition under this F value. However, a 50%-50% partition may not be good for traces with other F values.

Recall that DC adaptively updates its partition based on the publishing pattern and the request pattern over time. The intelligent storage partition may be another reason for the superiority of DC to DC-TA in Figure 2. To study the influence of the cache partition, DC-TA is tested under different partitions for each F. Based on our preliminary experimental results, the optimum storage fraction assigned to notification-based portion are 100%, 100%, 100%, and 75% for F of 1, 0.75, 0.5 and 0.25 respectively. The optimum fraction value of the notification-driven portion, denoted as OF, is positively correlated but not equal to F. Interestingly, OF is always larger than F when F is less than 1, which means giving higher importance to the notification-driven accesses and subscriptions benefits all users in reducing response time. Particularly, when at least half of the accesses are driven by notifications ($F \geq 0.5$), it is better to use the whole cache for the push-time placement and the access-time placement based on subscriptions and notification-driven accesses only.

DC-TA with Optimum Partition is called ODT in this paper. Figure 3 exhibits a dramatic improvement using the optimum partition over using fixed 50%-50% partition in DC-TA.

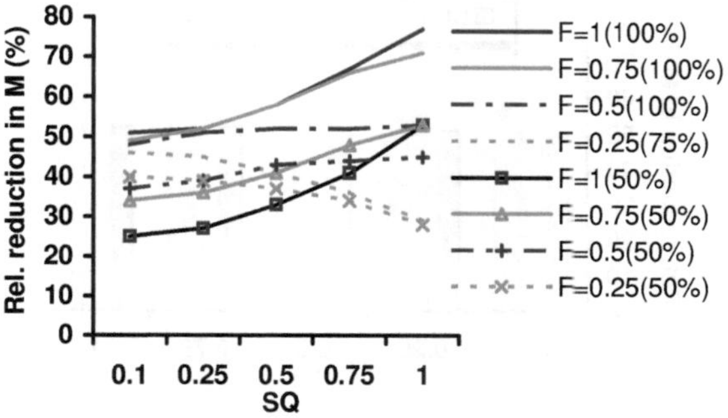

Figure 3. Performances of DC-TA with fixed or optimum partition

As shown in figure 4, using the optimum partitions, ODT has comparable performances to DC in many cases. When SQ is high, ODT yields the same or even better results than DC. The results in figures 3 and 4 support the hypothesis that adaptive partition based on access pattern is important to dual-caches approaches.

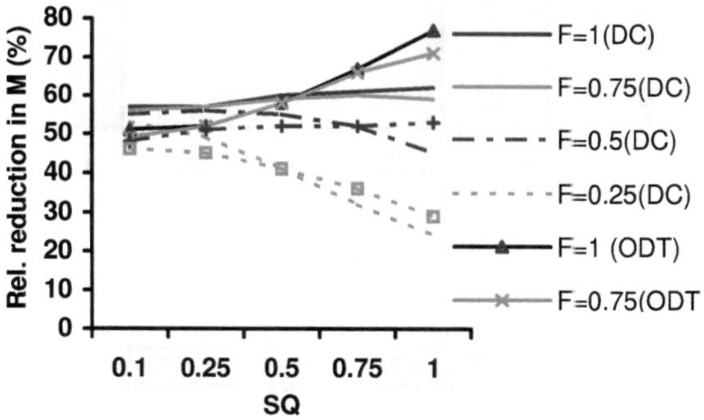

Figure 4. Performances of DC and ODT

4.4 Comparing SG1, HUG and DC

The experiments in sections 4.2 and 4.3 demonstrate that HUG outperforms RSG2, and DC outperforms DC-TA under most settings of F and SQ. The following figures compare the performances of HUG and DC to that of SG1, one of our previous approaches for pure notification-driven accesses.

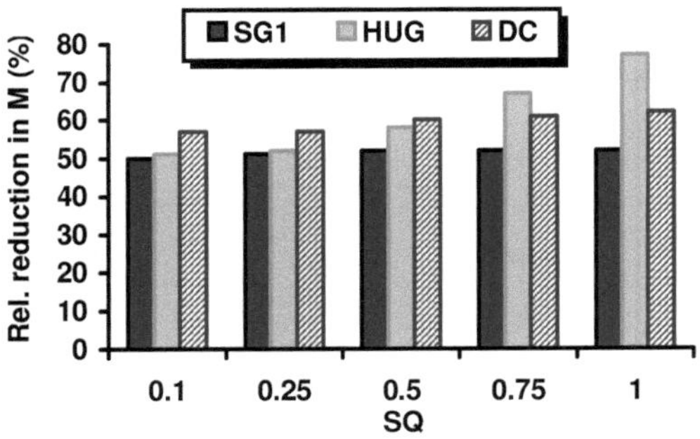

(a) F = 1

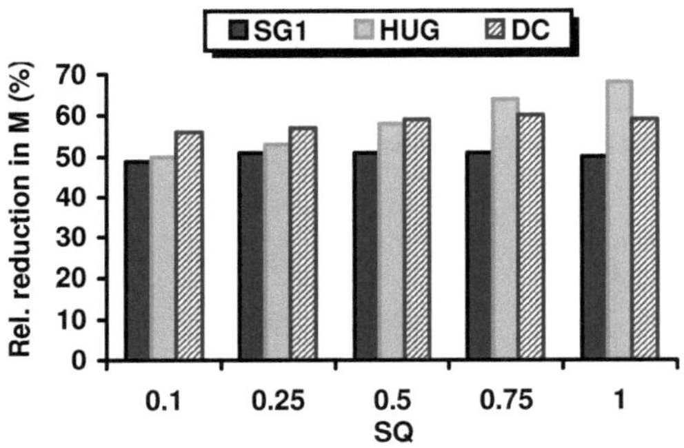

(b) F = 0.75

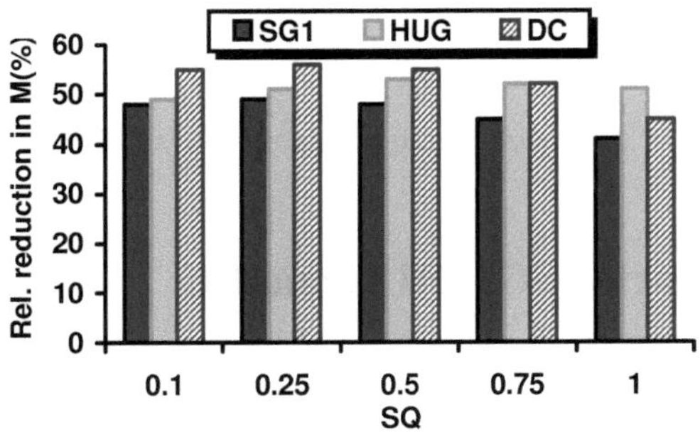

(c) F = 0.5

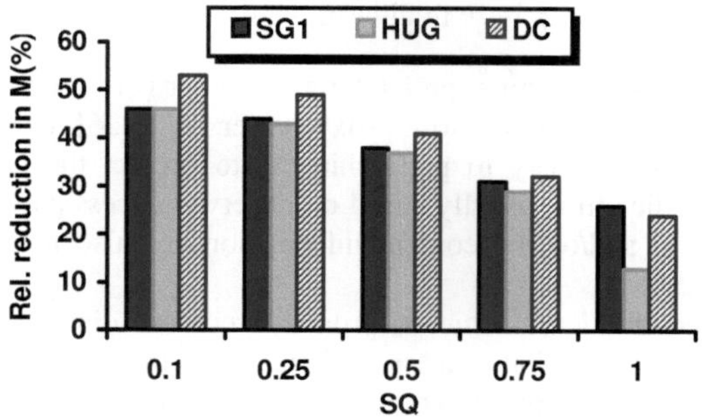

(d) F = 0.25

Figure 5. Performances of SG1, HUG and DC

All of the three approaches reduce the miss ratio over GD* under all settings of F and SQ. Table 1 summarizes the winners in different cases.

Table 1. Winners for traces at different F and SQ

SQ \ F	0.1	0.25	0.5	0.75	1
0.25	DC	DC	DC	DC	SG1
0.50	DC	DC	DC	HUG/DC	HUG
0.75	DC	DC	DC	HUG	HUG
1	DC	DC	DC	HUG	HUG

When both F and SQ are high (F ≥ 0.5, SQ > 0.5), HUG yields the highest relative gains over GD* among the three approaches, hence HUG is best when most requests follow the users' stated interest. DC is the winner for most settings of F and SQ when either F or SQ is low. As the simplest approach among the three, SG1 yields comparably good results in many cases except for when both F and SQ are high. Therefore, SG1 is suitable when computational cost is a major concern.

In application of the proposed content delivery methods, the choice of the three methods should be made based on the cost of cache misses, the observed user model in terms of F and SQ, the affordability of the algorithm complexity, and other application-specific factors.

5. Related work

Web caching has been widely applied to distribute content on the Internet. The GreedyDual-Size algorithm combines temporal locality, popularity, fetching cost, and page size into an evaluation function [3]. GreedyDual* generalizes GDS by balancing the effects of long-term popularity and short-term reference correlation in requests [14].

In contrast to passive caching, prefetching techniques are used to proactively pull information from an original site to a proxy server [7, 9, 26], or from a proxy to a browser cache [10]. The key to prefetching is to predict the pages that will be requested. The prediction is usually based on observed access patterns [10, 26]. The linkage information and/or the content information are also useful in prefetching techniques [4, 9, 20].

Server-initiated multicast achieves proactive content distribution on behalf of publishers. Based on global access information, the popular pages can be automatically pushed to the proxy servers that request the pages frequently [2, 13].

In existing content delivery systems, caching, prefetching, and pushing are mainly based on inferred user interest typically from access patterns. Our work is distinct in exploiting the stated user interest in addition to inferred user interest.

Recently, there has been increasing interest in the placement problem for content delivery networks (CDN). Kangasharju et al. [15] formulate an optimum content replication problem as the problem of minimizing the average number of network hops to fetch a copy, assuming the locating algorithm always gets a copy from the nearest server holding a copy. The optimum placement in an overlay network with a graph topology has been proved to be NP-hard, while there exist polynomial solutions for other topologies such as tree [19, 21]. Cidon et al. [6] present an optimum solution in a distribution tree, but the approach assumes a static environment in which all requests are given precisely in advance and the network condition does not change.

Kangasharju et al. [15] propose four heuristics for content placement, but the best one needs to exploit global knowledge about the network topology, the global reference distribution and the global content image at different times. Qiu et al. [23] propose several heuristics to choose M replica sites from N candidates for a given site, also assuming a relative stable reference pattern at the candidate sites.

In this paper, each cache server is autonomous in managing its storage using local subscription and access information. Therefore, our approaches are scalable in a globally distributed system that faces dynamic content publishing and request patterns, and network conditions.

Karlsson and Mahalingam [17] argue that caching works at least as well as replica placement algorithms, if the cache storage is ample and the replacement algorithm runs periodically rather than after each access. For combining placement and caching, Korupolu and Dahlin propose a "static partition" approach that divides a server's cache into two portions and runs placement and replacement algorithms on the two portions separately [18]. The approach is analogous to our DC algorithm, but several major distinctions exist between the two. DC takes into account stated user interest rather than being purely based on request distribution. For the placement portion, our push-time placement is triggered by content generation and matching the content to users' subscriptions, while placement is called periodically in [18]. Moreover, DC re-partitions a server's cache adaptively to the access and publishing pattern.

An important component in CDN is load balancing. Load balance is usually achieved by request redirection [1, 8] using different hashing schemes [16, 25].

Caching can be used as complimentary to request redirection-based CDN. To the best of our knowledge, the existing CDNs that work with conventional caching systems still rely on request patterns only. Gadde et al. [11] suggest a natural limit to the marginal benefits of redirection-based CDNs, since the hit ratio in proxy caches increases dramatically as ISPs serve larger user communities. A recent experimental study on content delivery systems also points out the importance of widely deployed proxy caches as compared to a separate CDN [24].

6. Conclusion

We proposed several content delivery approaches for an environment where a publish-subscribe service exists, and user may request contents based on notifications that satisfy their subscriptions or by general browsing. In addition to three approaches from our earlier study that assumed all accesses are based purely on notifications (with necessary adaptation to the new scenario), we presented two new methods that exploit information about access type, access patterns, and subscriptions. Our performance metric is the miss rate in local caching servers, assuming a lower miss rate can generally be translated into a smaller response time.

We find that all of our approaches that use subscription information outperform an access-based caching-only algorithm. Interestingly, this is true even when only a small fraction of accesses is notification-driven and when subscriptions do not predict subscribers' accesses very well. Especially, when users' stated interest matches users' access patterns fairly well, HUG, the new approach considering the access type, is the best approach; otherwise, our previous dual-caches approach DC outperforms our other approaches. As the best algorithm in our earlier study on pure notification-driven accesses, SG2 is not the best approach in the new scenario any more, even after necessary adaptation.

A major direction for our future work is to examine the use of subscription information in cooperative placement and locating schemes for a distributed system.

References

[1] Akamai. http://www.akamai.com.

[2] Besravros, A. Demand-based document dissemination to reduce traffic and balance load. In *Proceedings of the 7th IEEE Symposium on Parallel and Distributed Processing (SPDP'95)*, 1995.

[3] Cao, P. and Irani, S. Cost-aware WWW proxy caching algorithms. In *Proc. of USENIX Symp. on Internet Technology and Systems*, 1997.

[4] Chi, E. H., Pirolli, P., Chen, K., and Pitkow, J. Using information scent to model user information needs and actions on the Web. *Proc. CHI*, 2001, pp. 490 – 497.

[5] Chen, M, LaPaugh, A. and Singh, J. P. Content distribution for publish/subscribe services. In *Proc. of ACM/IFIP/USENIX Middleware*, 2003, pp. 83 – 102.

[6] Cidon, I., Kutten, S., and Soffer, R. Optimal allocation of electronic content. In *Proceedings of IEEE Infocom* 2001.

[7] Deolasee, P., Katkar, A., Panchbudhe, A., Ramamritham, K., and Shenoy, P. Adaptive push-pull: Disseminating dynamic Web data. In *Proceedings of WWW10*, 2001.

[8] Digital Island. http://www.digitalisland.com/.

[9] Duchamp, D. Prefetching hyperlinks. In *Proc. of USENIX Symp. on Internet Technologies and Systems*, 1999.

[10] Fan, L., Cao, P., Lin, W., and Jacobson, Q. Web prefetching between low-bandwidth clients and proxies: Potential and performance. In *Proc. of ACM SIGMETRICS*, 1999.

[11] Gadde, S., Chase, J., and Rabinovich, M. Web caching and content distribution: A view from the interior. *Proc. 5th International Web Caching and Content Delivery Workshop (WCW)*, 2000.

[12] GT-ITM: Georgia Tech Internetwork Topology Models. http://www.cc.gatech.edu/fac/Ellen.Zegura/graphs.html

[13] Gwertzman, J. and Seltzer, M. An analysis of geographical push-caching. 1997.

[14] Jin, S. and Bestavrous, A. GreedyDual* Web caching algorithm: Exploiting the two sources of temporal locality in Web request streams. *Computer Comm.*, 24(2):174-183, Feb. 2001.

[15] Kangasharju, J., Roberts, J., and Ross, K. W. Object replication strategies in content distribution networks. In *Proceedings of WCW'01: Web Caching and Content Distribution Workshop*, June 2001.

[16] Karger, D., Lehman, E., Leighton, F. T., Levine, M., Lewin, D., and Panigrahy, R. Consistent hashing and random trees: Distributed caching protocols for relieving hot spots on the World Wide Web. In *Proceedings of the 29th Annual ACM Symposium on Theory of Computing*, pp. 654-663, 1997.

[17] Karlsson, M. and Mahalingam, M. Do we need replica placement Algorithms in content delivery networks. In *Proc. of WCW '02: Web Caching and Content Distribution Workshop*, 2002.

[18] Korupolu, M. R. and Dahlin, M. Coordinated placement and replacement for large-scale distributed caches. *IEEE Transactions on Knowledge and Data Engineering*, 14(6):1317-1329, 2002.

[19] Krishnan, P., Raz, D., and Shavitt, Y. The cache location problem. *IEEE/ACM Transactions on Networking*, 8(5):568-582, October 2000.

[20] Lieberman, H. Letizia: An agent That assists Web browsing. *Proceedings of the International Joint Conference on Artificial Intelligent*, 1995.

[21] Li, B., Golin, M. J., Italiano, G. F., and Deng, X. On the optimal placement of Web proxies in the Internet. In *Proc. of IEEE INFOCOM*, 1999.

[22] Padmanabhan, V. N. and Qiu, L.-L. The content and access dynamics of a busy Web site: Findings and implications. In *Proc. of ACM SIGCOMM* 2000.

[23] Qiu, L., Padmanabham, V. N., and Voelker, G. M. On the placement of web server replicas. In *Proceedings of 20th IEEE INFOCOM*, 2001.

[24] Saroiu, S., Gummadi, K. P., Dunn, R. J., Gribble, S. D., and Levy, H. M. An analysis of Internet content delivery systems. In *Proc. of OSDI*, 2002.

[25] Wang, L.-M., Pai, V., and Peterson, L. The effectiveness of request redirection on CDN Robustness. In *Proceedings of the Fifth Symposium on Operating Systems Design and Implementation (OSDI'02)*, 2002.

[26] Venkataramani, A., Yalagandula, P., Kokku, R., Sharif, S., and Dahlin, M. The potential costs and benefits of long-term prefetching for content distribution. Technical Report TR-01-13, UT, Austin, 2001.

COOPERATIVE ARCHITECTURES AND ALGORITHMS FOR DISCOVERY AND TRANSCODING OF MULTI-VERSION CONTENT

Claudia Canali[1], Valeria Cardellini[2], Michele Colajanni[3],
Riccardo Lancellotti[2], and Philip S. Yu[4]
[1]*University of Parma,* [2]*University of Roma "Tor Vergata",* [3]*University of Modena and Reggio Emilia,* [4]*IBM T.J. Watson Research Center*

Abstract A clear trend of the Web is that a variety of new consumer devices with diverse processing powers, display capabilities, and network connections is gaining access to the Internet. Tailoring Web content to match the device characteristics requires functionalities for content transformation, namely *transcoding*, that are typically carried out by the content provider or by some proxy server at the edge. In this paper, we propose an alternative solution consisting of an intermediate infrastructure of distributed servers which collaborate in discovering, transcoding, and delivering multiple versions of Web resources to the clients. We investigate different algorithms for cooperative discovery and transcoding in the context of this intermediate infrastructure where the servers are organized in hierarchical and flat peer-to-peer topologies. We compare the performance of the proposed schemes through a flexible prototype that implements all proposed mechanisms.

1. Introduction

The Web is rapidly evolving towards a highly heterogeneous accessed environment, due to the variety of consumer devices that are increasingly gaining access to the Internet. The emerging Web-connected devices, such as handheld computers, PDAs, mobile phones, differ considerably in network connectivity, processing power, storage, display, and format handling capabilities. Hence, there is a growing demand for solutions that enable the transformation of Web content for adapting and delivering it to these diverse destination devices.

The content adaptation mechanism, called *transcoding*, can be applied to transformations within media types (e.g., reducing the color depth of an image), across media types (e.g., video clip to image set) or to both of them. The existing approaches to deploy Web content adaptation fall into three broad categories depending on the entity that performs the adaptation process [1, 12]: *client-based*, *edge-based* (also called *proxy-based*), and *server-based* adaptation. A comprehensive analysis on related work

F. Douglis and B.D. Davison (eds.), Web Content Caching and Distribution, 205-221.

is in Section 2. The edge-based approach uses the proxy server to analyze and adapt the content on-the-fly, before delivering the result to the user. This component is often called *edge server* as in the delivery chain between the client device and the content server it is generally located close to the client. So far, the edge-based approach has been typically carried out by some edge server that directly connects to the clients. In this paper, we explore a different alternative that considers a distributed system of cooperative servers which collaborate in discovering, transcoding, and delivering Web content. Minimizing the user response time and bounding its variability are the main goals of this distributed cooperative architecture. Indeed, the computational cost of transcoding can be notably reduced by discovering the desired resource in other servers and also by involving some other servers to perform the task of content adaptation. Although the cooperative transcoding is based on distributed schemes, it is not a simple extension to cooperative caching because two main issues have to be addressed to achieve a suitable solution. First, the presence of diverse consumer devices requires to recognize, discover, and cache the multiple variants of the same resource obtained by transcoding operations. Moreover, the transcoding process is expensive in terms of server computing resources. Issues related to workload distribution, which are not usually considered in Web caching, can become of fundamental importance in the case of cooperative transcoding.

In this paper, we propose and investigate some architectures and algorithms for cooperative discovery, transcoding, and delivery, including servers organized in hierarchical and flat topologies. We compare their performance through a prototype called ColTrES (Collaborative Transcoder Edge Services), which is a flexible testbed based on Squid [19]. ColTrES implements all considered mechanisms by extending the traditional cooperative caching systems to an environment characterized by heterogeneous client devices. The first extension transforms a traditional cache server into an active intermediary, which not only caches Web objects but also transcodes them, stores the results, and allows multi-version lookup operations [7, 18]. The second novel extension allows the cooperation of the active intermediaries for caching and transcoding. We take advantage of the scalability opportunities provided by cooperation to reduce the response time experienced by the users of a heterogeneous client environment.

We are not aware of any other research work dealing with the study and implementation of cooperative transcoding and caching systems with both hierarchical and flat topologies. Through our prototypes, we demonstrate that all proposed algorithms and cooperative architecture are immediately applicable to the Web infrastructure. The real testbed allows us to evaluate the reduction of the user response time achievable by different cooperative discovery and transcoding schemes. Moreover, we clearly demonstrate the advantages of cooperative transcoding through flat topologies over hierarchical schemes.

The rest of this paper is organized as follows. Section 2 analyzes related work. Section 3 discusses the main features of a distributed system for cooperative transcoding. Sections 4 and 5 explore different topologies and protocols for cooperative discovery, transcoding, and delivery. Section 6 describes the workload model used to exercise the prototype. Section 7 presents the experimental results. Section 8 concludes the paper with some final remarks.

2. Related work

The client-based approach to content adaptation seems not suitable for all the cases in which clients offer limited processing power and connection bandwidth. The server-based approach, that adds content adaptation services to traditional Web server functionalities [14], increases the complexity of the server platform and software, but remains a valid alternative. However, in this paper we will focus on content adaptation carried out by an intermediate architecture of distributed servers. The advantages of proxy-based content adaptation have been explored by many recent studies [4, 3, 5, 7, 9, 10, 13, 18]. This scheme can use one [10, 13, 18] or more [4, 3, 9] servers to analyze and adapt the content on-the-fly,up to fully distributed peer-to-peer networks as in [17] (although the study of Shi *et al.* is more focused on personalized contents). An intermediate adaptation can shift load from content-providing servers and simplify their design. Moreover, this solution is also viable because the large majority of consumer devices requires some proxy to access the Web. A transcoding service located at intermediary points of the network can also tailor resources coming from different content servers. The intermediate server plays also another important role that is, it can cache the results of content adaptation, thus avoiding some round-trips to the content server and costly transcoding operations when resources can be served from the cache [9, 18].

Most research has focused on handling the variations in client bandwidth and display capabilities (e.g., [5, 9, 10]), without focusing on caching aspects. In these proposals, the edge server that directly connects to the clients typically reduces the object size (thus reducing bandwidth consumption), apart from providing a version that fits the client device capabilities. A limited number of recent proposals have also exploited techniques to combine both adaptation and caching to reduce the resource usage at the edge server [7, 13, 18].

The large majority of research efforts have been devoted to the investigation of solutions in which the adaptation and caching functionalities are provided on stand-alone edge servers that do not cooperate among them. The main motivation that leaded us to study distributed architectures for intermediate services of caching and adaptation is the limited scalability of a single proxy-based approach because of significant computational costs of adaptation operations [12]. Fox *et al.* [9] address this scalability issue by proposing a cluster of locally distributed edge servers. This approach may solve the CPU-resource constraint, but it tends to move the system bottleneck from the server CPU to the interconnection of the cluster. On the other hand, the proposes infrastructure is designed to be distributed over a wide area network thus preventing network bottlenecks.

In recent works we have started to examine how to extend traditional caching architectures to the active support of cooperative transcoding. In [4] the authors have obtained some preliminary simulation results that demonstrated the importance of distributing the computational load of transcoding in a cooperative hierarchical scheme. Some preliminary experimental results on flat topologies have been presented in [3], in which the authors demonstrate that a cooperative distributed system consistently outperforms in terms of user response times a system of non-cooperative servers.

3. Main features of the intermediate infrastructure

The servers of a distributed system can be organized and cooperate through a large set of alternatives. Each node of the intermediate infrastructure may have one or more functionalities that is, it can act as a *transcoder*, a *cache* or an *edge* server. *Edge* servers receive requests directly from the clients and deliver the requested resources. *Cache* servers provide caching functionalities for both original and transcoded resources. *Transcoder* servers perform content adaptation. In this paper we consider hierarchical and flat topologies. In flat topologies all nodes are peers and provide all functions, while in hierarchical topologies the nodes may provide different functionalities. In this section we outline the main common operations that characterize an intermediate infrastructure for cooperative transcoding, while the specific features of hierarchical and flat organizations are described in Sections 4 and 5, respectively.

We identify three main phases that may require some cooperation among the nodes of the intermediate infrastructure, namely *discovery*, *transcoding*, and *delivery* phases. Even the traditional phases differ from the corresponding ones of a standard cooperative caching scheme. We describe the three phases in a reverse order.

Once the desired version of the requested object is found (or generated), the *delivery* phase transfers the resource to the client. The final delivery is always carried out by the edge server first contacted by the client. Hence, if the resource is found in another node, the delivery phase includes its transmission to the edge server. Although for some applications a request can be satisfied with a lower quality resource than that specified by the client, we do not consider such possibility in this paper.

The *transcoding* phase is specific to the problem here considered. We assume that any server of the cooperative system is equipped with software that can perform the transcoding operations required by any type of client device that contacts an edge server. The features of client devices vary widely in screen size and colors, processing power, storage, user interface, software, and network connections. Recently, the WAP Forum and the W3C have also proposed the standards CC/PP and UAProf for describing the client capabilities [1]. The client may also include the resource type it can consume as a meta-information in the HTTP request header. Hereafter, we will refer to the information describing the capabilities of the requesting client as the *requester-specific capability information* (RCI). An object which has been previously transcoded may be further adapted to yield a lower quality object. In particular, each version may be transcoded from a subset of the higher quality versions. Different versions of the same object (and the allowed transcoding operations among them) can be represented through a *transcoding relation graph* [4].

It is worth to observe that we consider a generic infrastructure that does not involve the content-providing server in the transcoding process as the proposal of server-directed transcoding [11]. Hence, we assume that the content server always returns the original version of the requested resource. Our cooperative architectures for transcoding and caching can be integrated with content server decisions or not, without altering main performance considerations and conclusions of this paper.

During the *discovery* phase, the servers may cooperate to search for the version of the Web object requested by the client. Since multiple versions of the same object typically exist in the caches, in this phase it is necessary to carry out a multi-version lookup process that may require cooperation among the servers. The discovery phase includes a local lookup and may include an external lookup. Once the edge server has determined the client capabilities, it looks for a copy of the requested resource in its cache. The local lookup may generate one of the following three events. (1) *Local exact hit*: the cache contains the exact version of the requested object, that can be immediately delivered to the client. (2) *Local useful hit*: the cache contains a more detailed and transcodable version of the object that can be transformed to match the client request. Depending on the transcoding cooperation scheme, the edge server can decide either to perform the transcoding task locally or to activate an external lookup, which is carried out through some cooperative discovery protocol. (3) *Local miss*: the cache does not contain any valid copy of the requested object. The edge server must activate an external lookup to fulfill the request.

When both exact and useful hits are found in the local cache, the former is preferred because it does not require any adaptation task, and no external lookup is necessary. We recognize that our architectures opens many novel possibilities for push caching and object replacement [7], that we do not consider in this paper.

In the case of local miss and sometimes of useful hit, the edge server may activate some cooperative discovery mechanism to locate a version on some other server. The external lookup may provide one of the following results. (1) *Remote exact hit*: a remote server holds the exact version of the requested object, which is transferred to the requesting server. (2) *Remote useful hit*: a remote server contains a more detailed and transcodable version of the requested object that can be transformed to meet the request. Depending on the transcoding cooperation scheme, the cooperating server can decide either to perform the transcoding task locally or to provide the useful version to the requesting server, which will execute the transcoding process. (3) *Remote miss*: no remote server contains any valid copy of the object, that is a *global cache miss* occurs. The original version of the requested resource must be fetched from the content server.

4. Hierarchical topologies

In this paper, we consider a pure hierarchical architecture where sibling servers do not cooperate, and only the bottom level nodes (called *leaf nodes*) are *edge* servers that can be contacted by the clients [15]. We use the commonly adopted three-level tree from leaf nodes to the root node, because hierarchical architectures follow the idea of hierarchical Internet organization, with local, regional, and international network providers.

Some schemes for distributing the transcoding load among the servers organized in a hierarchy have been described in [4]. In this paper we consider two approaches, called *Hierarchical root* and *Hierarchical leaf* cooperation schemes. In the Hierarchical root scheme, each node is both a *transcoder* and a *cache* server. In the case of local miss, the request is forwarded by the edge server up to the hierarchy, until it is satisfied with either an exact or useful hit. In the case of global miss (that is, no level holds a valid copy of the requested resource), the root node retrieves the original resource from the content provider, if necessary adapts it, and sends the exact version of the object to the lower-level server. Each node experiencing a local exact hit responds by sending the resource to the requesting entity, which can be a client or a lower-level server. In the case of useful hit, the contacted server performs locally the content adaptation before sending the exact version of the resource downwards the hierarchy. A copy of the object is stored in the caches of all the nodes along the request path.

As the root node must perform the transcoding service for every global miss and content adaptation may involve costly operations, there is a great risk of overloading this server. Indeed, different studies have shown that pure hierarchical architectures, even when applied to traditional cooperative caching, may suffer from scalability and coverage problems, especially when the number of nodes is large (e.g., [8, 20]). This situation can dramatically worsen in the case of cooperative transcoding. For this reason, we propose the *Hierarchical leaf* scheme, that differentiates the roles of the nodes in the intermediate infrastructure. The leaf nodes maintain the roles of edge, cache and transcoding servers, while the upper-level nodes provide just cache services for original versions of the resources. When necessary, content adaptation is performed locally by the leaf nodes.

5. Flat topologies

An alternative to hierarchical topology is the flat organization, in which all nodes are *peers* and provide the functionalities of *transcoder*, *cache*, and *edge* server. This flat organization allows us to explore various algorithms for cooperative discovery and cooperative transcoding, which are discussed in the following sections.

5.1 Cooperative discovery

Although the discovery phase in a flat topology can be based on different protocols, we limit the research space of alternatives to the most interesting and widely used systems. It is worth to remark that the cooperation protocols for object discovery and delivery considered in this section differ from the traditional ones because multiple versions of the same object may be present in the caches of the cooperative edge servers. Moreover, there are three possible results of the external lookup process: miss, exact hit, and useful hit.

Cooperative lookup among distributed servers requires a protocol to exchange local state information, which basically refers to the cache content, although when we consider a CPU-bound task such as transcoding, other data can be useful (e.g., server load). Cooperative resource discovery has been studied for a while and many mechanisms have been proposed to address the related issues [15]. Most of those mechanisms can be adapted to the lookup of multiple versions. The two main and opposite approaches for disseminating state information are well defined in the literature on distributed systems: *query-based protocols* in which exchanges of state information occur only in response to an explicit request by a peer, and *directory-based protocols* in which state information is exchanged among the peers in a periodic way or at the occurrence of a significant event, with many possible variants in between. In the following, we consider a query-based protocol and a summary-based protocol (a simplified version of the directory-based protocols).

Query-based protocols are conceptually simple. When an edge server experiences a local miss or even a useful hit (depending on the cooperative transcoding algorithm), it sends a query message to all the peers to discover whether any of them caches a copy of the requested resource. In the positive case, the recipient edge server replies with an exact hit or with a useful hit message; otherwise, it may reply with a miss message or not reply at all. In the case of useful hit, the response message should provide some information about the available version of the resource to allow its retrieval. As the protocol for our query-based cooperation, we use the popular ICP adopted in NetCache and Squid [19]. In our ColTrES prototype we added the support for multi-version lookup into the Squid version of ICP by including the version identifier to the URL contained into the messages.

Directory-based protocols are conceptually more complex than query-based schemes, especially because they include a large class of alternatives, being the the two most important ones the presence of one centralized directory vs. multiple directories disseminated over the peers, and the frequency for communicating a local change to the directory/ies. It is impossible to discuss here all the alternatives that have been the topics of many studies. We consider distributed directory-based schemes because it is a common view that in a geographically distributed system any centralized solution does not scale, the central directory server may represent a bottleneck and a single point of failure, and it does not avoid the query delays during the lookup process.

In a distributed directory-based scheme, each edge server keeps a directory of the resources that are cached in every other peer, and uses the directory as a filter to reduce the number of queries. Distributing the directory among all the cooperating peers avoids the polling of multiple edge servers during the discovery phase, and, in the ideal case, makes object lookup extremely efficient. However, the ideal case is affected by large traffic overheads to keep the directories up-to-date. Hence, real implementations use multiple relaxations, such as compressed directories (namely, *summary*) and less frequent information exchanges for saving memory space and network bandwidth. Examples of compression that reduce the message size are the Bloom filters, used by Summary Cache [8] and Cache Digests [16], that compress the cache indexes so that a certain amount of false hits is allowed. For our experiments, we choose Cache Digests as a representative of the summary-based architectures, because of its popularity and its implementation in the Squid software. Support for caching and discovery of multiple versions has been added to our ColTrES prototype into the summary-based lookup process through URL-encoding the resource version identifier. Therefore, the basic mechanism of Cache Digests cooperation is preserved. However, the lookup process becomes more expensive because it has to carry out a search for every possible useful version.

5.2 Cooperative transcoding algorithms

Cooperative transcoding is necessary only when a local or a remote useful hit occurs, while misses and exact hits are handled as described in Section3 and they are unrelated to the cooperative transcoding algorithms. We can identify some alterna-

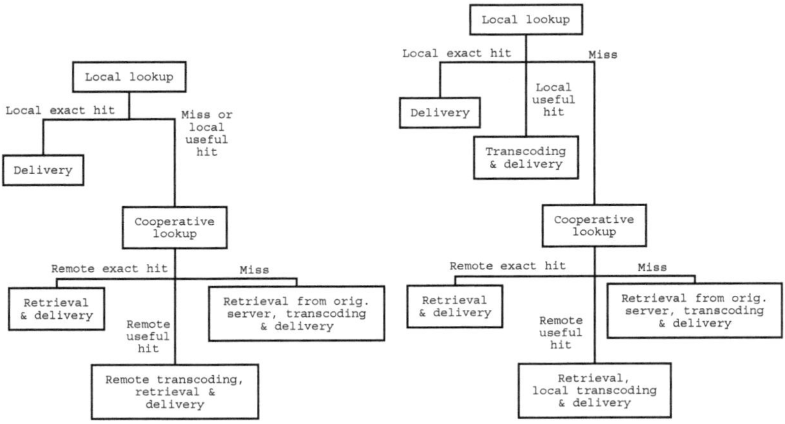

(a) Blind-lazy algorithm. (b) Blind-active algorithm.

Figure 1. Load-blind algorithms.

tives in the case of local and remote useful hits. Since transcoding a useful hit may be computationally expensive, several load-balancing algorithms can be used. In particular, we distinguish between **load-blind** algorithms that do not take into account any load state information and **local load-aware** algorithms, that use load information about the local server itself to decide which node will perform the transcoding task. We propose two load-blind algorithms and a local load-aware algorithm.

The two load-blind algorithms are called *blind-lazy* and *blind-active*. The **blind-lazy** algorithm, whose flow diagram is shown in Figure 1(a), tends to limit the computational costs of transcoding by taking most advantage of the cooperative peers. In the case of a local useful hit, the edge server continues the discovery phase by activating an external lookup process to look for an exact version of the requested object in some peer proxy. In case of a remote useful hit, the edge server always delegates the transcoding task to the peer server that reported the useful hit. The rational behind this approach is to exploit as much as possible the remote exact hits and to distribute in a nearly random way the transcoding process. The price of the external lookup process is worth when the remote exact hit is found and the network links are not saturated; otherwise, a (guaranteed) local useful hit may be preferable to a (possible) remote exact hit.

The **blind-active** algorithm, shown in Figure 1(b), follows an approach opposite to its blind-lazy counterpart. Whenever possible, it saves network usage for the external lookup at the price of local computation. In the case of a local useful hit, the edge server transcodes the useful version found in its cache without continuing the discovery phase. In the case of a remote useful hit, the resource is retrieved from the peer and transcoded locally.

The **load-aware** algorithm we propose in this paper is based on load information at the local server. When a local or a remote useful hit occurs, the node decides whether to perform locally the transcoding operation or to continue the discovery phase on the basis of its current load. This is a typical threshold-based algorithm that follows one of the two previous algorithms depending on the server load (i.e., CPU utilization). When the CPU utilization of the edge server surpasses a certain threshold, it behaves in a *lazy* mode, as the lazy approach tends to save local CPU resources. Otherwise, the edge server adopts the *active* approach, because there is enough spare CPU power to perform transcoding.

6. Workload model

In this section we describe the client and workload models used to evaluate the performance of the cooperative schemes. We consider a classification of the client devices on the basis of their capabilities of displaying different objects and connecting to the assigned edge server [4, 7]. The classes of devices range from high-end workstations/PCs which can consume every object in its original form, to cellular phones with very limited bandwidth and display capabilities. We introduced six classes of clients; the description of the devices capabilities and the values of their popularity can be found in [2]. In this paper, we consider that most transcoding operations are applied to image objects (GIF, JPEG, and BMP formats), as more than 70% of the files requested in the Web still belong to this class [6]. In our experiments we also consider a workload scenario where the transcoding operations have higher costs. This may be found in a near future when the Web will provide a larger percentage of multimedia resources.

The first workload, namely **light trans-load**, aims at capturing a realistic Web scenario with a reduced transcoding load. The set of resources used in this workload are based on proxy traces belonging to the nodes of the IRCache infrastructure. Some characterizations performed on the images of this workload, such as file size, JPEG quality factor, and colors of GIF images, evidenced that they are very close to the characteristics reported in [6]. The measured costs of transcoding operations required by this set of resources on the machines used for our experiments gave the following results: 0.04 and 0.22 seconds for the median and the 90-percentile service time, respectively.

The second workload model (called **heavy trans-load**) aims at denoting a scenario where the transcoding process has a major cost. As the trend of the Web is towards a growing demand for multimedia resources, this workload can represent a situation with a large amount of multimedia objects, such as video and audio. In this scenario, the costs for transcoding operations are 0.27 and 1.72 seconds for the median and the 90-percentile service time, respectively. In both workload models, the client request distribution among the edge servers is uniform, with each node receiving the same number of client requests. However, the types of requests in each trace can differ substantially, because the file size follows a heavy-tailed distribution, especially for the light trans-load working set.

From the file list of each workload model, we obtained 80 different traces that were used in parallel during the experiments. Each trace consists of 1000 requests with a random delay that elapses between two consecutive requests. The total size of the original resources for all workloads is similar. It is 10% higher than the sum of the cache sizes of the nodes used in our experiments. On the other hand, the mean file size of the two workloads differs considerably. Hence, the light workload determines higher cache hit rates than the heavy one. We have also introduced a popularity resource distribution by defining a set of hot resources (corresponding to 1% of the working set): 10% of requests refers to this hot set.

To study the performance of the cooperative transcoding algorithms, we had to consider a scenario where the servers of the intermediate infrastructure are under heavy stress. To this purpose, we used two workload models (called **uniform** and **bimodal**) which are based on the heavy trans-load, but are characterized by different client request distributions. In the uniform scenario each edge server receives the same number of requests, while the bimodal scenario is characterized by an uneven request distribution among the edge servers, where 50% of the servers receive 90% of the client requests and the remaining half of the nodes handle only 10% of the traffic.

7. Experimental results

In this section we first outline the performance metrics and the server-side experimental setup and then discuss the experimental results. As main performance metrics we consider the *cache hit rates* (local, global, exact, useful), the *CPU utilization* of the servers, and the *system response time* that corresponds to the interval elapsed between the instant in which the client sends a request to the edge server and the instant in which the client receives all the response.

As our main target is to enable heterogeneous devices to access Web content, the servers transcode the object to best fit the client capabilities, while we do not explore object compression to reduce transmission time as done in [10]. We also consider only complete transcoding relation graphs, where each version can be obtained from any higher quality version [4]. In our experiments we set up a system of 16 servers. The servers are equipped with ColTrES and configured to cooperate through different architectures and discovery protocols.

7.1 Comparison of the architectures

In this section we compare the performance of the hierarchical and flat architectures of servers that collaborate in discovering, transcoding, and delivering Web objects. We set up a scenario where all servers are well connected among them and with the clients. The content servers are placed in a remote location, connected through a geographic link with 14 hops in between, a mean round-trip time of 60 ms, and a maximum bandwidth of 2Mb/sec. We verified that in this scenario the network path to the content servers (reached in case of global miss) was a possible system bottleneck. Hence, the global cache hit rate may impact the response time.

We consider the **Hierarchical leaf** and **Hierarchical root** schemes for the hierarchical architecture, the query-based (**Flat query-based**) and summary-based (**Flat summary-based**) for the flat architecture.

The hierarchical architectures are configured on the basis of a three-level hierarchy with 12 leaves, 3 intermediate servers (with a nodal out-degree of 4), and one root node. The client are redistributed to let only the leave nodes receive their requests. The configuration for Flat query-based and Flat summary-based are based on ICP and Cache Digests protocols, respectively, and a flat cooperation scheme, where all edge servers have sibling relationships among them. For a fair comparison, in this set of experiments the flat schemes use the blind-active algorithm as the hierarchical schemes.

In these experiments we use both light trans-load and heavy trans-load workloads. First, we evaluate the cache hit rates, and then we focus on the system response time, which is the crucial performance metric to the end users.

Tables 1 and 2 show the cache hit rates for light trans-load and heavy trans-load workloads, respectively. For each cooperation scheme, we report the local exact and useful hit rates (columns 2 and 3, respectively) as well as the remote hit rates (columns 4 and 5). The last column shows the global hit rate, which is the sum of the various hit rates. For the hierarchical leaf scheme, we do not report the remote useful hits, because the requests to the parent nodes refer only to the original version of the resources.

Table 1. Cache hit rates (*light trans-load*).

	Local exact	**Local useful**	**Remote exact**	**Remote useful**	**Global**
Flat query-based	19.4%	16.9%	13.8%	19.3%	69.4%
Flat summary-based	21.2%	11.9%	11.5%	11.5%	56.1%
Hierarchical root	17.9%	6.8%	7.1%	7.7%	39.5%
Hierarchical leaf	10.2%	8.2%	19.6%	n/a	38.0%

We describe Table 1 and use the results in Table 2 to confirm our observations or to evidence differences. From the last column of Table 1 we can observe that there are some significant differences in the global hit rates, depending on the used cooperation mechanism. In particular, Flat query-based provides the best results, while Flat summary-based turns out to be less effective in finding hits. Flat summary-based performance deteriorates because the Cache Digests protocol tends to become imprecise (i.e., the accuracy of the exchanged cache digests decreases) and its remote hit rates diminish. This is particularly evident for the heavy workload, but it can be also observed for the light workload: the presence of larger objects causes faster changes in the caches, having as a consequence a reduction of the accuracy of the exchanged digests. Columns 4 and 5 in Table 1 show that the reduction in the global hit rate is caused by a reduction of the remote hit rate.

The two hierarchical schemes achieve similar global hit rates (last column of Table 1). However, their global hit rates are lower than those of flat architectures. The most evident and expected result observed from comparing Tables 1 and 2 is the higher hit rates obtained under the light trans-load model, because the object sizes in the heavy trans-load model are much larger. The lower hit rate of the heavy trans-load model increases the replacement activity, thus reducing the hit rate of Flat summary-based. For this reason, the reduction in remote hit rates of this scheme, which has been

Table 2. Cache hit rates (*heavy trans-load*).

	Local exact	**Local useful**	**Remote exact**	**Remote useful**	**Global**
Flat query-based	5.1%	4.7%	20.3%	22.1%	52.2%
Flat summary-based	5.3%	4.6%	10.3%	8.9%	29.1%
Hierarchical root	6.3%	4.7%	5.2 %	4.4 %	20.6%
Hierarchical leaf	6.1%	4.3%	11.6%	n/a	22.0%

already observed for the light trans-load model, is even more evident from columns 4 and 5 of Table 2.

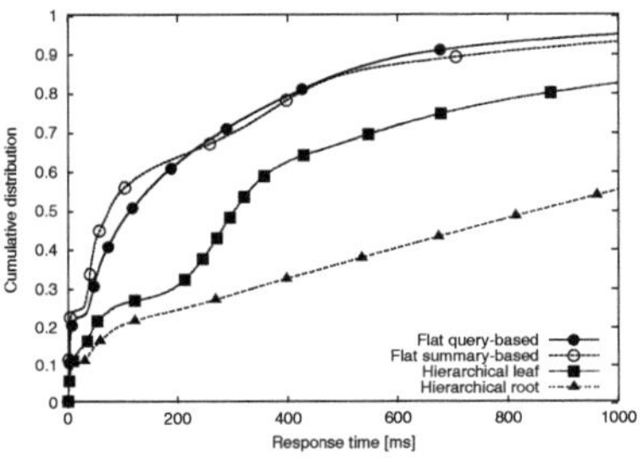

Figure 2. Cumulative distributions of system response times (*light trans-load*).

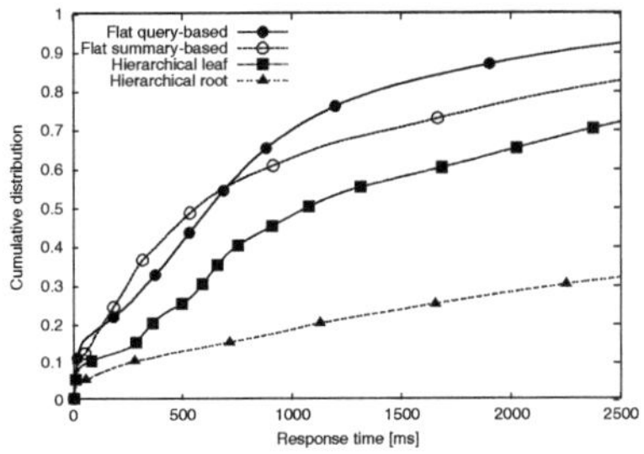

Figure 3. Cumulative distributions of system response times (*heavy trans-load*).

We now pass to consider the response time. Figures 2 and 3 show the cumulative distribution of system response time for the considered schemes under the light trans-load and heavy trans-load workloads, respectively. Most of the curves shows several steps as a consequence of the different kinds of cache hit (that is, remote vs local, and useful vs exact) during the discovery phase. All the schemes present a first step (located on the left side of each graph) due to local exact hits that are nearly instantaneous with respect to other hits and misses. Useful local hits have longer response times, which are typically comparable to the ones of remote exact hits. Remote useful hits have even longer response times, but do not show evident steps on the response time curve because of the larger variance in the response time. Misses generate the highest response times, hence they are typically located in the right side of each curve.

The two figures confirm that the hierarchical leaf scheme clearly outperforms the hierarchical root architecture. However, none of the hierarchical schemes can compete with flat architectures. The expected bad performance of the hierarchical root scheme is due to the bottleneck of the root node. We observed that the higher levels of the hierarchy are often overloaded because they have to handle transcoding operations of all misses from the lower levels. Measurements on the CPU load show that the mean load of the root node is nearly 0.90 and 0.99 for light trans-load and heavy trans-load model, respectively, as this node has to process every miss occurred in the lower levels.

On the other hand, leaf edge servers are often idle (the corresponding mean CPU load is less than 0.02 for both workload models), thus waiting for the upper-level nodes to process their requests.

The hierarchical leaf scheme achieves better performance: the response times in Figures 2 and 3 are much lower than those obtained by the hierarchical root. However, even the hierarchical leaf scheme is penalized with respect to the flat schemes. There are two reasons for this result. In hierarchical leaf scheme, the upper hierarchy levels can only act as pure cache servers (in our testbed prototypes, 4 nodes over 16 do not contribute in transcoding operations). Moreover, as shown in the Tables 1 and 2, the flat cooperation schemes achieve the highest cache hit rates.

Flat architectures offer the best results. A preliminary performance comparison between Flat query-based and Flat summary-based is in [3]. With the experiments carried out in this paper we confirm the previous observations: the higher global hit rates of Flat query-based tend to reduce the response time of the resources that are found in the nodes of the intermediate architecture. On the other hand, due to the faster lookup mechanism of Flat summary-based, remote hits are typically served faster than those of Flat query-based. For this reason, it seems interesting to analyze the cumulative distribution of the response time. Table 3 provides a summary of data in Figures 2 and 3. It shows the median (50-percentile) and 90-percentile of the response time for each cooperation scheme and both workload models.

Table 3. Median and 90-percentile of system response times [sec].

	Light trans-load		**Heavy trans-load**	
	median	**90-perc.**	**median**	**90-perc.**
Flat query-based	0.11	0.64	0.62	2.24
Flat summary-based	0.07	0.78	0.56	3.76
Hierarchical root	0.86	2.82	5.52	14.57
Hierarchical leaf	0.30	1.74	1.07	5.11

Figure 2 shows that the difference between the two curves of Flat query-based and Flat summary-based is slight, with the former only slightly superior to the latter on the right side of the graph. This result occurs even if the global hit rate of the two flat schemes differs significantly (69.4% vs. 56.1%). Moreover, if we analyze the median response time, we can see that Flat summary-based is faster than Flat query-based (also shown in the column 2 of Table 3). This can be explained by the high lookup time required by the Flat query-based scheme. On the other hand, under the heavy trans-load model (Figure 3) the curves of response times are more differentiated, with Flat query-based outperforming Flat summary-based. This result is due to the higher difference in their cache hit rates (52.2% vs. 29.1%) that cannot be compensated by the faster lookup of Flat summary-based. However, even in this case the median response time for Flat summary-based is the lowest.

Table 3 summarizes the results that can be get from the previous figures: the hierarchical root scheme is the slowest; flat architectures outperform hierarchical schemes; Flat query-based is the fastest scheme to serve the large majority of the requests, even if Flat summary-based can be faster than Flat query-based to serve half of the requests for both workloads.

7.2 Cooperative transcoding algorithms

In this section we compare the performance of the cooperative transcoding algorithms for flat architectures described in Section 5.2. For this set of experiments we choose the Flat flat-based scheme, because it typically offers the highest cache hit rates and lowest response times. Indeed, the flat-based scheme performs well for a wide range of workload models, at least until the system is under heavy stress, as noted in [3]. Under low and medium load, the difference between the various transcoding algorithms is very small. Therefore, it is more interesting to explore the performance gain achievable with the proposed cooperative transcoding algorithms when the server CPUs are nearly always busy due to transcoding operations. To this purpose, we used the bimodal and uniform workloads described in Section 6.

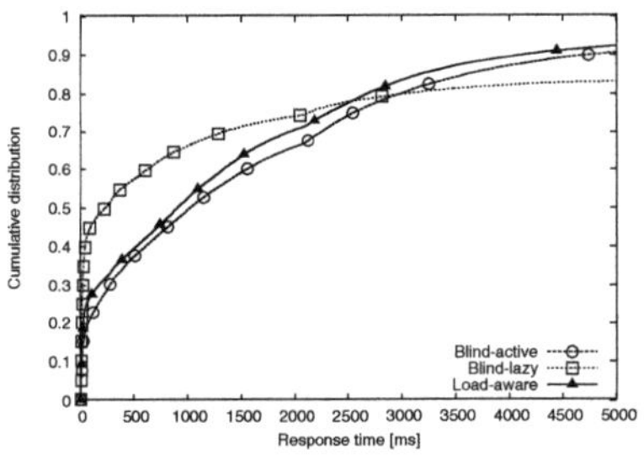

Figure 4. Cumulative distributions of system response times (*bimodal workload*).

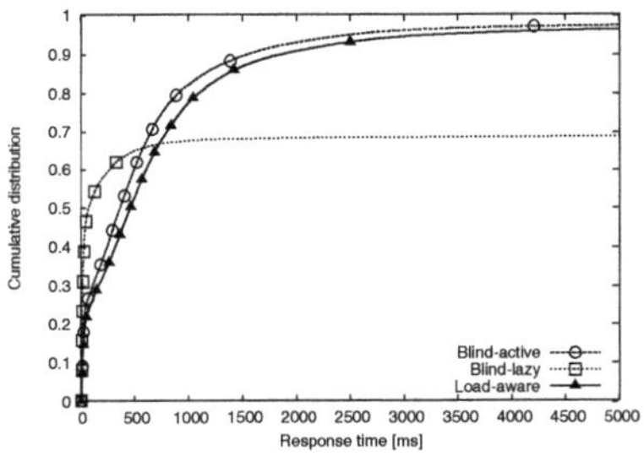

Figure 5. Cumulative distributions of system response times (*uniform workload*).

Figure 4 shows the cumulative distribution of response time for the load-blind and load-aware algorithms with the bimodal workload, while Figure 5 refers to the uniform workload. It is worth to note that the load-aware algorithm is a typical threshold-based policy that uses the CPU utilization as the activation parameter. We performed experiments for different thresholds ranging from 0.1 to 0.9 and found that for bimodal workload the typical common-sense value of 0.66 for the threshold offers the most stable performance in terms of the 90-percentile of response time. Therefore, Figure 4 shows only the curve related to this threshold value. On the other hand, for the uniform workload we found that no "best" threshold value exists: the 90-percentile of the response time grows monotonically as the threshold value decreases from 0.9 to 0.1. In Figure 5 the curve of the load-aware algorithm corresponds to the same load threshold value (0.66) used in Figure 4. Under the uniform workload, the curve corresponding to the best threshold value (0.9) is in between the curve of the blind-active algorithm and the one for threshold equal to 0.66.

Response time curves achieved by blind-active and load-aware algorithms are similar, with the load-aware algorithm providing better response times in the case of bimodal workload and the blind-active algorithm being faster in the case of uniform workload. On the other hand, the blind-lazy algorithm shows a different behavior. It is able to reduce the response time for most requests, but it becomes unacceptably slow for up 30% of the requests, depending on the workload.

To better present the performance differences, in Table 4 we report the median and the 90-percentile of the response time.

Table 4. System response times for load-blind and load-aware algorithms [sec].

	Bimodal workload		**Uniform workload**	
	median	**90-percentile**	**median**	**90-percentile**
Blind-active	1.03	4.88	0.36	1.56
Blind-lazy	0.25	121.12	0.07	239.79
Load-aware	0.89	3.98	0.46	1.90

The blind-lazy algorithm bets on finding either a useful hit or an exact remote hit on a less loaded peer. If it succeeds, it can reduce the response time. However, if no remote hit is found or, even worse, if the peer having a remote useful hit is overloaded, the response time can increase significantly. This explain why the blind-lazy algorithm can successfully reduce the median response time (as shown in columns 2 and 4 of Table 4), but it tends to have poor performance when considering 90-percentile due to the high number of pathological cases (columns 3 and 5 of Table 4). The problem is more evident when the load is evenly distributed (column 5 of Table 4), because in this case there is a higher probability of finding a peer with a heavier load. On the other hand, the blind-active algorithm seems to offer better performance because the transcoding load is only related to the client requests being served and not to the requests directed to other peers. The response time has a more definite upper bound, thus reducing the 90-percentile of the response time with respect to the blind-lazy algorithm. On the other hand, the median response is higher than that of the lazy policy.

The load-aware algorithm offers some performance gains when the load is unevenly distributed (bimodal workload), because it can act smarter than the load-blind algorithms. In particular, it reduces of about 22% the 90-percentile (as shown in column 3 of Table 4) and about 14% the median response time (column 2 of Table 4) with respect to the blind-active algorithm. On the other hand, in the case of uniform workload the load-aware algorithm is ineffective in reducing the response time, and there is a performance loss on both 90-percentile and median response time. Indeed, when the skew of the workload is low, we need a more sophisticate algorithm, possibly based on information on the load of a larger (maybe entire) set of servers of the intermediate architecture.

8. Conclusions

In this paper, we have proposed an intermediate distributed architecture for cooperative caching and transcoding that can be implemented in the existing Web infrastructure. We have investigated various schemes that use different server organizations (hierarchical, flat), and different cooperation mechanisms for resource discovery (query-based, summary-based) and transcoding (load-blind, load aware). We have compared their performance through ColTrES, a flexible prototype testbed based on Squid that implements all proposed mechanisms. From the performance evaluation, we have found that flat peer-to-peer topologies are always better than hierarchical schemes, because of bottleneck risks in the higher levels of the hierarchy combined with limited cache hit rates. Among the flat cooperation schemes, we evaluated multi-version lookup extensions of Cache Digests and ICP and found that, ICP tends to have better performance due to the lower cache hit rates of Cache Digests. As a further contribution of this paper, we verified that the proposed load-aware algorithm can achieve some performance gains only when the client load is unevenly distributed among the edge servers of the intermediate infrastructure. On the other hand, in the case of rather uniform load distribution, the load-aware algorithm does not seem to achieve any significant improvement.

An intermediate infrastructure of distributed servers that cooperate in multi-version content caching, discovery, and transcoding opens many research topics. A limited number of issues have been investigated in this work, that to the best of our knowledge represents the first implementation of cooperative transcoding and caching systems for both hierarchical and flat topologies. There are other research issues that this paper opens up, such as cooperative cache replacement policies for multi-version content, transcoding policies based on global information on server load and available network bandwidths, and integration with server-direct transcoding to preserve the end-to-end content semantics.

Acknowledgements

We acknowledge the support of MIUR-FIRB "Wide-scale, Broadband, Middleware for Network Distributed Services".

References

[1] M. Butler, F. Giannetti, R. Gimson, and T. Wiley. Device independence and the Web. *IEEE Internet Computing*, 6(5):81–86, Sept./Oct. 2002.

[2] C. Canali, V. Cardellini, and R. Lancellotti. Squid-based proxy server for content adaptation. Technical Report TR-2003-03, Dept. of Comp. Eng., Univ. of Roma "Tor Vergata", Jan. 2003. `http://weblab.ing.unimo.it/research/trans_caching.shtml`.

[3] V. Cardellini, M. Colajanni, R. Lancellotti, and P. S. Yu. A distributed architecture of edge proxy servers for cooperative transcoding. In *Proc. of 3rd IEEE Workshop on Internet Applications*, pages 66–70, June 2003.

[4] V. Cardellini, P. S. Yu, and Y. W. Huang. Collaborative proxy system for distributed Web content transcoding. In *Proc. of 9th ACM Int'l Conf. on Information and Knowledge Management*, pages 520–527, Nov. 2000.

[5] S. Chandra, C. S. Ellis, and A. Vahdat. Application-level differentiated multimedia Web services using quality aware transcoding. *IEEE J. on Selected Areas in Communication*, 18(12):2544–2465, Dec. 2000.

[6] S. Chandra, A. Gehani, C. S. Ellis, and A. Vahdat. Transcoding characteristics of Web images. In *Proc. of Multimedia Computing and Net. Conf.*, Jan. 2001.

[7] C.-Y. Chang and M.-S. Chen. On exploring aggregate effect for efficient cache replacement in transcoding proxies. *IEEE Trans. on Parallel and Distributed Systems*, 14(6):611–624, June 2003.

[8] L. Fan, P. Cao, J. Almeida, and A. Z. Broder. Summary cache: A scalable wide-area Web cache sharing protocol. *IEEE/ACM Trans. on Networking*, 8(3):281–293, June 2000.

[9] A. Fox, S. D. Gribble, Y. Chawathe, E. A. Brewer, and P. Gauthier. Cluster-based scalable network services. In *Proc. of 16th ACM Symp. on Operating Systems Princ.*, pages 78–91, Oct. 1997.

[10] R. Han, P. Bhagwat, R. LaMaire, T. Mummert, V. Perret, and J. Rubas. Dynamic adaptation in an image transcoding proxy for mobile Web browsing. *IEEE Personal Communications*, 5(6):8–17, Dec. 1998.

[11] B. Knutsson, H. Lu, and J. Mogul. Architectures and pragmatics of server-directed transcoding. In *Proc. of 7th Int'l Workshop on Web Content Caching and Distribution*, Aug. 2002.

[12] W. Y. Lum and F. C. M. Lau. On balancing between transcoding overhead and spatial consumption in content adaptation. In *Proc. of ACM Mobicom 2002*, pages 239–250, Sept. 2002.

[13] A. Maheshwari, A. Sharma, K. Ramamritham, and P. Shenoy. TransSquid: Transcoding and caching proxy for heterogeneous e-commerce environments. In *Proc. of 12th IEEE Int'l Workshop on Research Issues in Data Engineering*, pages 50–59, Feb. 2002.

[14] R. Mohan, J. R. Smith, and C.-S. Li. Adapting multimedia Internet content for universal access. *IEEE Trans. on Multimedia*, 1(1):104–114, Mar. 1999.

[15] M. Rabinovich and O. Spatscheck. *Web Caching and Replication*. Addison Wesley, 2002.

[16] A. Rousskov and D. Wessels. Cache Digests. *Computer Networks*, 30(22-23):2155–2168, 1998.

[17] W. Shi, K. Shah, Y. Mao, and V. Chaudhary. Tuxedo: a peer-to-peer caching system. In *Proc. of the 2003 Int'l Conf. on Parallel and Distributed Processing Techniques and Applications (PDPTA'03)*, Las Vegas, NV, June 2003.

[18] A. Singh, A. Trivedi, K. Ramamritham, and P. Shenoy. PTC: Proxies that transcode and cache in heterogeneous Web client environments. *World Wide Web*, 2003.

[19] Squid Internet Object Cache. `http://www.squid-cache.org`.

[20] A. Wolman, G. M. Voelker, N. Sharma, N. Cardwell, A. Karlin, and H. M. Levy. On the scale and performance of cooperative Web proxy caching. In *Proc. of 17th ACM Symp. On Operating Systems Princ.*, Dec. 1999.

USER SPECIFIC REQUEST REDIRECTION IN A CONTENT DELIVERY NETWORK

Sampath Rangarajan[1], Sarit Mukherjee[1], and Pablo Rodriguez[2]
[1]*Lucent Technologies Bell Laboratories,* [2]*Microsoft Research, Cambridge*

Abstract This paper discusses user specific request redirection for personalizing responses to user requests in Content Delivery Networks (CDN). User specific request redirection refers to the process of redirecting user requests to a server on the content delivery network based on user specific information carried in the user request. Current redirection schemes in CDNs are either based on the authoritative DNS model or the URL rewrite model. The authoritative DNS model is not flexible to support user specific redirection as user specific information is not available to the DNS server; the URL rewrite model cannot support user specific redirection in practice because of the cost of on-the-fly URL rewrites required on a per user request basis. We discuss a technique that allows for flexible user specific request redirection. The technique is simple enough to be implemented at wire-speed in a switch.

1. Introduction

Content Delivery Service Providers (CDSP) such as [1, 8] distribute content from origin sites to cache servers on the edge of the network and deliver content to the users from these edge servers (referred to as Content Delivery Servers or CDS). The distribution mechanism could be based both on push technologies such as multicasting the data to all the edge servers through terrestrial or satellite links or pull technologies such as those used by proxies. The goal is to decrease the latency of user access to the objects by delivering the objects from an edge server closest to the user.

A CDN consists of a set of CDSs across the Internet as well as a DNS infrastructure which is used to route user requests to the nearest CDS. For this to work, the DNS requests sent from the user browser now needs to be directed to the DNS of the CDSP. There are two ways to accomplish this.

Authoritative DNS: One way is for the CDSP to "takeover" the DNS functionality of the origin site and become the authoritative DNS for the origin site [9]. This is an easy approach to implement but the problem with this approach is that all the objects from the domain that has been taken over needs to be served from the CDSs. This approach will not work for example, if it is required that all html pages be served

F. Douglis and B.D. Davison (eds.), Web Content Caching and Distribution, 223-232.

from the origin site but all the images (e.g., gifs and jpgs) be served from the CDSs.

URL Rewrite: In this approach, the authoritative DNS functionality still stays with the origin site's DNS. Any top level page requested by a user will be served from the origin server. But before the page is served, all the embedded links found in the top level page are rewritten to point to the CDSP DNS so that requests to embedded objects can be redirected by the CDSP DNS to the closest CDS. This mechanism has been popularized by Akamai [1].

In this paper, we consider User specific request redirection. This refers to the process of redirecting requests to the same set of embedded objects in a top level page, arriving from different users, to different CDSs, based on information contained in the user request. The information found in the user requests that could be used include the user IP address, cookies present in the user request, user browser information, etc. Other information that is not contained in the user request could be used to make redirection decisions as well. These include time of day, priority of the user, cost of accessing the CDSP network and performance of the CDSP network.

The next section discusses and motivates the need for user specific request redirection. It is further argued that the two redirection mechanisms discussed above, namely Authoritative DNS and URL rewriting, cannot, in practice support user specific redirection. In the subsequent section, we propose a technique based on modifying HTTP response headers. This can be used to perform user specific redirection very efficiently in practice.

2. User Specific Request Redirection

The main objective of performing user specific request redirection is to provide service differentiation based on user specific information. In addition, by performing such redirection to specific server IP addresses, inaccuracies due to DNS based redirection can be eliminated and performance can be improved by bypassing DNS. These issues are discussed below. In the following discussion, URL rewriting is used only to illustrate user specific redirection. In a latter section it is argued that this is not a very practical approach.

2.1 Service differentiation based on user specific information

Service differentiation based on user specific information is very important for service providers and is very much tied to a user and the price the user is willing to pay. There are several different ways to ensure service differentiation with user specific redirection.

Service differentiation using DNS hierarchy: With user specific redirection, it becomes possible for the content provider to instruct the CDSP to provide different service levels to different users. For example, if two users carrying cookies "priority = high" and "priority = low" access a top-level page, the CDSP could provide the page rewritten in two different ways so that the service level provided to these two users, when they access the embedded objects in the top-level page, is different. One way to provide different service levels is to redirect the embedded object URLs to two different CDSP DNS hierarchies where one DNS hierarchy resolves the DNS query to one of a set of fast servers close to the user and the other DNS hierarchy resolves the query to one of a set of slower servers or a server which is on a site with limited bandwidth. In addition, requests from two different users could be redirected to two different service provider networks; for example, a high priority user may be directed to www.cdsp1.com DNS hierarchy which provides fast service but costs more and the low priority user may be directed to www.cdsp2.com which provides slower service but is cheaper.

Service differentiation using URL prefix: If URL rewriting is used to redirect user requests, when URLs are rewritten, different prefixes can be added to the URL path to provide indication to the CDS about the service level that should be provided to the user. For example, if embedded objects are prefixed with `level1` when the top level page is served to one user and with `level2` when it is served to another user, this could indicate to the CDS that if both requests arrive at the CDS, the request with a prefix of `level1` should be provided better service than the one with the prefix `level2`.

When redirection is performed, IP addresses could be used instead of host names. User specific redirection with IP addresses has advantages that are discussed in the next two sections.

2.2 IP address based redirection

Performing IP address resolution using the DNS of the CDSP has certain drawbacks as described below.

Source IP address inaccuracy: When a DNS request is received, the CDSP DNS checks the source IP address on the request, and based on this, returns the IP address of the CDS "closest" to the source IP address. This decision is made based on the assumption that the source IP address on the request is either the IP address of the user or one "close" to the user. But this may not be the case in practice. The source address on the DNS request is the IP address of the entity that sends the DNS request to the CDSP DNS. Normally, this is the local DNS server on the user site. Depending on how DNS requests are forwarded, it could also be a DNS server further along the hierarchy from the local DNS. Therefore, the selected CDS is "closest" to the entity that sends the DNS request but not necessarily "closest" to the user. Server selection based on local DNS server IP addresses can result in a non-optimal server selection since users are frequently distant from their local DNS severs [7].

Inaccuracy due to DNS caching: When the CDSP DNS returns the IP address of the "closest" CDS, this IP address is cached by the browser and subsequently used to resolve domain names to IP addresses locally. This means that subsequent DNS queries to the same domain name will not even be sent to the CDSP DNS until the cached information is flushed. A non-optimal CDS may be used for this period of time if the network conditions change. Similarly, the local DNS or one of the DNS servers upstream towards the CDSP DNS could also cache DNS information. This type of DNS caching may lead to inaccurate server selection. One way to address this issue is to specify a DNS timeout (TTL) which is very small. There are, however, two problems with this approach. The first problem is that the DNS caches do not need to obey the timeouts. The second problem is that it is difficult to select this timeout. The timeout needs to be small enough so that dynamic server selection is possible; on the other hand, a DNS timeout value that is too small will lead to very frequent DNS lookups at the CDSP DNS server. In addition to the drawbacks discussed above, DNS requests themselves add to the response time when content is retrieved from a CDN. The work in [4, 7] shows that DNS requests for rewritten URLs account for a significant overhead and clearly reduce the benefits of having content replicated at the network edge.

The aforementioned drawbacks can be eliminated by performing user specific redirection using IP addresses instead of host names. For example, with dynamic URL rewrite, when a URL www.xyz.com/fashion/style.gif is rewritten in a top level page, instead of pointing the URL to the CDSP DNS, the URL could be rewritten as 192.1.1.87 /www.xyz.com/fashion/style.gif where 192.1.1.87 is the IP address of the server "closest" to the user to whom the top level page is being served. The server IP address can be chosen accurately based on the source IP address of the client itself (as this is now visible to the URL rewriter). In addition, rewriting links with IP addresses will eliminate the need for a DNS lookup and hence make the download of the page faster. Using IP addresses for URL rewrite is especially beneficial to wireless users. DNS lookups across wireless links is a performance bottleneck. This is more of a problem when the TTL for the DNS responses are kept very small. By using IP addresses thereby eliminating DNS lookups, the latency for wireless users will be significantly decreased.

3. Implementing user specific request redirection

We now turn to the issue of implementation. Of the two redirection mechanisms discussed in an earlier section, the authoritative DNS is not flexible enough to provide user specific request redirection because DNS requests do not carry any user specific information. URL rewrite is a possible option for this purpose. We discuss URL rewrite in more detail below and argue that it is not a viable option for user specific redirection in practice.

The URL rewrite process involves rewriting the embedded links inside a html page to point to the CDSP DNS. For the most part, embedded objects that are automatically fetched by the browser are images that are large and will benefit from being served from the edge. During the content delivery process, the request to the top level page is sent to the content provider's origin server. DNS requests to resolve the rewritten URLs inside the top level page will then be sent to the CDSP's DNS. The CDSP DNS server returns the IP address for the CDS which is "closest" to the user. The request for these objects are then sent to this CDS and retrieved from that CDS.

Normally, the process of URL rewrite is performed statically using some software tool. Pages that need to be rewritten are parsed and the embedded URLs rewritten using this tool. This technique does not provide the flexibility to rewrite the same page in multiple different ways and provide different versions of the rewritten page to different users in order to personalize downloads of embedded objects based on information found in user requests (which is what is needed for user specific redirection). Thus, static URL rewriting is not a valid option for user specific redirection.

CDSP companies have also partnered with caching and switching vendors to provide what is referred to as *dynamic URL rewrite* [1, 6, 3, 5]. In this technique, a reverse proxy cache [6] or a load balancing switch [3] placed in front of the content provider's servers performs URL rewrite on the objects. The transformations need to be performed are downloaded into the device. When a user request is received at the device for the first time, the request is sent to a server and the object is fetched. This object is then (URL-)rewritten at the proxy/switch and then sent to the user.

There are two obvious ways in which Dynamic URL rewrite can be used to implement User specific redirection: **a)** Every time a user request is received, the html page is parsed and the embedded URLs are transformed appropriately based on the requesting user. For example, if the transformation is performed based on IP addresses, each URL will be transformed to point to the IP address of the best server for that user, and **b)** All the embedded URLs in the page are rewritten a priori in all possible ways in which it needs to be delivered to the users and all these copies are cached. When a user request is received, the appropriate page is delivered.

The first technique may not be practical as the html page needs to be parsed every time a user request is received. The second technique may be possible if the number of ways in which URL rewrite needs to be performed on a page is small. But if the transformation is performed using IP addresses, the number of ways in which URL rewrite can be performed on a page equals the number of server IP addresses. This will become impractical even for a reasonably large number of servers.

In the next section, we present a technique based on HTTP response header modification that enables user specific redirection to be implemented very efficiently. We refer to this technique as Request Deflection. The attractiveness of the technique is that it could be implemented in a switch at the packet level at wire-speed. A switch level implementation of our technique is not described in detail in this paper, but a brief discussion is included that shows the viability of such an implementation.

4. Request Deflection

The Request Deflection technique enables user specific redirection without the need for parsing the html page every time a user request is received or performing a URL rewrite on the page a priori in all possible ways and caching these page copies. There are two steps to the Request Deflection process as discussed below.

4.1 Step 1: A priori transformation of embedded URLs

In the first step, the page on which URL rewrite is to be performed is parsed and the embedded URLs are transformed as follows. This transformation is performed only once and the transformed page is cached.

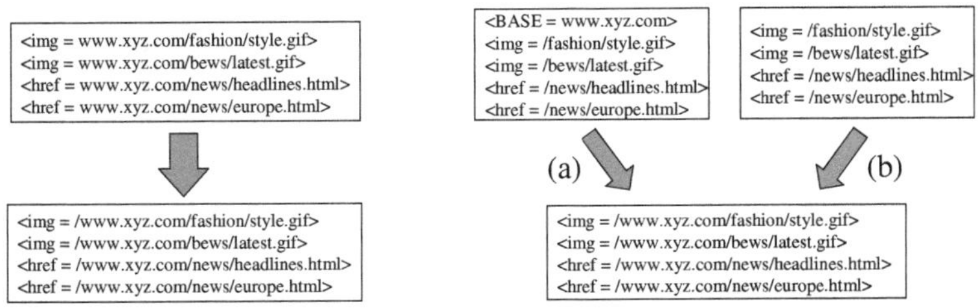

(A) Transformation of absolute URLs (B) Transformation of relative URLs

Figure 1. URL Transformation

Handling absolute URLs. All the embedded absolute URLs that are to be redirected to CDSs are converted into relative URLs. This process is shown in Figure 1(A).

Note that the domain name of the CDSP is not prefixed to the embedded URLs but the URLs have only been changed to relative URLs by prefixing a /.

Handling relative URLs. Note that if a relative URL exists in the page (see Figure 1(B)), there must either be (a) a `<BASE=>` command present at the top of the html document that indicates the server from which this object needs to be retrieved, or (b) a `<BASE=>` command is not present at the top of the html page but the expectation is that the server will add a `Content-Base` token to the HTTP header which points to the server name (or IP address) of the main page that contains these embedded objects.

For all relative URLs that are to be redirected to CDSs, in case of (a), the `<BASE=>` command should be removed and the server name found in the base command should be prefixed to the relative URL (after prefixing this relative URL, the embedded URL should still be relative; i.e., a / should be prefixed). In case of (b), the server name from where the main page was fetched should be prefixed to the relative URL (again leaving the URL still relative). In Figure 1(B) assume that the server name where the main page is present is www.xyz.com. In Figure 1(B), (a) shows a page with the `<BASE=>` command. The server name found in the command is prefixed to the embedded relative URLs and the URLs are left in relative form with a / in front. In Figure 1(B), (b) shows a page with no `<BASE=>` command but with the expectation that the server will send a `Content-Base` token as part of the HTTP header which will point to www.xyz.com, which is the server name where the main page is located. This page is similarly transformed by prefixing the server name of the main page leaving it intact as a relative URL with a / in the front.

4.2 Step 2: On-the-fly HTTP response header modification

The second step is performed every time a user request is received. The page transformed using Step 1 is sent to the user, but the server includes a `Content-Base` token to the HTTP response header that contains the server name (or server name +

URL prefix path) or the IP address (or IP address + URL prefix path) from which the embedded objects in the page need to be retrieved. This is a really fast operation in that the server does not have to parse the html page every time a user request is received, but can rewrite the page differently for each user with respect to the server name or server IP address from which the object should be fetched.

As an example, consider requests from four different users C1, C2, C3 and C4. Assume that to provide service differentiation, users C1 and C2 need to be redirected to www.cdsp1.com, user C3 needs to be redirected to www.cdsp2.com and user user C4 needs to be redirected to www.cdsp3.com. For each of these requests, the server will serve a transformed object (using Step 1) as shown in Figure 1(B), but will include a `Content-Base` token as part of the HTTP header which will be `Content-Base:` http://www.cdsp1.com for C1 and C2, `Content-Base:` http://www.cdsp2.com for C3 and `Content-Base:` http://www.cdsp3.com for C4. The user browsers at C1 and C2 will construct the embedded URLs as http://www.cdsp1.com/www.xyz.com/fashion /style.gif, etc, and fetch these objects from www.cdsp1.com. Similarly, C3 and C4 will fetch the objects from www.cdsp2.com and www.cdsp3.com respectively. If IP addresses are used for redirection and assuming that the "closest" servers to C1,C2,C3 and C4 are IP1,IP2,IP3 and IP4, the content-bases token used will be `Content-Base:` IP1 for C1, `Content-Base:` IP2 for C2, `Content-Base:` IP3 for C3, and `Content-Base:` IP4 for C4.

It is easy to see that Step 2 required for each user request requires very little overhead and can be implemented very easily in a reverse proxy. A more important observation is that adding a `Content-Base` token could be performed at wire-speed in a switch. We briefly discuss this issue in the next section.

4.3 Wire-speed request deflection

As described in the previous section, Step 1 of the request deflection process can be performed a priori and the html page can be cached. Assume that the server stores html pages after performing Step 1. The goal then is to perform Step 2, which is performed on a per user request basis, at wire-speed, using a switch in front of the server.

Let us consider performing Step 2 using IP addresses. When a main page is sent from the server to the user, if a `Content-Base` token is found in the HTTP response header, the value of the `Content-Base` token should be changed to http://IP where IP is the IP address of the server from which the embedded objects should be retrieved. If a `Content-Base` token does not exist, the string `Content-Base:` http://IP should be introduced into the HTTP response header. Let us consider the case where the `Content-Base` token does not exist and should be introduced into the HTTP response header. The case where a `Content-Base` token already exists can be considered as a situation where the existing token is removed and a new token that contains the IP address is introduced.

For this discussion, assume that the `Content-Base` token will be included into IP packets after re-assembly of IP fragments. That is, IP packets that are received at the switch are re-assembled to remove IP fragmentation and only then the `Content-Base` token will be added. At the switch, we cannot introduce new TCP segments into the flow. This means, when the `Content-Base` token is introduced into an IP packet, we need to make sure that the resulting IP packet does not go beyond a TCP segment.

The maximum size of the string that will be added to the IP packets will be strlen(`Content-Base:`) + strlen(http://) + strlen(maximum number of characters in an IP address) + an EOL character. The maximum number of characters in an IP address is 15. Thus, the maximum number of characters added will be 36. In order to accommodate these 36 bytes in a TCP segment without overflowing the segment, we need to make the server send TCP segments whose maximum length is 36 bytes less than the maximum segment size (MSS) that has been announced by the user. This way, the extra 36 bytes can be added to a TCP segment without exceeding the client requirement.

When the SYN packet for the TCP connection from the client (which will download the main page) is received at the switch, the value of the MSS TCP option is decreased by 36. For example, if the original MSS sent by the user is 1460, it is changed to 1424. This way, the server is forced to send TCP segments whose size does not exceed 1424. When the HTTP response is received from the server, the `Content-Base` token is added to the HTTP response header. Given that different tokens of the HTTP header can appear anywhere in the header, it is possible to add this header to the first response segment that is received from the server. Even after this addition, the TCP segment size will not exceed 1460 and so is acceptable to the client. Of course, to accommodate the change in the size of the TCP segment, the sequence numbers on all packets that follow this modified packet from the server to the client and the ack packet for this modified packet as well as subsequent packets from the client to the server need to be adjusted. Assume that after the addition of the `Content-Base` token, the length of the segment is increased by x, where $x \leq 36$. The ack sequence number on all packets from the server to the client that follow the modified packet should be decreased by x and the sequence number on the ack for the modified packet as well as subsequent packets from the client to the server should be increased by x. The length field of the modified packet should be adjusted as well.

The drawback of decreasing the MSS indicated to the server is that the server is forced to send TCP segments smaller than what the user can accept. This means, the server may end up sending more TCP segments. The justification for this is that main pages are mostly of small size and should fit in a few TCP segments. Because of this, the number of extra TCP segments sent should be very minimal.

A Request Deflection device (either a reverse proxy or a switch) needs a mechanism to decide the mapping between a user request and the IP address of the server that should be chosen to serve the embedded objects. The next section discusses two possible approaches for this.

4.4 Resolving server IP addresses

Downloading the mapping table into the Request Deflection device: Currently the CDSP DNS server serves two main functions: a) collect information about the network and the status of the CDSs and runs a proprietary algorithm based on this information to construct a table of user IP address to CDS IP address mapping, and b) reply to DNS requests from the users with the IP address of the appropriate CDS.

With Request Deflection using IP addresses, the CDSP still will compute a table of user IP address to CDS IP address mapping, but instead of serving as a DNS server, it will now download the table into Request Deflection device. There are two problems with this solution. Firstly, the CDSP DNS (which is now only an IP address mapper) needs to download this information into all the Request Deflection devices that may exist in the CDSP network. This problem is compounded by the fact that normally one CDSP DNS server does not compute all user to CDS mapping. This computation is performed by multiple DNS servers in the CDSP DNS hierarchy. This means that information needs to be downloaded into all the Request Deflection devices from all the DNS servers that compute the mapping. Secondly, the table size could be large; storing this table at all the Request Deflection devices could be difficult. This approach may be feasible only with small CDSP networks like the ones that will be found within an enterprise (enterprise level CDSP networks) [2].

Performing address lookup from the Request Deflection device: Instead of downloading the mapping table, the Request Deflection device could perform an operation similar to a DNS lookup at the CDSP DNS whenever a user to CDS mapping entry is needed. The device could send the IP address of the user to the CDSP DNS which will return the IP address of the optimal CDS for that user.

When such an approach is followed, there is no need to maintain and update a table at the device and the scalability of the CDSP network is not a problem anymore. Still, this solution requires that the CDSP DNS be changed so that it understands queries sent by the Request Deflection device. The CDSP DNS could let the devices cache this information by providing a timeout value. It is expected that the devices will obey this value. Also note that unlike the DNS lookups performed in current CDSP implementations, the user IP address that is sent to the CDSP DNS is the IP address of the real user. It is also possible to send more detailed information to the CDSP DNS about the user. For example, information such as the cookies contained in the request and the user browser information could be sent in addition to the user IP address. This way, the CDSP DNS could return an IP address for a CDS whose identity depends not only on the "closeness" to the user but also on other user characteristics. In fact, multiple users making requests from the same user machine could be identified and redirected to different CDSs, if needed. Also note that the value returned by the CDSP DNS does not just have to be the IP address. It could return a URL prefix that not only contains the IP address, but also information about the customer making the request for billing purposes (for example, the customer code as used by Akamai [1]), serial numbers and other such information that will enable the CDSs to understand this request in a proprietary way.

5. Conclusions

User specific redirection is a useful technique for response personalization. This gives the ability for a content provider to differentiate users based on their IP addresses as well as other user identifiers such as cookies and browser type and provide service differentiation to these users although the objects are served to these users from a CDSP network. We have proposed a technique which makes user specific redirection practical. The advantage of the technique is that it could be implemented at wire-speed in a switch.

There is a limitation to the proposed technique. With Request Deflection, all objects referred to by the embedded URLs in a given page must be retrieved from the same CDS, thus, not permitting for different objects to be delivery from different CDSs. The reason for this is that `Content-Base` token can only specify one single server name or IP address to be used for all relative embedded objects in a given page.

Fetching all embedded objects from the same CDS is normally recommended to optimize end-user performance due to the following considerations: **a)** When different embedded objects are fetched from different CDSs, client browsers generate new DNS lookups and new TCP connection setups for each CDS. These new DNS requests create a significant overhead, thus, reducing the benefits of replicating content at the network edge and **b)** With the advent of HTTP/1.1 that uses persistent connections to download objects, fetching embedded objects from the same CDS enables the embedded objects to be fetched over a single (or a few) TCP connections. This means, the TCP connection setup overhead is minimized.

However, despite the fact that fetching all objects from the same CDS improves end-user experience, there are cases where different objects must be retrieved from different CDSs. For instance, different CDSs may be required to handle streaming content and images. One solution to overcome this limitation is to host a set of CDSs needed for all content types (e.g. streaming, WAP content, images) under the same virtual IP address. These CDSs are then front-ended by a Layer 4 or Layer 7 switch. The switch can then be used to redirect different requests for different content types towards the right CDS based on the destination port number or on application specific information.

References

[1] Akamai Technologies. http://www.akamai.com/en/html/services/content_delivery.html.

[2] Cisco. http://www.cisco.com/univercd/cc/td/doc/product/webscale/content/cdnent/encdn302/rn_en302.htm.

[3] F5 Networks. http://secure.f5.com/news/articles/article052300b.html.

[4] J. Kangasharju, K. W. Ross, and J. W. Roberts. Performance evaluation of redirection schemes in content distribution networks. In *Proceedings of the 5th Web Caching Workshop*, Lisbon, 2000.

[5] Nortel Networks. http://www142.nortelnetworks.com/bvdoc/alteon/whitepapers/Accel_Delivery.pdf.

[6] Novell. http://www.novell.com/news/press/archive/2000/10/pr00103.html, 2000.

[7] A. Shaikh, R. Tewari, and M. Agrawal. On the effectiveness of DNS-based server selection. In *Proceedings of the IEEE Infocom conference*, Apr. 2001.

[8] Speedera Network. http://www.speedera.com/technology/technology_implement.html.

[9] Unitech Networks. http://www.sw.unitechnetworks.com/en/products/idns/.

FRIENDSHIPS THAT LAST: PEER LIFESPAN AND ITS ROLE IN P2P PROTOCOLS

Fabian E. Bustamante and Yi Qiao
Department of Computer Science, Northwestern University

Abstract We consider the problem of choosing who to "befriend" among a collection of known peers in distributed P2P systems. In particular, our work explores a number of P2P protocols that, by considering peers' lifespan distribution a key attribute, can yield systems with performance characteristics more resilient to the natural instability of their environments.

This article presents results from our initial efforts, focusing on currently deployed decentralized P2P systems. We measure the observed lifespan of more than 500,000 peers in a popular P2P system for over a week and propose a functional form that fits the distribution well. We consider a number of P2P protocols based on this distribution, and use a trace-driven simulator to compare them against alternative protocols for decentralized and unstructured or loosely-structured P2P systems. We find that simple lifespan-based protocols can reduce the ratio of connection breakdowns and their associated costs by over 42%.

1. Introduction

Peer-to-peer computing has been defined as the sharing of computer resources and services by direct exchange between the participating nodes. Since Napster's [18] introduction in 1999, the area has received increasing attention from the research community and the general public, as the model's many advantages have being recognized.

In its purest form, the P2P model has no concept of a server, but rather considers all participants as equals, regardless of resource capacity, connectivity or "commitment" to the common good. While this assumption of equality enables very simple protocols, it could also translate into the loss of some of the model's most appealing attributes [1].

Part of the problem stems from the clash between this equality assumption and the degree of heterogeneity and transiency found in recent studies of current P2P systems. Far from being equal, peers' populations have been shown to exhibit significant variations in attributes such as storage, bandwidth, latency and their degree of sharing. Peers' commitment to the system, in particular, has been found to differ by more than four orders of magnitude [25].

Peers in P2P systems typically define an overlay network topology by keeping a number of connections to other peers, their "friends," and implementing a maintenance protocol that continuously repairs the overlay as new members join and others

F. Douglis and B.D. Davison (eds.), Web Content Caching and Distribution, 233-246.

leave the system. The implication of the degree of peer transiency on the overall system's performance is directly related to the degree of peers' investment in their friends. At the very least, the amount of maintenance-related messages processed by any node would be directly related to the degree of stability of the node's neighboring set. Beyond this, and in the context of content distribution P2P systems, the degree of replication, the effectiveness of caches, and the spread and satisfaction level of queries will all be affected by how dynamic the peers' population ultimately is.

We consider the problem of selecting who to "befriend," among a collection of known peers in distributed P2P systems. Our work explores new protocols that, by considering peers' lifespan distribution a key attribute, can yield systems with performance characteristics more resilient to the natural instability of their environments. This article presents results from our initial efforts, focusing on currently-deployed unstructured and loosely structured P2P systems.

We have measured the observed lifespan of peers in a widely-deployed P2P system and identified a functional form that fits the distribution well. We have designed a number of P2P protocols based on insights gain from this distribution and used a trace-driven simulator to compare them against alternative protocols for decentralized and unstructured or loosely-structured (superpeer-based) P2P systems. We found that P2P protocols based on simple heuristics that prioritize long lived peers when selecting a peer's new "friends" can significantly reduce (up to 42%) the ratio of connection breakdowns and their associated costs.

2. Background

P2P computing has experienced an explosive growth in the last few years and a number of widely-deployed and research-oriented protocols have become available. Although the goal of most P2P systems is to provide general distributed resource sharing among participating peers [27, 9, 10, 12], one of the most popular applications is content distribution.

In general, a set of participant nodes in a P2P system carries the system traffic consisting of functionality as well as control-related messages. P2P systems, or more precisely the protocols they implement, can be classified based on the participating nodes' reliance on centralized servers and the set's degree of structure [15].

In protocols adopting a centralized architecture, e.g. Napster, a central server is approached first to obtain meta-information, such as the identity of the peer in which some information is stored, and all subsequent communication is done directly between the peers themselves. In order to improve query performance and/or reduce control traffic, a number of protocols with more structured but decentralized architectures have been proposed. In loosely structured protocols the location of objects could be more or less controlled [7], or some degree of hierarchy may be imposed among peers [13, 14]. Within highly-structured protocols, both the network topology and the placement of resources are precisely determined [20, 23, 30, 31]. Decentralized and unstructured protocols such as early versions of Gnutella [8] neither rely on centralized directories nor enforce any precise control over the network topology or object placement, resulting in systems that are highly resilient to the transient nature of P2P populations.

In most unstructured and loosely-structured P2P protocols nodes join the network by first contacting a set of peers already in the system (whose contact information may have been obtained through a well-known web site). Connected peers interact with each other exchanging various types of messages, most of which are broadcasted or back-propagated. Broadcasted messages are sent on to all other peers to which the sender has open connections. Back-propagated messages are forwarded on a specific connection on the reverse of the path taken by an associated broadcasted message. A user wishing to find a given resource issues a query to its own peer. Queries are forwarded among peers for as long as they are alive, determined by a Time-To-Live field associated with the query itself and decremented after each forward. Besides queries and replies, other types of messages include object transfer and group membership messages such as *ping*, *pong* and *bye*. *Pings* are used to discover hosts on the network. Ping messages are replied with *pongs* containing information on the responding peer and all others this peer knows about. Information on neighbor nodes can be provided either by creating pongs on their behalf or by forwarding the ping to them and back-propagating the replies. Pong messages include the contact point of a peer as well as information on what resources it makes available. *Byes* are optional messages used to report the closing of connections.

2.1 Related work

There have been a number of studies reporting on experimental data collected from currently deployed P2P systems [29, 21, 25, 6, 16, 26, 4]. Ripenau et al. [21] identify a mismatch between the topologies of the Gnutella application-level network and that of the underlying Internet that leads to ineffective use of the physical networking infrastructure. More relevant to our work, the authors found that, by November 2000, only 36% of the total traffic (in bytes) was user-generated (query), while 55% of the remaining traffic was used to maintain group membership. While these numbers have significantly improved with the last modifications to the protocol, they are still a good indication of some of the potential effects of instability.

A few of these studies have looked at peers' participation in P2P systems. Saroiu et al. [25] examine node uptime and a range of other attributes such as reported bandwidth, latency, and degree of sharing, and found large degrees of heterogeneity among peers in the systems, with variations of three and up to five orders of magnitudes in the characteristics sampled. For their lifetime study, they recorded the uptime of 17,125 peers during 60 hours. Chu et al. [6] present results from a considerably longer experiment (over six weeks) on a smaller number of peers (5,000 IP:port pairs), focusing on node availability and object transfer. Their experiments results show serious fluctuations in the number of available nodes, a highly transient population and significant time-of-day effects. The authors also found a high level of locality in the stored and transferred objects, suggesting that the use of caches may significantly reduce network traffic and improve the user experience. Sen and Wang [26] analyze P2P traffic collected passively at multiple border routers across a large ISP network and report similar high-level system dynamics. The node availability measurements in both Saroiu et al. [25] and Chu et al. [6] were gathered by actively probing previously collected TCP/IP addresses of peers. Due to this methodology, their probes can only determine with certainty if a node is or is not accepting TCP connections in the requested port without distinguishing what application is connected to it. In our experiments we collected around 1 million lifespan entries for over a half-million peers; to avoid potential errors in our measurements, we tried to set application-level connections (checking for the specific packet header in Gnutella messages). In their study on availability [4], Bhagwan et al. discuss the potential effects of aliasing on modeling host availability.[1] The authors rightly point out that, in trying to accurately capture the availability of hosts, IP address aliasing can result in great overestimation of the number of hosts in the systems and underestimation of their availability. By comparison with the authors' study, our work aims at characterizing the lifespan distribution of individual sessions, during which a peer's IP:port tuple will not change.

Our research is partially motivated by the seminal work of Harchol-Balter and Downey [11] on process lifetime distribution and its implications on load-balancing techniques. The authors measured the distribution of Unix processes and propose a UBNE (used-better-than-new-in-expectation) distribution that fits it well. Based on their finding, Harchol-Balter and Downey present a new policy for preemptive process migration in clusters of workstations.

We consider the problem of selecting who to "befriend" so as to yield systems with performance characteristics more resilient to the dynamic nature of their environment. Bernstein et al. [3] propose the use of machine learning for the selection of peers as sources from which to download. Banerjee et al. [2] introduce a scalable unicast-based technique to locate nearby peers. While efficiently choosing nearest peers is an important problem for many applications, such as overlay multicast and content distribution networks, selecting among similarly near peers using our proposed lifespan-based heuristics could significantly improve system stability, further reducing network load.

3. Peer Lifespan Distribution

Because of the potential implications of high degrees of transiency in P2P populations, we performed an independent study of peers' lifespans in a current and widely deployed P2P network with the intention of developing a model that accurately describes its distribution. In the remainder of this section we describe our methodology and discuss our findings.

3.1 Collecting observed peers' lifespans

To actively measure the lifespan of peers in Gnutella, we modified an open source Gnutella client [17] to both keep track of every peer found and periodically check its availability. Our monitoring peer maintains a hash table, initially empty, of peers it has seen so far. Each entry in the hash table includes fields for (1) IP:port of peer, (2) node type (leaf- or ultra-peer), (3) time of birth (TOB), (4) time when found (TWF), and (5) time of death (TOD).

[1] Aliasing effects could be due, for example, to the use of DHCP and NATs, as well as the sharing of a host by multiple users.

On each iteration the monitoring peer updates the existing entries and inserts new ones as it finds new peers. Since it only knows with certainty the TOB of previously known and reborn peers, first time found (live) peers are included in the table with only the TWF field set to the current time. A peer is considered dead when a connection attempt fails (i.e. a third try times out, using the default timeout value of 10 seconds) or an unexpected response is received. Table 1 summarized the strategy used for updating peer lifespan information.

A single monitoring peer scanning the whole table will clearly be too slow, resulting in too coarse a granularity for our lifespan measurements. To avoid this we evenly distribute the peer table (based on the hash values of peers) over 20 monitoring peers running across 17 hosts. This approach allows us to achieve a granularity of 1,300 seconds (about 21 minutes), when scanning over 30k to 40k entries per client.

3.2 Peer lifespan distribution

We measured the lifespans of more than 500,000 peers for over 7 consecutive days between March 1st and 8th, 2003. To account for the fact that sessions may be active (or inactive) for times longer than our sampling duration, we resort to the *create-based method* [22, 25]: we divide the captured trace into two halves and report lifespans only for sessions started in the first half. If a session ended during either the first or second half, we can obtain its lifespan by subtracting the starting time from the ending time; if a session was still active at the very end of the trace, we get a lower bound for its lifespan, which is larger than half the trace length, i.e. 3.5 days. This method provides accurate information about the distribution of lifespans for sessions that are shorter than half the trace, as well as percentage of sessions that are even longer. In addition, due to the granularity of our measurement, we could only give lifespan distribution for sessions longer than 1,300 seconds.

Figure 1 presents the Reverse Cumulative Distribution Function (RCDF) of peers' observed lifespans shorter than 3.5 days (and longer than 1,300 seconds). Lifespan distribution is presented in both normal axes (a) and log-log scale (b). The distribution in the log-log scale plot can be approximated by a straight line, indicating that the peer lifespan distribution can be modeled by a Pareto distribution of the form λT^k $(k < 0)$.

Table 1. Strategy used for updating the peer table in each iteration (*T:* Current Time). As described in the fourth case in the table (in italics), if a peer was found dead in the previous scan and alive in the current one, its Time Of Birth (TOB) is set to the current time. A peer is considered dead when a connection attempt fails or an unexpected response is received.

Last Scan	**Current Scan**	**Action**
Unknown	Dead	None
Unknown	Alive	TWF = T
Dead	Dead	None
Dead	*Alive*	*TOB* = T
Alive	Dead	TOD = T
Alive	Alive	None

More precisely, the probability of a session exceeding T seconds is λT^k. The R^2 value higher than 0.99 verifies the very high goodness of fit of the model. In contrast, the exponential curve fails to model the observed data with a R^2 value of only 0.80.

The Pareto distribution belongs to the UBNE class of distributions. In our context, this means that the expected remaining lifetime of a peer is directly proportional to its current age: the older a peer is, the longer we can expect it to remain in the system. In the remainder of this article we introduce a number of P2P protocols that take advantage of this observation. This set of protocols is not meant to be an exhaustive one, but is only used to illustrate the potential advantages of the proposed approach.

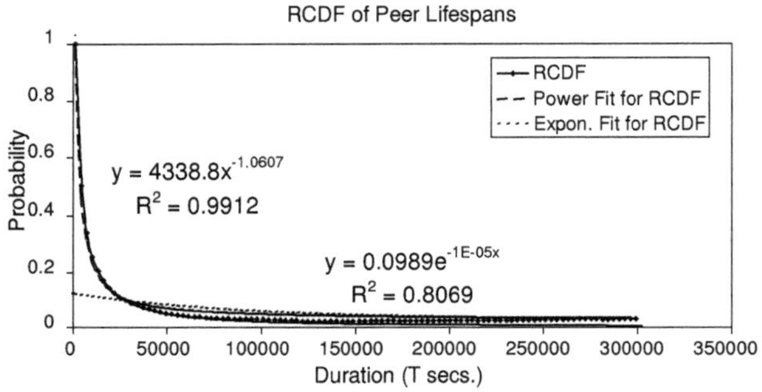

(a) Normal plot.

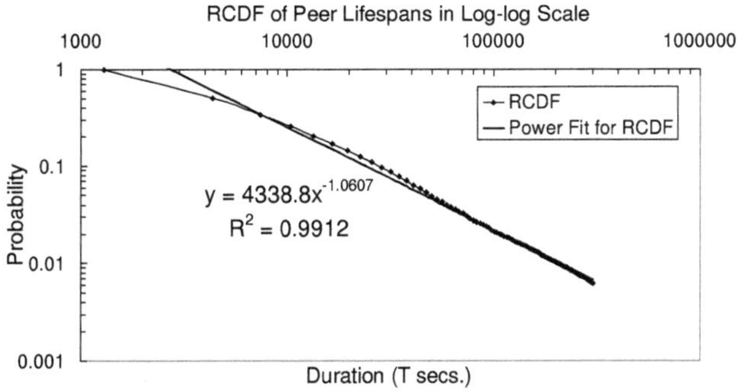

(b) Log-Log plot.

Figure 1. Distribution of lifetimes in a current P2P system (Gnutella) over a period of 7 days. The two additional lines in subfigure (a) show two attempts to fit a curve to these data: one a Pareto distribution and the other one an exponential curve. Subfigure (b) shows the same distribution on a log-log scale; the straight line in the log-log space indicates that the distribution can be modeled by λT^k, where the constant k is less than zero and proportional to the slope of the line.

4. Peer Lifespan and P2P Protocols

In most P2P protocols, there are at least two instances where peers need to choose among "acquaintances": (1) when deciding who to befriend and (2) when needing to respond to a third-party's request for references. In Gnutella-like protocols, the first group would be contacted for connection requests and the second one would be included in *pong* replies to *ping* messages.

Peers normally keep a number of other peers as close friends by accepting an upper-bounded number of incoming connections and trying to maintain a lower-bounded number of outgoing ones. To cope with the dynamic changes in P2P user population, most systems implement some kind of maintenance protocol that continuously repairs the overlay as nodes join and leave the network. Nodes joining the network use a number of control messages to let others know of their arrival. The departure of a node is noticed by its neighbors through periodic monitoring.

It is clear that the amount of control messages processed by any node would be in direct relation to the degree of stability of their neighboring set. Beyond this, the implications of peers' transiency on the overall system's performance would be directly related to the degree of investment peers place on their friends. In the context of content distribution networks, the degree of replication, the effectiveness of caches, and the spread and hit ratio of queries would all be affected by how dynamic the peers' population ends up being.

It is surprising, then, to find that peers in most P2P protocols are hardly selective when choosing friends; consistent with the pure P2P model assumption on peers, the most widely-used strategy to select whom to befriend and/or choose which acquaintances to recommend is just random.

The basic idea behind the proposed protocols is to dynamically increase the system's degree of dependency on a node as the node's commitment to the community becomes clear. One way of achieving this is to give preference to peers with longer expected lives. Given the UBNE nature of peers' observed lifespan distribution, a fair estimate for a peer's remaining lifetime can be derived from its current age.[2]

The rest of this section describes three lifespan-based protocols. Table 2 highlights their main aspects.

4.1 Lifespan-based friend selection

A very simple protocol using this heuristic, *LSPAN-1*, takes peers' observed lifespans into consideration only when deciding with whom to open a connection. Peers piggy-back their own birth time in their ping messages and propagate other peers' birth times with their replies. When a peer needs to open a new connection, after the departure of a friend for example, it simply selects the oldest known peer as its new partner.

Notice that the selection process incorporates some degree of randomness. While a peer chooses the oldest peer(s) from among those it knows of, this group is made from

[2]This approach could also be seen as an *incentive* based system, where long standing members of the peer population have a higher degree of connectivity than new-comers [25].

the random set of recommendations forwarded by other peers in the network (through the ping/pong message exchange already described).

4.2 Lifespan-based friend selection and recommendation

A more selective scheme, *LSPAN-2*, uses lifespan in both opportunities: when selecting who to connect to and when generating a response to a third-party's request for references. From the perspective of the peer trying to open a new connection, this insures that the set of potential friends is made of long-lived peers.

The two protocols introduced so far would blindly favor older peers and will naturally result in an increase in the number of connection attempts made to them. Since the actual number of incoming connections that a peer can accept is typically bounded by its maximum number of incoming connections, our last protocol considers the estimated number of available incoming connections of a peer when selecting who to connect to.

4.3 Taking available connections into consideration

LSPAN-3 uses a weighted credit selection scheme that incorporates both criteria: the peer's current age and the estimated number of available incoming connections. The estimated number of available incoming connections is the difference between the optimal and current number. The optimal number of incoming connections is upper-bounded, and its value at a given point in time lies between a half and three-fourths of the maximum number of incoming connections, depending on the peer's age (the older it is, the larger the optimal incoming connection number).

In deployed P2P systems we expect to find a positive correlation between the lifespan of a peer and its maximum number of connections: peers behind a modem can only support very limited connections to others, and tend to remain online for very short times, while peers using T1/T3 connections will have a larger maximum of connections and often stay active for several days [25]. Correspondingly, the number of maximum connections allowed by a given peer in our protocols is related to the peer's current lifespan. For our experiments this number ranged between 5 and 50, with an average value of 20.[3]

The following section presents evaluation results of the three protocols and compares them with two alternative ones that do not rely on lifespan information.

5. Evaluation

To explore the role of peers' lifespan distribution in P2P protocols, we have implemented a trace-driven simulator for P2P systems. Using a subset of the collected trace-data we evaluate the proposed lifespan-based protocols and compare them with

[3]For completeness, we evaluated the performance of our protocols following the node capacity distribution model suggested by Chawathe et al. [5], where each node belongs to one of five capacity levels and has a maximum connection number directly proportional to its level. Nodes' capacities are assigned arbitrarily and have no correlation with the nodes' lifespans. The results, available upon request, are comparable to those included in the article.

Table 2. Lifespan-based protocols and the different strategies used to select which acquaintances should a peer "befriend" and which ones it should recommend.

Protocol	**Connect?**	**Recommend?**
LSPAN-1	Oldest	Random
LSPAN-2	Oldest	Oldest
LSPAN-3	Oldest & more avail. connections	Random

two others that represent decentralized unstructured and loosely structured P2P systems.

The remainder of this section describes our experimental setup and the different strategies evaluated. We then present results showing that P2P systems more resilient to the transient nature of their environments are possible with simple heuristics that prioritize long lived peers when selecting new "friends."

5.1 Experimental setup

Our trace-driven, event-based simulator for P2P systems consists of about 3,500 lines of commented C++ code, implementing all membership management related messages. We are currently extending it to include different protocols for object searching and replication.

We ran our simulation (on a cluster of Linux PCs) using one of the 20 traces collected, with a total simulation period of 510,000 seconds (or about six days), capturing the lifespan of 36,577 peers (using the remainder traces yield similar results). The simulation starts "cold," i.e. without any peer. The number of peers in the system increases during the first day and stabilizes for the remaining time, varying between 700 and 1,000 at any given point. The results reported in this section exclude this warm-up period (80,000 sec.).

Strategies. We evaluate three different protocols based on lifespan distribution and compare them with two decentralized protocols based on currently used systems. The lifespan-based protocols were introduced above; the remainder of this subsection describes the two alternative protocols.

The alternative protocols used for comparison are closely based on Gnutella- and Kazaa-like protocols: *Unstructured Decentralized Protocol (UDP)* is based on an improved version of Gnutella v0.4 [8] and *Hybrid Decentralized Protocol (HDP)* is modeled after hybrid protocols that rely on ultra- or super-peers [28] such as Kazaa [13] and Gnutella v0.6 [14]. We heavily rely on the specifications and available RFCs. When the specifications are vague or unavailable, we resort to our understanding of (open-source) clients currently in use and other publically available documents.

Both UDP and HDP utilize separate pools for cached pongs, one pool per connection. Upon receiving a ping message, the protocols *randomly* choose a specified number of pong entries from their caches (currently 10) and respond to the request.

HDP distinguishes between leaf-peers and ultra-peers: ultra-peers can connect to any other peer (ultra or leaf), while leaf peers can only connect to ultra-peers. The scheme basically creates a two-level hierarchy among participating nodes, where more powerful, faster ultra-peers take over much of the load from slower ones. For our experiments we mark each peer as either leaf- or ultra-peer as indicated by our trace information.

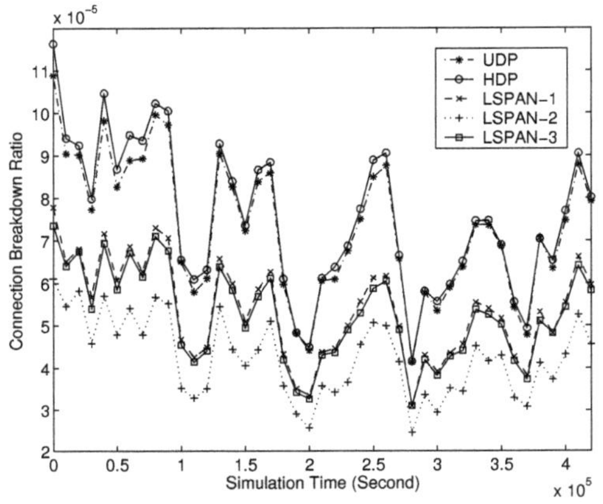

Connection Breakdown Ratio – Main Statistics

Protocol	Avg	Std	Min	Max
UDP	7.318	1.632	4.142	10.894
HDP	7.533	1.756	4.159	11.645
LSPAN-1	5.385	1.155	3.101	7.793
LSPAN-2	4.262	0.951	2.443	6.104
LSPAN-3	5.235	1.103	3.017	7.360

* Values are in 10E–5

Figure 2. Ratio of connection breakdowns to number of effective connections over time (aggregated over 10,000 sec.). The associated table shows some basic statistics including average, standard deviation as well as minimum and maximum observed values.

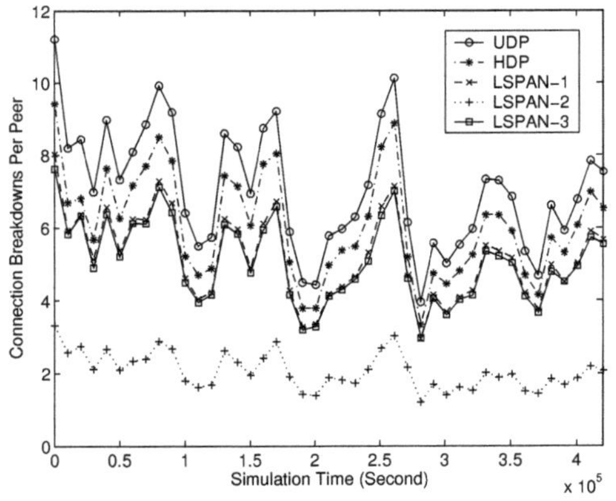

Connection Breakdowns per Peer – Main Statistics

Protocol	Avg	Std	Min	Max
UDP	7.067	1.694	3.395	11.199
HDP	6.100	1.462	3.331	9.420
LSPAN-1	5.199	1.184	2.954	8.023
LSPAN-2	2.072	0.504	1.205	3.309
LSPAN-3	5.064	1.137	2.961	7.602

Figure 3. Connection breakdowns per peer over time (aggregated over 10,000 sec.). The associated table shows some basic statistics including average, standard deviation as well as minimum and maximum observed values.

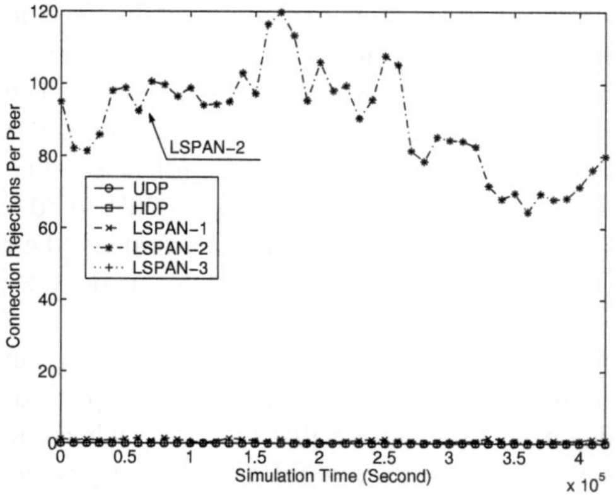

Connection Rejections per Peer – Main Statistics

Protocol	Avg	Std	Min	Max
UDP	3.46E–5	2.27E–4	0	0.002
HDP	0.003	0.002	0	0.007
LSPAN–1	0.859	0.348	0.343	1.561
LSPAN–2	89.89	14.03	64.57	1.2E+2
LSPAN–3	0.158	0.158	0	0.621

Figure 4. Connection rejections per peer over time (aggregated over 10,000 sec.). The associated table shows some basic statistics including average, standard deviation as well as minimum and maximum observed values.

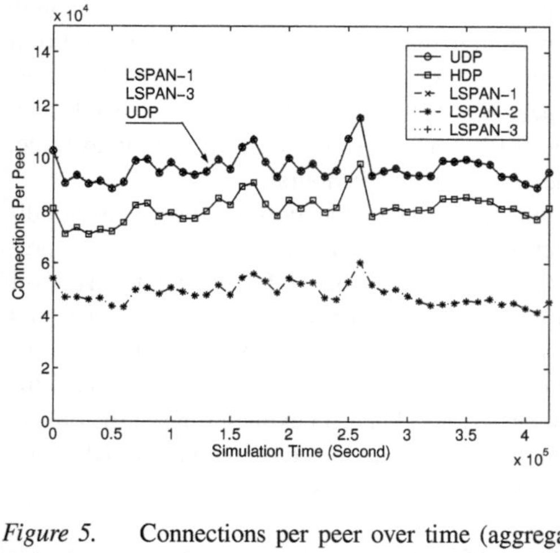

Connections per Peer – Main Statistics

Protocol	Avg	Std	Min	Max
UDP	0.964	0.053	0.886	1.154
HDP	0.811	0.054	0.709	0.980
LSPAN–1	0.965	0.053	0.884	1.157
LSPAN–2	0.486	0.040	0.415	0.605
LSPAN–3	0.966	0.053	0.888	1.158

* Values are in 10E+5

Figure 5. Connections per peer over time (aggregated over 10,000 sec.). The associated table shows some basic statistics including average, standard deviation as well as minimum and maximum observed values.

5.2 Comparison

A good indicator of the effect of a protocol on system stability is the ratio of connection breakdowns to the number of effective connections. Figures 2 shows this ratio for every one of the protocols discussed, while the associated table presents some basic statistics including the average and standard deviation. As can be observed, all lifespan-based protocols yield much lower ratios of connection breakdowns than random-based protocols (UDP and HDP), a natural result of the UBNE property of peers' lifespan distribution and the former protocols' preference for older peers. The most selective lifespan-based protocol, LSPAN-2, naturally gives the lowest ratio of connection breakdowns over time, with reductions of 42-43% by comparison with that of UDP and HDP. LSPAN-1 and LSPAN-3 yield comparative savings of 26% to 30% (by contrast with UDP) in the ratio of connection breakdowns in the system and their associated costs. Figure 3 show the closely related connection breakdowns per peer over time.

Both graphs show a clear *sawtooth* shape resulting from time-of-day patterns in our Gnutella-originated traces, something also observed in other studies of peer-to-peer systems [6, 26]. These time-of-day patterns are especially interesting when one considers the expected independence of Gnutella logical topology from geographic location [6].

It would be expected that the lifespan-based protocols' preference for long-lived peers will lead to higher numbers of connection requests to these nodes and, consequently, a higher overall number of connection rejections. This higher rejection number may be seen as a reasonable price to pay for longer-lasting "friendships," and its cost would become comparatively less important as peers increase their degree of investment in their friends. While our most selective protocol (LSPAN-2) has, indeed, a high rejection number, the rejection number of LSPAN-1 and LSPAN-3, although higher than that of UDP or HDP, is low enough that it can be ignored (Figure 4). From the associated table of statistics in Figure 4, for example, we could see that the average connection rejection number per peer for LSPAN-1 is only 0.859 for every 10,000 seconds. This number further drops to 0.158 for LSPAN-3, meaning that, in average, a peer will only face a connection rejection every 17.58 hours!

It is interesting to note that the number of actual connections per peer does not directly follow from the number of rejections experienced by it. While all LSPANs protocols result in higher, if only minor, rejection numbers than UDP and HDP, LSPAN-1 and LSPAN-3 have even higher ratios of connections per peer than both UDP and HDP (Figure 5). Recall that LSPAN-1 and LSPAN-3 select long lasting peers from a random set of candidates and, in the case of LSPAN-3, with some knowledge of the candidate peers' available incoming connections. This can explain their high connection numbers, resulting from both long lasting connections (compared with UDP and HDP) and low rejection numbers (by contrast with LSPAN-2).

6. Conclusions and Future Work

We have presented trace-driven simulation results that illustrate the potential advantages of considering peers' age as a key system attribute in the design of P2P protocols. From independent measurements of peer lifespan in a current P2P system we developed a functional form that fits the distribution well. We use this distribution as the basis for a number of illustrative P2P protocols which we compare against two alternative ones for decentralized and unstructured or loosely-structured (superpeer-based) P2P systems. We find that even simple lifespan-based protocols can achieve up to 42% reductions in the ratio of connection breakdowns and their associated costs.

We are currently exploring the effect of lifespan-based protocols in different query and caching algorithms for P2P systems. While decentralized and unstructured protocols could yield more resilient systems, na"ve implementations of query mechanisms in these systems have been shown to scale poorly [15]. As part of our future work, we are investigating the benefits of the higher resilience of lifespan-based systems on the hit rates and response times of more scalable search mechanisms [15, 28]. Similarly, while potentially highly beneficial [29, 24], the effectiveness of caching in P2P systems will be directly connected to the lifespan of the caching peer. We have also started to experiment with these and similar ideas in the context of highly structured protocols [30, 31, 23, 19].

Acknowledgments

We would like to thank the Northwestern University Information Technology group and the Computing Support Group of the Department of Computer Science, in particular Roger A. Safian, Scott Hoover and Geoffrey Pagel, for their aid and understanding while obtaining the trace data used for our experiments. We are also grateful to Peter Dinda for stimulating conversations and to Jeanine M. Casler, Stefan Birrer and the anonymous reviewers for their helpful comments on early drafts of this article.

References

[1] E. Adar and B. A. Huberman. Free riding on Gnutella. *First Monday*, 5(10), September 2000.

[2] S. Banerjee, C. Kommareddy, and B. Bhattacharjee. Scalable peer finding on the Internet. In *Proc. of Globecom*, Taipei, Taiwan, R.O.C., November 2002.

[3] D. S. Bernstein, Z. Feng, B. N. Levine, and S. Zilberstein. Adaptive peer selection. In *Proc. of the 2nd IPTPS*, Berkeley, CA, USA, February 2003.

[4] R. Bhagwan, S. Savage, and G. M. Voelker. Understanding availability. In *Proc. of the 2nd IPTPS*, Berkeley, CA, USA, February 2003.

[5] Y. Chawathe, S. Ratnasamy, L. Breslau, and S. Shenker. Making Gnutella-like P2P systems scalable. In *Proc. of SIGCOMM*, Karlsruhe, Germany, August 2003.

[6] J. Chu, K. Labonte, and B. N. Levine. Availability and locality measurements of peer-to-peer file systems. In *Proc. of ITCom*, July 2002.

[7] I. Clarke, O. Sandberg, B. Wiley, and T. W. Hong. Freenet: A distributed anonymous information storage and retrieval system. In *Proc. ICSI Workshop on Design Issues in Anonymity and Unobservability*, pages 43–58, Berkeley, CA, USA, July 2000.

[8] Clip2. The Gnutella protocol specification v0.4. RFC, The Gnutella RFC, 2000.

[9] Entropia, Inc. http://www.entropia.com. 2003.

[10] Groove Networks, Inc. http://www.groove.net. 2003.

[11] M. Harchol-Balter and A. B. Downey. Exploting process lifetime distribution for dynamic load balancing. *ACM Transactions on Computer Systems*, 15(3):253–285, August 1997.

[12] S. Iyer, A. Rowstron, and P. Druschel. SQUIRREL: A decentralized, peer-to-peer web cache. In *Proc. of the ACM PODC*, Monterey, CA, USA, July 2002.

[13] Kazaa. http://www.kazaa.com. 2001.

[14] T. Klingberg and R. Manfredi. Gnutella 0.6. RFC, The Gnutella RFC, June 2002.

[15] Q. Lv, P. Cao, E. Cohen, K. Li, and S. Shenker. Search and replication in unstructured peer-to-peer networks. In *Proc. of ICS*, New York, NY, June 2002.

[16] E. P. Markatos. Tracing a large-scale peer to peer system: an hour in the life of Gnutella. In *Proc. of CCGrid*, Berlin, Germany, May 2002.

[17] Mutella. http://mutella.sourceforge.net. 2003.

[18] Napster. http://www.napster.com. 2003.

[19] Overnet. http://www.overnet.com. 2003.

[20] S. Ratnasamy, P. Francis, M. Handley, R. Karp, and S. Shenker. A scalable content-addressable network. In *Proc. of SIGCOMM*, pages 161–172, San Diego, CA, USA, August 2001.

[21] M. Ripeanu, I. Foster, and A. Iamnitchi. Mapping the Gnutella network: Properties of large-scale peer-to-peer systems and implications for system design. *IEEE Internet Computing Journal*, 6(1):50–57, January/February 2002.

[22] D. Roselli, J. R. Lorch, and T. E. Anderson. A comparison of file systems workloads. In *Proc. of the USENIX Annual Technical Conference*, San Diego, CA, June 2000.

[23] A. Rowstron and P. Druschel. Pastry: Scalable, distributed object location and routing for large-scale peer-to-peer systems. In *Proc. of the IFIP/ACM Middleware*, Heidelberg, Germany, November 2001.

[24] S. Saroiu, K. P. Gummadi, R. J. Dunn, S. D. Gribble, and H. M. Levy. An analysis of Internet content delivery systems. In *Proc. of the 5th USENIX Symposium on Operating Systems Design and Implementation*, Boston, MA, USA, December 2002.

[25] S. Saroiu, P. K. Gummadi, and S. D. Gribble. A measurement study of peer-to-peer file sharing systems. In *Proc. of Multimedia Computing and Networking*, San Jose, CA, USA, January 2002.

[26] S. Sen and J. Wang. Analyzing peer-to-peer traffic across large networks. In *ACM SIGCOMM Internet Measurement Workshop*, Marseille, France, November 2002.

[27] SETI@Home. http://setiathome.ssl.berkeley.edu/. 2003.

[28] A. Singla and C. Rohrs. Ultrapeers: Another step towards Gnutella scalability. Working draft, Lime Wire LLC, December 2001.

[29] K. Sripanidkulchai. The popularity of Gnutella queries and its implications on scalability. In *O'Reilly's http://www.openp2p.com*, February 2001.

[30] I. Stoica, R. Morris, D. Karger, F. Kaashoek, and H. Balakrishnan. Chord: A scalable peer-to-peer lookup service for internet applications. In *Proc. of SIGCOMM*, pages 149–160, San Diego, CA, USA, August 2001.

[31] B. Y. Zhao, J. Kubiatowicz, and A. D. Joseph. Tapestry: An infrastructure for fault-tolerant wide-area location and routing. Technical Report UCV/CSD-01-1141, Computer Science Division, University of California, Berkeley, CA, USA, April 2001.

A FINE-GRAINED PEER SHARING TECHNIQUE FOR DELIVERING LARGE MEDIA FILES OVER THE INTERNET*

Mengkun Yang and Zongming Fei
Department of Computer Science, University of Kentucky

Abstract In this paper we propose a fine-grained peer sharing technique for delivering large media files in content distribution networks. The replica servers are divided into groups, and those in the same group cooperate with each other. The key difference of the technique from conventional peer to peer systems is that the unit of peer sharing is not a complete media file, but at a finer granularity, in order to increase the flexibility of replica servers for handling client requests. We design a protocol for peers to exchange the information about available resources with each other, and a scheduling algorithm to coordinate the delivery process from multiple replica servers to a client. Our simulations show that the fine-grained peer sharing approach can reduce the initial latency of clients and the rejection rate of the system significantly over a simple peer sharing method.

1. Introduction

Content distribution networks (CDNs) have been successful in delivering conventional web documents to clients by distributing content from an origin server to replica servers located at the edges of the network and directing clients to a nearby replica. They usually distribute the whole file of a document to replica servers. The upcoming streaming media files tend to be much larger and the number of files a replica can store is limited, even with the increasing disk space. The result is that either we have to store only a small portion of contents at the replica servers and the rest will be delivered directly from the origin server, or we have to frequently remove previously stored files from the storage to accommodate new files. This consumes quite a lot of network bandwidth and can also overload the origin server.

To improve the service capabilities of resource-constrained replica servers, we employ the peer sharing technique of peer-to-peer systems at the replica servers. Peer sharing among clients is an attractive approach, as proposed by most peer-to-peer systems. However, the difficulty lies in that clients are more willing to get the files than

*This work was supported in part by the National Science Foundation under Grants CCR-0204304 and EIA-0101242.

F. Douglis and B.D. Davison (eds.), Web Content Caching and Distribution, 247-256.

serving other clients. Studies showed that almost 70% of users share no files at all and nearly 50% of all responses are returned by the top 1% of sharing hosts [1]. In this paper we explore the possibility of peer sharing among replica servers, which content service providers have direct control and can equip the new technique if proved useful. In a typical CDN, every replica server achieves the best performance when it stores the most popular files in the vicinity. The result is that files stored at different replica servers have a very large overlap. This is hard for them to help each other to serve client requests. Instead, we can let cooperating replica servers store different files. By careful planning, the overlap between different replica servers can be reduced and the total number of media files stored in a group of replica servers can be increased. Thus their capacities of serving clients are improved.

We go one step further in this paper by proposing a fine-grained peer sharing technique. The unit of peer sharing is not a complete media file. Rather, a media file is divided into segments, which will be distributed to different replica servers. A client makes a request to its nearest replica server, which will act as a *coordinator* to schedule which segment will be streamed from which cooperating replica server to the client at what time. We design a protocol for peers to exchange the information about available resources with each other, and a scheduling algorithm to coordinate the delivery process from multiple replica servers to a client. A nice feature of the scheme is that the coordinator only needs to contact with those peers having the segments involved.

2. Fine-Grained Peer Sharing in CDNs

2.1 Peer sharing in CDNs

In a CDN, a client is normally directed to the nearest replica server. Under the limited storage constraint, we assume that a replica server can store n media files. Usually it will store the most popular or most recently accessed objects. If a media file is available at the replica server, the client can get it directly. For those files that are not stored in the replica server, the client has to get them from the origin server, or the replica server can fetch them from the origin server on its behalf. This may impose a tremendous load on the origin server, because of the large number of potential clients and the long streaming duration.

One approach to dealing with this problem is *peer sharing*, i.e., cooperating replica servers store different objects and they will serve the requests made to their peer servers. If we have P servers peering with each other, we can store $P * n$ different objects, instead of n in the non-sharing case. We can decrease the load on the origin server substantially and better serve client requests. An observation is that a larger P will result in more objects being served from replica servers. However P cannot be too large. One reason is that grouping more servers together means a longer distance from one of them to the client making the request. This may affect the quality (delay, delay jitter, bandwidth) of the streaming path. The other reason is that the management cost will be higher because of the increased complexity of communication between replica servers. We envision a two-tier model for organizing the replica servers in the CDN. The replica servers are divided into groups and those in the same group will cooperate with each other.

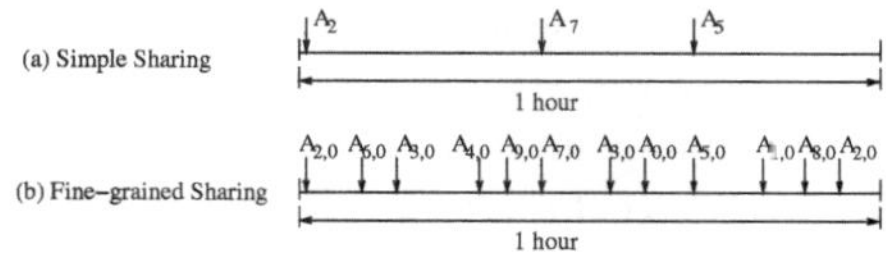

Figure 1. Comparison between simple sharing and fine-grained sharing

2.2 Fine-grained peer sharing

The problem with the simple peer-sharing technique described above is that it is possible that the load on one of the replica servers may be extremely high because it stores one or several hot media files. In the non-sharing case, the requests to these hot media files are distributed to all replica servers. In the sharing case, all these requests are now concentrated on a particular replica server that stores the hot files. Therefore, clients may experience excessive delay in obtaining the media file requested. We propose a fine-grained peer sharing technique to solve the problem.

Instead of storing the whole media file in one server, we divide each media file into segments. Generally one replica server will only store one segment of a media file. A client still makes the request to the nearest replica server, which will communicate with those peering servers having segments of the requested file and schedule streaming of each segment to the client.

We use a simplified example to illustrate the idea of the fine-grained peer sharing technique. Assume four replica servers A, B, C and D peer with each other, and each of them can stream 3 media files simultaneously at the playout rate. Each of them can hold 10 media files that last 1 hour. First consider the *simple peer sharing* case. As a result of careful placement, each of them holds different files. Specifically we assume that A stores files $A_0, A_1, \cdots, A_9$, B stores files $B_0, B_1, \cdots, B_9$, etc. Because each file lasts 1 hour and a server can stream 3 media files at the same time, the total number of requests for files $A_0, A_1, \cdots, A_9$ that we can start to serve within one hour is 3. The same is true of files stored at servers B, C and D.

Now consider the *fine-grained peer sharing* case. We divide each file into 4 segments and store them into different servers. For example, file A_0 is divided into $A_{0,0}, A_{0,1}, A_{0,2}$ and $A_{0,3}$, which will be stored to servers A, B, C and D, respectively. All other files are divided and distributed in a similar way, except that $B_{k,0}$ will be kept in server B, $C_{k,0}$ in C, $D_{k,0}$ in D ($k = 0, 1, 2, \cdots, 9$) and the rest of segments are distributed in a round robin fashion. Now consider the total number of requests for files $A_0, A_1, \cdots, A_9$ that we can start to serve within one hour. Since each segment only lasts 1/4 hour, in each 1/4 hour we can start 3 streams. Over 1 hour period of time, we can start 3*4 = 12 streams for those requests for $A_0, A_1, \cdots, A_9$. As an example, Figure 1(a) shows that the simple sharing method can only start three video files (A_2, A_7, A_5) within one hour, while Figure 1(b) shows that the fine-grained peer sharing method can start 12 video files within one hour.

It is true that the total number of requests that can be served by a given set of cooperating replica servers within a certain period of time is not necessarily increased.

However, because the load for serving a given file is evenly distributed, the number of requests for *a specific file* or *a specific subset of files* that can be served within a certain period of time by the fine-grained peer sharing scheme is increased. Therefore the capacity of these replica servers to serve a given media file or a given set of media files is improved. In general, if media files are divided into N segments, the capability of servicing a given set of files will be increased up to N times.

The idea of fine-grained peer sharing is similar to the disk striping technique [2] in balancing the load, but there are two key differences. First, the scheduler in disk striping schemes has the knowledge of all the requests and states of all the disks, while the scheduler in the fine-grained peer sharing scheme needs to communicate with other replica servers to know their status to make a decision for individual clients. Second, one goal in disk striping is to improve throughput from the disk to CPU by accessing different parts of a file from different disks at the same time, while the goal of the fine-grained peer sharing technique is to improve the availability of a given set of files.

In the next section, we will describe a protocol for peers to share information and redirect requests to others, and design a scheduling algorithm for a replica server to decide when to get each segment from which sharing peer.

3. Communication Protocol and Scheduling Algorithm

Accessing different segments from different servers needs coordination between these servers for several reasons. First, the load and the streaming schedule at replica servers will change dynamically. We do not want to have a central server to maintain all the information. When a client requests a media file, those replica servers holding segments of the file need to be coordinated to schedule when to send the segments to the client based on the current load. Second, the client can only receive from a limited number of servers at the same time. At any point of time, the number of servers streaming to the client cannot exceed this limit. Third, it is possible that several initial segments are received by the client at an earlier time and the client starts to play back, but later segments cannot be streamed fast enough to catch up with their supposed playout time. This can lead to a gap in the playout by the client. One way to eliminate the gap is to let the client postpone playout with an initial delay. We want to minimize this delay and guarantee that there is no gap in the playout. These factors motivate us to design the communication protocol and the scheduling algorithm described in this section.

When a media file is delivered to a group of P peering replica servers, it is divided into N (usually $N \leq P$) segments, which can be of different size [1]. The first segment will be stored at the replica server that has the smallest number of first segments of other files already there. Assume the server number is i. The second segment will be distributed to server $(i + 1) \bmod P$. We continue the distribution of the rest of segments in the round-robin fashion until we finish distributing all the segments. The

[1]In the simulations of this paper, we use segments of equal size. However, the protocol and the algorithm described in this paper can deal with segments of unequal size as well. The effects of segment size on the performance were discussed in another paper.

information about locations of each segment of each file will be distributed to all replica servers in the group, so that they know where to ask for help when needed.

The server and client capacities are abstracted as channels. By a *channel*, we mean the bandwidth used to stream data at the playout rate of media files. If the server bandwidth is B and the playout rate is b, then the number of channels this server has is $B_s = \lfloor B/b \rfloor$. Similarly, we has a parameter B_c, which represents the number of streams (at the playout rate) that a client can receive from replica servers at the same time.

3.1 Communication protocol

Streaming a media file from replica servers to a client involves several steps of communications among the client, the nearest replica server and other peer replica servers. The communication protocol can be described as the following steps.

First, the client sends its request for a file to its nearest replica server, which acts as the *coordinator* for this request.

Second, the coordinator looks up in the table about the locations of segments of this file, and forwards *segment requests* to those peering servers having segments of this file.

Third, after receiving segment requests, the peering replica servers will try to find available slots that can be used to stream the segment. The available slots are in the form of a list of pairs (a, b), which means there is a channel that is free between time a and time b. Note b can be ∞. These slots will be reserved for a period of time τ for the coordinator. It is a policy of the replica server that determines how many slots and how long these slots will be reserved for the coordinator.

Fourth, for segment j, a list of time slots $(a_{j0}, b_{j0}), (a_{j1}, b_{j1}), \ldots, (a_{jk}, b_{jk})$, will be received by the coordinator, where $0 \leq j \leq N - 1$. Upon receiving replies from these peers, the coordinator uses the *scheduling algorithm* described in the next subsection to choose an appropriate time slot for streaming each segment, under the assumption that a client can receive from B_c channels at the same time. The goal is to minimize the initial latency and guarantee there is no jitter between playing out different segments.

Fifth, confirmation messages are sent back to peer servers from the coordinator. They will inform each peer which slot (or part of it) has been scheduled to stream the requested segment. At the same time, the coordinator sends the client a message about when to start playing out and where to get each segment. It is up to the client to make the actual requests. We understand that the client may need a certain size of buffer to hold those segments that will be streamed before their playout time.

Finally, after receiving the reservation messages, each peer server confirms the reservation of the selected time slot for transmitting the segment of the file and releases all other channel slots previously reserved for this request. If a certain server does not receive the reservation message from the coordinator before the lifetime τ expires, it will cancel all channel slots reserved for the segment. It also informs the coordinator, which may cancel channel reservations and transmissions for all other segments of this file accordingly.

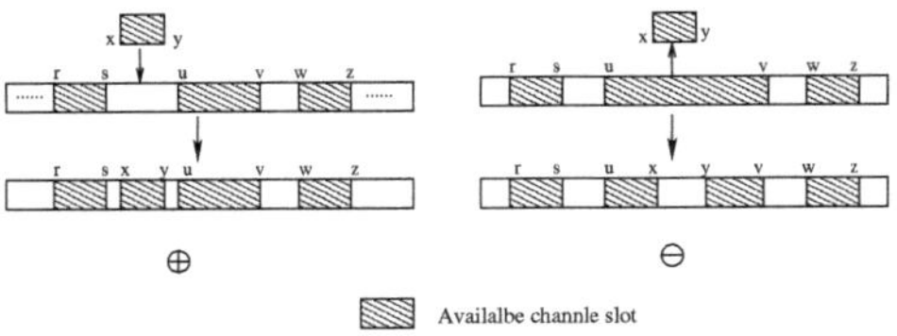

Figure 2. Operations on Ψ_l: $\Psi_l \oplus (x, y)$ and $\Psi_l \ominus (x, y)$

One issue worth further discussion is how to deal with concurrent requests from multiple peers. In the third step, a replica server may receive segment requests from multiple coordinators, which may ask for the same segment or different segments. Obviously reserving all available slots for one coordinator will block requests from all other coordinators. Our strategy is to count the number of requests every τ time period. Assume the number in the most recent period is Γ_c and the variable used to represent the load on this server is Γ. We calculate the current load as weighted sum, i.e., $\Gamma = w * \Gamma_c + (1 - w) * \Gamma$, where $w = 0.3$. The proportion of slots that will be reserved for each coordinator will be $1/\Gamma$ of all available slots.

3.2 Scheduling algorithm

Upon receiving information about available time slots for streaming each segment, the scheduling algorithm of the coordinator will determine the time slot that will be actually used. Assume that the client makes the request at time t_{rqst}. The length of the media file is L, which is divided into N segments. Assume the length of segment j is s_j. (If all segments are of equal size, we have $s_j = s = L/N$.) Ideally, there will be no delay if the client can start downloading segment j $(0 \leq j \leq N-1)$ at time $t_j = t_{rqst} + \sum_{k=0}^{j-1} s_k$. However we may have to download segment j before or after t_j because the time slots starting at time t_j on all channels of the replica server may have already been reserved for other requests. While an actual downloading time before t_j will not cause any problem, a downloading time after t_j will cause the client to start playing out the whole media file after some initial latency. Actually the final initial latency is the maximum of the latencies caused by all segments. Our algorithm will try to reduce this latency. We start with some notations used in the algorithm.

- For any request being processed, the available server channel slots returned from the peer replica server holding segment j $(j = 0, \cdots, N-1)$ is $\Omega_j = \{(a_{j0}, b_{j0}), (a_{j1}, b_{j1}), (a_{j2}, b_{j2}), \cdots\}$, where $a_{j0} \leq a_{j1} \leq a_{j2} \leq \cdots$.

- We assume that the client making the request can receive from B_c streams at the same time. The capability for downloading each stream is modeled as a channel. At one point of time, the available slots on the l-th channel of the client is $\Psi_l = \{(x_{l0}, y_{l0}), (x_{l1}, y_{l1}), (x_{l2}, y_{l2}), \ldots (x_{lh_l}, \infty)\}$, where $l = 0, \cdots, B_c - 1$ and $x_{l0} < x_{l1} < x_{l2} < \cdots < x_{lh_l}$. Initially, all channels are (t_{rqst}, ∞).

Problem Formulation: For any request, given server slots $\Omega_0, \Omega_1, \cdots, \Omega_{N-1}$ together with client channels $\Psi_l = \{(t_{rqst}, \infty)\}$ for $l = 0, \cdots, B_c - 1$, we need to find the streaming time u_j for segment j $(0 \leq j \leq N-1)$, such that $(u_j, u_j + s_j)$ is a time slot that does not collide with time slots for downloading other segments in the same client channel, and there exists a k such that $(u_j, u_j + s_j) \subseteq (a_{jk}, b_{jk})$. The latency caused by segment j is $d_j = \max(u_j - t_j, 0)$, and the final initial latency of the client will be $d = \max\{d_j | j = 0, \cdots, N-1\}$. The client can start playing out the media file at time $t_{rqst} + d$ without jitter for each segment [2]. The goal of the algorithm is to minimize d.

The basic idea of the algorithm is to reserve the *earliest possible slot* on the client channels for each segment first, then try to make adjustment to minimize the initial latency d. Given the current available slots Ω_j from peer servers and client channels $\Psi_0, \cdots, \Psi_{B_c-1}$, the current *earliest possible slot* for segment j is $(u_j, u_j + s_j)$ satisfying: 1) $u_j \geq t_{rqst}$; 2) $(u_j, u_j + s_j) \subseteq (a_{jk}, b_{jk}) \cap (x_{lg}\ y_{lg})$ for some k, l, g; and 3) for any e, such that $e \geq t_{rqst} \wedge (e, e + s_j) \subseteq (a_{jk'}, b_{jk'}) \cap (x_{l'g'}, y_{l'g'})$ for some k', l', g', we have $u_j \leq e$. We call such a slot satisfying the *earliest condition* in the algorithm.

The greedy reservation for the earliest possible slots favors early segments and may use up the only slots that can be used to prefetch some later segments that have a larger latency. So the next step in the algorithm is to gradually reduce the highest segment latency by moving a segment with zero or very low latency from its current reserved slot to a later free slot, and let the vacated slot on a client channel to accommodate the segment with the highest latency.

Two operations, $\oplus$ and $\ominus$, are defined on Ψ_l, as illustrated in Figure 2. Basically, $\oplus$ inserts a slot (x, y) into the channel Ψ_l, and $\ominus$ removes (x, y) from the channel. If inserting a slot makes some slots on a channel contiguous to each other, we further combine them to one larger slot. The priority queue $Q[0 \ldots N-1]$ stores segment indices such that $d_{Q[0]} \geq d_{Q[1]} \geq \cdots \geq d_{Q[N-1]}$.

The scheduling algorithm is outlined in Figure 3. Lines 1-5 reserve slots on client channels for segments as early as possible. So at the time we schedule for segment j, $(u_j, u_j + s_j)$ on the channel l_j is the earliest possible slot. The priority for reservation decreases as the segment index increases. Some segments may get high latencies since their residing servers cannot stream them on time, and the only possible slots able to prefetch them have already been taken up by other segments. To solve this problem, lines 6-21 move the segment with the highest latency to an earlier slot previously reserved for another segment with lower latency. In this way, the initial latency of the client can be decreased. The process is continued until no such moving is possible.

4. Performance Evaluation

We evaluate the performance of the *fine-grained* peer sharing and its protocol and scheduling algorithm by simulation. Each client is assumed to have B_c (with the

[2]To accommodate the latency from the replica servers to the client, we may have to add an extra to d. This extra can be the maximum value of one way delay from the replica servers to the client. For the clarity of presentation, we will not mention this extra in the rest of the paper.

Scheduling Algorithm

for ($j = 0$ to $N - 1$) **do**
  find $(u_j, u_j + s_j)$ satisfying *earliest condition* for segment j on a client channel and record that the slot is on l_j-th client channel.
  $\Psi_{l_j} \leftarrow \Psi_{l_j} \ominus (u_j, u_j + s_j)$
  $d_j \leftarrow \max(u_j - t_j, 0)$
Sort d_j and fill in $Q[0 \ldots N-1]$ accordingly.
$j \leftarrow N - 1$
while $j > 0$ **do**
  $i \leftarrow Q[0]$, $i' \leftarrow Q[j]$
  $\Psi_{l_i} \leftarrow \Psi_{l_i} \oplus (u_i, u_i + s_i)$
  $\Psi_{l_{i'}} \leftarrow \Psi_{l_{i'}} \oplus (u_{i'}, u_{i'} + s_{i'})$
  find $(v, v + s_i)$ satisfying *earliest condition* for segment i on a client channel and record that the slot is on l-th client channel.
  find $(v', v' + s_{i'})$ satisfying *earliest condition* for segment i' on client channel and record that the slot is on l'-th client channel.
  if $v < u_i$ and $\max(v' - t_{i'}, 0) < d_i$ **then**
    $u_i \leftarrow v$, $l_i \leftarrow l$
    $u_{i'} \leftarrow v'$, $l_{i'} \leftarrow l'$
    Update d_i, $d_{i'}$, and $Q[0 \ldots N-1]$
    $j \leftarrow N - 1$
  else
    $\Psi_{l_i} \leftarrow \Psi_{l_i} \ominus (u_i, u_i + s_i)$
    $\Psi_{l_{i'}} \leftarrow \Psi_{l_{i'}} \ominus (u_{i'}, u_{i'} + s_{i'})$
    $j \leftarrow j - 1$

Figure 3. The scheduling algorithm

default value 4 unless specified otherwise) channels for receiving media files, and every replica server has B_s (default value 10) channels to transmit data. The file size is $L = 4000$ second MPEG video and each server can store 50 files. Each peering group consists of P (default value 20) replica servers. So the total storage size of the group is $P * 50$ files. In the *fine-grained* scheme, a file is divided into N (default value 10) segments, and they are placed onto servers using the *round-robin placement*. In the *simple sharing* case, the whole file is always distributed to the server currently having the smallest sum of popularity rankings of files already on it, so long as the server has enough available storage. Request arrivals are modeled as a Poisson process and the file popularities are approximated by a Zipf-like distribution with $\alpha = 0.733$. We assume requests are equally likely originated from any client. The highest possible request arrival rate the system can support is $\lambda_{max} = \frac{B_s \times P}{L}$. That is the case in which *all* channels of *all* servers are kept busy *all the time*. The actual arrival rate λ in our simulation is set as a percentage of λ_{max}. Every experiment simulates a time period of L*25 second long.

The performance measures we study include: 1) *Initial latency*. It is the average interval between the time a client makes a request and the starting time of playout. 2)

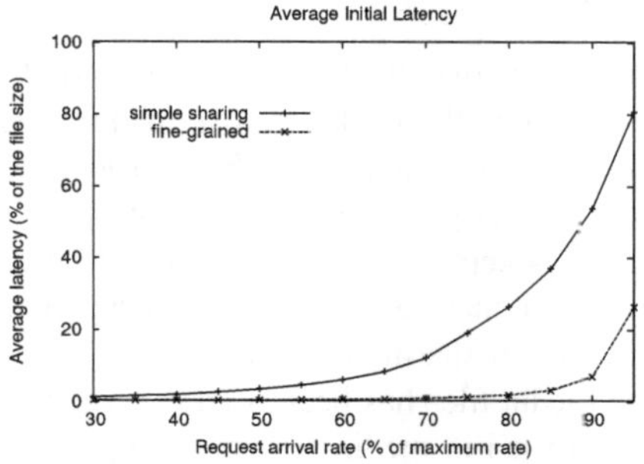

Figure 4. Average latency using different approaches

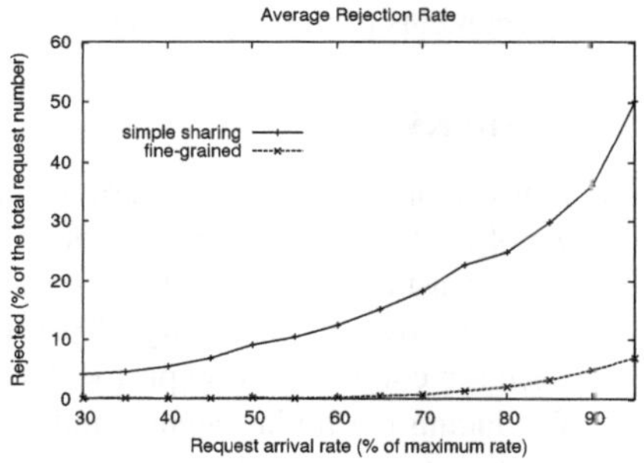

Figure 5. Rejection rate using different approaches

Rejection rate. It is the ratio of the number of rejected requests over the total number of requests.

Figure 4 shows the initial latency of the *simple sharing* and the *fine-grained* schemes, expressed as the percentage of the video length. As the volume of traffic increases, the difference between the initial latencies of the *simple sharing* and the *fine-grained* schemes becomes much larger. When the request rate is 80% of the system capacity, the initial latency of the simple sharing is almost 25% of the video length, i.e., more than 15 minutes, while the initial latency of the fine-grained scheme is still very close to 0.

Figure 5 plots the rejection rate when the admission control is applied. Specifically, if a coordinator finds that the initial latency is greater than a threshold $T_{thresh} = L/20$, the request will be rejected. We note that much lower rejection rate can be achieved in the *fine-grained* scheme. For example, when the request rate is 80% of the system capacity, the rejection rate of the *fine-grained* scheme is less than 2%, while the rejection rate of the *simple sharing* scheme is as high as 27%. Even though the request rate is only 30% of the system capacity, the *simple sharing* scheme has to reject about 4% of the requests.

5. Related Work

Streaming multimedia content over the Internet has been studied extensively. Proxy caches have been used to reduce the backbone bandwidth requirement for scalable video delivery [3]. Recently, prefix caching has been proposed to reduce the storage requirement for the proxy servers [4, 5]. Only the prefix part is delivered to replica servers while the suffix part is kept at the origin server and usually depends on multicast for its delivery. The peer sharing technique was originally used in peer-to-peer systems [6]. However, the unit of sharing is a complete file.

Segmentation of large media files has been used in large video servers. The data striping technique has been used in media servers to improve the disk access speed and reliability [2]. It allows multiple disk transfers to occur in parallel. However, the scheduling task is easier in the video server because it has knowledge of all files and load on all disks. Our scheduling algorithm is more complicated because it requires communication between replica servers. It is executed by each replica server and only involves those replicas having segments of the requested file.

6. Concluding Remarks

In this paper we present a novel fine-grained peer sharing technique for streaming large media files in the context of content distribution networks. Our scheme does not depend on the availability of multicast deployment. Our design employs peer-to-peer sharing at a finer granularity to improve the flexibility of peering replica servers to handle client requests. We design a communication protocol and a scheduling algorithm to achieve the goal of reducing the initial latency and the rejection rate. The simulations demonstrate a significant performance gain of the fine-grained peer sharing technique.

References

[1] E. Adar and B. Huberman. Free riding on Gnutella. *First Monday*, 5(10), 2000. http://www.firstmonday.dk/issues/issue5_10/adar/.

[2] A. L. Drapeau, P. M. Chen, J. H. Hartman, E. K. Lee, E. L. Miller, K. Shirriff, S. Seshan, R. H. Katz, G. A. Gibson, and D. A. Patterson. RAID-II: A high-bandwidth network file server. In *Proceedings of 21st International Symposium on Computer Architecture*, April 1994. Chicago, IL.

[3] Z. li Zhang, Y. Wang, D. H. Du, and D. Su. Video staging: A proxy-server-based approach to end-to-end video delivery over wide-area networks. *IEEE/ACM Transactions on Networking*, 4(8):429–442, August 2000.

[4] S. Ramesh, I. Rhee, and K. Guo. Multicast with cache (mcache): an adaptive zero-delay video-on-demand service. In *Proceedings of IEEE Infocom'01*, 2001.

[5] S. Sen, J. Rexford, and D. Towsley. Proxy prefix caching for multimedia streams. In *Proceedings of IEEE Infocom'99*, April 1999.

[6] I. Stoica, R. Morris, D. Karger, M. Kaashoek, and H. Balakrishnan. Chord: A scalable peer-to-peer lookup service for Internet applications. In *Proceedings of ACM SIGCOMM 2001*, pages 149–160, August 2001. San Diego, CA.

PROXY-CACHE AWARE OBJECT BUNDLING FOR WEB ACCESS ACCELERATION

Chi-Hung Chi, HongGuang Wang, and William Ku
National University of Singapore

Abstract In this paper, we address the acceleration of web page access through a novel proxy-cache aware object bundling technique. Richer content for a web page is always the result of increasing the number of embedded objects inside. However, due to the significant setup costs of a `TCP` connection, `HTTP` is known to be inefficient for transfers of small objects. As a result, request bundling techniques such as `MGET` and `N-to-1 Bundle` were proposed to eliminate the need for multiple requests by packing all associated embedded objects into a single bundle. However, these proposals all suffer from the lack of support from proxy caching. To address this problem, we firstly present the traffic characteristics of embedded object retrieval. This is then used as an argument upon which our proxy-cache aware **`PC-Bundle`** mechanism is built. The unique feature of our mechanism is that the advantages of all previously proposed bundle techniques can be achieved without any redundant object retrieval in the presence of proxy cache. Compared to previously proposed bundle requests such as `N-to-1 Bundle`, our `PC-Bundle` mechanism can achieve an improvement ranging from 21.2% to 34.8% for page latency reduction and 17.2% to 36.4% for page network traffic, which are very significant.

1. Introduction

We expect that multimedia and data traffic will surpass the traditional voice traffic in mobile networks by the year 2005 [18]. High quality streaming multimedia content is likely to form a significant portion of this traffic. It is therefore important that large mobile networks find ways to manage client traffic and data efficiently.

Page retrieval latency is always a key metric that content delivery and distribution network service providers would like to improve. Besides the potential reuse of static web objects by proxy cache [17] [21], researchers are actively investigating mechanisms to accelerate the downloading process of page retrieval. This direction has huge potentials because it covers all pages and objects, independent of their cacheability.

F. Douglis and B.D. Davison (eds.), Web Content Caching and Distribution, 257-273.

When a user requests for a web page, its page container object, usually in the form of a HTML file, would be returned. Then the content would be parsed chunk-by-chunk and individual requests for the embedded objects that are used by the container object will be requested. Hence, the retrieval of a web page can be broken into two stages: the retrieval of the page container object followed by its associated embedded objects. While the page container object has to be retrieved in only one manner, the retrieval of its embedded objects can be accelerated in various ways. Examples include persistent connection [6] [10], parallel connection [8] [14] [18] [19], and bundling [7] [10] [16] [22]. Note that while these mechanisms try to address the same latency problem, their approaches and the targets for performance optimization are quite different [11]. Hence, they should be viewed as complementary (or alternative) to each other [22].

In this paper, we would like to focus on the direction of bundling web objects together into a single request for more efficient content delivery. Earlier development in this direction includes the proposal for `GETALL` and `GETLIST` [16] and `MGET` [7]; the latest one is the `N-to-1 bundle` [22]. While all these mechanisms achieve very good results, they share one important limitation. Nothing about the impact of the presence of proxy cache to their schemes is mentioned. This is a valid concern because when the bundle request arrives at the content server, all objects described in the bundle request will be sent out as one package, independent of the potential hits of some of the bundle objects in the proxy caches that are along the data transmission path. This will cause substantial performance loss, as proxy caching has been proven to be an important technology to reduce web access latency [10] [17].

To address the above problem, we propose a proxy-cache aware version of the bundle request, which we call **`PC-Bundle`**. The unique feature of this mechanism is that the advantages of all previously proposed bundle techniques can be achieved without any redundant object transfer in the presence of proxy cache. Compared to previously proposed bundle requests such as `N-to-1 Bundle`, our `PC-Bundle` mechanism can achieve an improvement ranging from 21.2% to 34.8% for page latency reduction and 17.2% to 36.4% for page network traffic, which are very significant.

In our study, the object retrieval latency is simply the time taken required to completely transfer the object while the page retrieval latency is the aggregate of the retrieval latency of the page container object and all of its associated embedded objects. The page retrieval latency needs not necessarily be the simple sum of the respective retrieval latency of the page container object and its embedded objects. This is due to the parallelism existed in web browsers for simultaneously object fetching. Currently, the parallelism width being used in Netscape and Microsoft IE browsers is four [5].

2. Existing Methods to Retrieve Embedded Objects

There are three main approaches used to accelerate the retrieval of embedded objects within a web page. They are: (i) parallel connections, (ii) persistent connection, and (iii) bundle transfer.

Some web browsers, after retrieving the page container object, establish parallel connections, one per embedded object, to retrieve the embedded objects. This is in part due to that the web browsers start parsing the initial portion of the container object that they receive and then make requests for the embedded objects, even when the retrieval of the container object is still underway. In addition, the browsers want to simultaneously render the embedded objects (especially images). This is possible with parallel connections. Thus the user would be able to view the web page in increasing quality and hence reduces the user-perceived page retrieval latency.

WebMUX [8] is an experimental multiplexing protocol that allows multiple application-layer sockets to transfer data fragments using a single transport-layer TCP connection. Paraloading [14] [18] [19] refers to the parallel segmented download (PSD) of an object using parallel connections to multiple mirrors (one connection per mirror). This is different from the convention of retrieving the object solely from one site. The user would first determine the mirrors from which the required object could be downloaded. Then, the user would select the mirrors to which a persistent connection would be established each.

The use of parallel connections, however, has some disadvantages such as additional overheads and slow transfer rates due to TCP slow-start. Also, if a container object has many embedded objects (for example more than 20), the use of parallel connections can have a negative impact on the performance and thereby increasing the page retrieval latency and not to mention user-perceived retrieval latency, which is definitely not desirable.

A persistent connection can be used, in lieu of parallel connections, for a serialized transfer of the embedded objects [6] [10]. Here, the server would return the embedded objects one-by-one, in the same sequence that the requests have been received. The client can also pipeline its requests so that it would reduce its waiting time for the transfers to take place. Nevertheless, the bulk of the retrieval latency comes from the transfer of the objects. The main advantage of using a persistent connection is that the client would bypass the overheads for the establishment of connections and the TCP slow-start stage that it would have incurred if parallel connections were used. It is possible that a serialized transfer of a group of embedded objects is faster than the use of parallel connections for the same purpose. However, it is reported in [4] [22] that support for persistent connections does not necessarily lead to reduced retrieval times for a set of objects from a server. This can be due to the inconsistent support from web servers, user agents, and intermediaries for persistent connections, particularly with pipelining.

For bundle request, the earlier proposals are MGET, GETALL, and GETLIST [7] [10] [16]. They are the precursor of pipelining in a persistent connection. GETALL specifies that the server returns the requested page container object as well as all the associated embedded objects (if any) to the client. As it is not cache-friendly and has performance and security issues, we would not discuss it further. MGET and

`GETLIST` allow a user to specify a list of requests (for objects) in a single request. The server would then parse through this list and return the objects (if applicable) in sequence. While this might mean that the client would have to consolidate all the requests to construct this list (and thereby incurring some waiting time), it would allow the server to know immediately what are the objects that the client requests at one instance.

`N-to-1 Bundle` is the latest version of the bundle request [22]. It suggests that the retrieval of objects would be shorter if they are bundled (and possibly compressed) together than that of their individual transfers. The server would advertise the availability of a bundle. This bundle would contain objects that have intra-page or inter-page relevance and in short, and most likely to be embedded objects. In addition, the bundle can be pre-determined or dynamically generated. In the case of a pre-determined bundle, the bundle might contain objects that the client does not want. On the other hand, the bundle may also lack the objects the client wants. Also, this form of bundling is not exactly cache-friendly. This is a trade-off from the associated costs and complexity of bundling on-the-fly and to the client's specifications.

3. Traffic Characteristics of Embedded Objects

In this section, we would like to examine some traffic characteristics of embedded objects based on a set of IRCACHE proxy traces [20] as our dataset. This should give us an insight on how to improve upon their retrieval latency.

3.1 Number of embedded objects per page

Table 1 illustrates the breakdown of the number of objects per page. This is derived from our IRCACHE dataset from which we have extracted 2725159 pages with 1934929 embedded objects from a total of 4660088 objects.

Table 1. Some Statistics Pertaining to Number of Objects per Page (from NLANR)

Number of Objects	1	≥ 2
Percentage of Entries	40.0%	60.0%
Percentage of Pages	68.4%	31.6%

While a large proportion of the web pages served is a single object by itself, about 32% have at least one embedded object. In addition, web pages with at least one embedded object amount to about 60% of the total number of entries. Hence, while there are fewer pages with at least one embedded object, they account for a larger portion of the NLANR dataset. This is consistent with the studies by [13].

3.2 Content of embedded objects

Table 2 shows that images are the dominant group of embedded objects. This is followed by text resources. GIFs, JPEGs and PNGs images form the majority of the images resources while HTML objects are the most numerous among the text resources.

Table 2. Content-Type of Embedded Objects

Content-Type of Embedded Object	Text	Images	Audio	Video	App.	Others
% of Total	28.0%	62.4%	0.14%	0.39%	7.93%	1.15%

Given the high percentage of image resources, it is understandably from the user's perspective on the desire to have simultaneous rendering of the embedded images. The GIF, JPEG and PNG support this form of progressive display whereby the quality of the image rendered improves as more of the image is retrieved. As such, the use of parallel connections is preferred to achieve this. A serialized transfer would not be able to achieve this type of simultaneous rendering. However, if a serialized transfer of a group of images is somehow faster than the parallel retrievals of the images, the faster retrieval might be chosen in preference over the simultaneous rendering.

3.3 Sizes of embedded objects

It can be observed that the embedded objects could be segmented into four main size classes namely:

- < 2KB
- 2KB - 5KB
- 5KB - 10KB
- 10KB

Table 3. Sizes of Embedded Objects

Size of Objects	< 2KB	2KB – 5KB	5KB – 10KB	> 10KB
% of Entries	56.6%	20.2%	11.24%	12.0%

Table 3 shows the various size categories of the embedded objects from our IRCACHE dataset. About 57% belongs to what we would refer to as small objects, accounting for more than half the total number of embedded objects that are in our dataset. The group of objects of sizes above 10KB represents about 12% of the total. We would refer to these objects as big objects. These big objects reflect the potential amount of the head-of-line blocking effect in transfers using a persistent connection.

3.4 Retrieval latency of embedded objects

Table 4 lists the mean retrieval latency of embedded objects classified in terms of size as in Table 3. We observe that the mean retrieval latency for objects in the groups < 2KB, 2KB to 5KB and 5KB to 10KB are rather similar. In addition, objects that are below 2KB in size, have a relatively lower transfer rate than that for the other objects.

Table 4. Mean Retrieval Latency of Embedded Objects

Object Size (bytes)	< 2KB	2KB – 5KB	5KB – 10KB	> 10KB
Mean Retrieval Latency (millisecond)	736	824	980	5255

A further analysis in the mean transfer rates is listed in Tables 5 and 6. They show that objects of sizes below 2KB have relatively lower transfer rates compared to that of other groups of objects. Given that from Table 3 that objects below 2KB in size forms the majority (more than half) of the embedded objects, this suggests that it would be beneficial to reduce the mean retrieval latency of this group of objects.

Table 5. Mean Transfer Rate of Embedded Objects for Sizes Below 5KB

Object Size (bytes)	< 512	512 – 1KB	1KB – 2KB	2KB – 3KB	3KB – 4KB	4KB – 5KB
Mean Transfer Rate (KB/s)	0.38	1.23	1.39	3.06	5.45	3.96

Table 6. Mean Transfer Rate of Embedded Objects of Sizes 5KB to 10KB

Object Size (bytes)	5KB – 6KB	6KB – 7KB	7KB – 8KB	8KB – 9KB	9KB - 10KB
Mean Transfer Rate (KB/s)	5.59	8.87	6.60	6.73	9.04

The probable reasons why the small embedded objects have a relatively lower transfer rates are the following:

- Overheads of connection establishment: as assumed, a fixed amount of resources is required for connection establishment. For larger objects, this cost is amortized with the relatively lengthy transfer.
- `TCP` slow-start: `TCP` slow-start confines the server to a low transfer rate to be increased gradually with more successful transmission. The transmission of small embedded objects would usually complete before the maturity of the slow-start stage.

4. Proxy-Cache Aware PC-Bundle Transfer of HTTP Messages

Bundle transfer, in abstraction, is a form of transfer coding except that it is applied to a group of HTTP messages rather than to a message. This group of HTTP messages could be a list of requests or a list of responses. Here, we formally propose a proxy-cache aware mechanism for the bundle transfer of HTTP messages. Changes to the HTTP specifications are required.

4.1 Basic mechanism

A client C would retrieve a web page with the container object Obj_0 from a server S along the data transfer path with proxy caches P_1, P_2, P_3, …, P_i, …, P_m. From Obj_0, C would determine the embedded objects that it would retrieve. Suppose that there are N such unique embedded objects. As such, C would have to make N number of HTTP requests to be denoted as Req_1, Req_2, ..., Req_n. Instead of sending out these requests (pipelining or otherwise), C would bundle-transfer them in a manner similar to that of MGET and GETLIST. C would encapsulate these requests as a list of requests inside a HTTP request. Let this new HTTP request be denoted as *REQ*. Hence, *REQ* would contain a list of HTTP requests as its message body and a slightly different set of HTTP request and entity headers that would provide some information about the list of requests.

When *REQ* arrives at each proxy cache P_i (where $1 \le i \le m$), the encapsulated object request list in the bundle *REQ* will be parsed. Each requested object will be checked against the proxy cache storage. For any object Obj_i (where $1 \le i \le N$) that is found in the local cache, the object content will be sent back to C immediately without waiting for the other object requests in *REQ*. At the same time, Obj_i will be deleted from *REQ* before the *updated* bundle request REQ' (which is the difference between *REQ* and all Obj_i found in proxy cache) is forwarded to the next level of the network. Note that all objects found in the local proxy cache can be sent back to C in one single bundle reply, just like the server S.

S receives the latest REQ' and parses it to obtain the individual requests. As S prepares the various responses, it determines whether to perform PC-bundle transfer and if so, which of the responses to be bundle-transferred. This would be a server-driven policy. In our case, S would only choose small responses. Small responses would usually refer to those responses containing small embedded objects or no entities at all. S will encapsulate these selected responses into a HTTP response to be denoted as *RES*. Thus *RES* will contain the list of selected responses as its message body and a slightly different set of HTTP entity headers that would provide some information about the list of responses. This list of responses is a subset of the responses to all the requests in *REQ* or *REQ'*. S will transfer *RES* first followed by the individual transfers of the other responses (to requests in *REQ*) which are not bundled in *RES*.

4.2 Assumptions

We make the following assumptions in the following three sub-sections.

Client-Specific Assumptions. The assumptions that are client-specific are:

- The client might issue a list of requests instead of individual requests. This is to be known as a request-list. The client suffers insignificant performance penalty in the construction of this list.
- The client will receive a bundle that is a collection of responses to some or all of its requests (in a request-list). The client will exert a fixed amount of resources to dissemble the bundle.
- There might be more than one bundle (although unlikely) received as response for each request-list. The client might receive a mixture of bundles and individual responses to a request-list.
- The client is prepared to receive responses not in the same order as that of its requests. In addition, the order of requests and responses is not important to the client.

Server-Specific Assumptions. The assumptions that are server-specific are:

- The server will determine whether or not to use the PC-Bundle mechanism whenever a request-list is received. The server incurs insignificant overhead for this action.
- The server will determine which of the responses to a request-list should be bundled. It will exert a fixed amount of resources to determine and assemble a bundle.
- The server should consider that only responses to requests from a request-list should be bundled.

Proxy-Specific Assumptions. The assumptions that are server-specific are:

- Upon receiving the request-list in a `PC-bundle`, the proxy has the ability to parse and understand all individual object requests.
- The proxy will remove from the request-list all objects that are found in its cache. It will form a new `PC-bundle`'s request-list to be forwarded to the server or next proxy.
- For all the cache-hit objects in the request-list, their content will be sent to the client using the same mechanism as the server does.

4.3 Detailed Specifications of Modifications to HTTP for `PC-Bundle`

Request. The motivation for a client to `PC-Bundle` its requests is that the server would be able to receive all these requests at one pass. This would allow the server to `PC-Bundle` all applicable responses back to the client, without waiting for potential additional requests from the client for whose responses could be piggybacked onto the bundle transfer. Encapsulating multiple requests in a single message would also mean a reduction in the number of packets in the network. This would help to reduce the load on the network. At the same time, the client must be able to tolerate some delay in the consolidation of all requests. In addition, it would take additional latency to transmit the encapsulation of requests since it would be considerably larger in size than the individual requests. (Note that the later situation is not very bad, as the encapsulated response can be interpreted in a chunk-by-chunk manner).

Here, we propose the following changes to the syntax of `HTTP` specifications in order to accommodate the capability of `PC-Bundle` of requests. The syntax of a `HTTP` request message looks like the following:

```
Request =  Request-Line
           *(General-Header|Request-Header|Entity-Header)
              CRLF
              [Message-Body]
Request-Line = Method SP Request-URI
                HTTP-Version CRLF
```

Request-Line. We propose that the field Request-URI is to be made optional and omitted only when the method to use `PC-Bundle` is invoked. Note that in a `MGET` request, the Request-URI would not be required. Thus the new Request-Line would look like the following.

```
Request-Line = Method SP [Request-URI]
                HTTP-Version CRLF
```

Thus a `PC-Bundle` request is different from a normal `HTTP` request in that the Request-URI is omitted and that it has a Message-Body which is simply the concatenation of multiple `HTTP` requests.

Request-Header. The client would want to specify that it is able to receive a `PC-Bundle` type of responses. It could only do so through the use of the `Accept` header for the recommended new media type `bundle/responses`.

```
Accept: bundle/responses
```

The inclusion of this Request-Header is mandatory for the responses of the `PC-Bundle` to take place.

Entity-Header. The client would have to specify the following Entity-Headers:

- Content-Length: the size of the Message-Body which should be the sum of all the encapsulated requests.
- Content-Type: This should be text/ASCII or a recommended new media type bundle/requests. This new media type should be only used to mean the PC-Bundle requests. The definition of the field media-type for Content-Type includes the optional input of an attribute. This option may be exercised to specify the number of requests that are encapsulated in the Message-Body of the PC-Bundle request.

Message-Body. The Message-Body of a PC-Bundle request would be simply the concatenation of the requests. These requests may be separated by a CRLF. In addition, these requests must be full-fledged requests on their own. The recipient of a PC-Bundle request should be able to parse through the Message-Body, extract the individual requests and resend them to another intermediary or origin server without (non-transparent) modifications. The Message-Header must not contain any other PC-Bundle requests.

Response. Only the recipient of a PC-Bundle request which includes an Accept: bundle/responses Request-Header would be in a position to use PC-Bundle responses back to the client. If the recipient does not receive a PC-Bundle request or receive a PC-Bundle request without an Accept: bundle/responses Request-Header, the recipient must not send any PC-Bundle responses back to the client. The subsequent decision whether to use PC-Bundle or not and which of the responses to use PC-Bundle is a server-driven policy.

The motivations for using PC-Bundle as the responses include the increase in the effective retrieval latency for small responses and a reduced network load since less number of packets would be transmitted. PC-Bundle response is preferred for non-dynamic responses.

A typical HTTP Response message has the following syntax:

```
Response = Status-Line
           * General-Header|
             Response-Header|Entity-Header)
           CRLF
           [Message-Body]
Status-Line =  HTTP-Version  SP  Status-Code  SP  Reason-
               Phrase CRLF
```

Status-Line. Two types of responses to a PC-Bundle request are possible. There would be either a PC-Bundle for some or all responses (to the list of requests) or none at all, meaning that the responses are returned as per normal. A "successful" PC-Bundle response would have a 2XX status code. The recommended status code is 207. The Reason-Phrase could be "Bundled

`Content"`. Hence the Status-Line of a `PC-Bundle` response would look like the following

```
HTTP/1.1 207 Bundled Content
```

Entity-Header. The `Content-Type` and `Content-Length` Entity-Headers are mandatory in a `PC-Bundle` response. The value to be assigned to `Content-Type` would be the media-type `bundle/responses`. As with the `Content-Type` header in the `PC-Bundle` Request, there is an optional input of an attribute for the media-type field. This option may be exercised to specify the number of responses that are encapsulated in the Message-Body.

Message-Body. The Message-Body is the concatenation of various responses that are selected for `PC-Bundle`. These responses may be separated using a `CRLF`. An additional Entity-Header URI must be included in every response that has a Message-Body. It will be used to identify the entity in the Message-Body. The value of URI must correspond to that of the Request-URI listed in the Request-Line of the request. The Message-Body must not contain any other `PC-Bundle` responses.

5. Experimental Results on PC-Bundle Mechanism

In this section, we would like to highlight the potentials of our `PC-Bundle` mechanism through trace-driven simulation. In particular, we would like to compare our mechanism with the latest `N-to-1 Bundle` proposal [22] in the presence of proxy cache. To make the study more realistic, the performance metric is chosen to be the total page latency instead of the object retrieval latency to reflect the effect of parallel object fetching to the actual performance. We also try to make our simulation environment and assumptions as close as possible to those in the latest `N-to-1 Bundle` proposal.

5.1 Environment of simulation

Our study is done in a trace-driven simulation using three sets of proxy traces from NLANR [9], Digital [15] and Berkeley [3]. The objects from the entries in these three proxy traces are grouped into pages and the traffic characteristics of these pages are examined. *Total page latency* refers to the total time required to retrieve the pages in each set of proxy traces while *total page traffic* refers to the effective bandwidth consumed in the retrieval of these pages. In our sensitivity study, we shall look at three cache parameters. They are: (i) cache size (as a percentage of total size of unique objects in the trace), (ii) replacement policy, and (iii) degree of parallelism for simultaneous object fetching.

To make a fair comparison, our simulation emulates the environment set in [22]. Bundling makes use of the best transfer rate among that of the objects in a page to be the effective transfer rate of the page. In the event of a partial page hit (the container page or some embedded objects are misses), the page is treated as a complete page miss. The effective transfer rate of the page used is then the best transfer rate among that of the object misses. Likewise, in the event of a complete page (the container page and the embedded objects are all hits), the best transfer rate among that of these hits is chosen as the effective transfer rate of the page. The page latency is then the time taken to transfer the amount of data of size equal to the sum of the sizes of the container page and its embedded object, using the effective transfer rate. In our `PC-Bundle` mechanism, it has two transfer rates. It uses the best transfer rate among that of the object hits and the best transfer rate among that of the object misses. The page latency is then the time taken to retrieve the bundle of hits and the bundle of misses respectively, using their respective effective transfer rates.

Page traffic effectively refers to the amount of data transferred for the retrieval of a page. For bundling in the case of a partial page hit or a complete page miss, the page traffic of this page is effectively the sum of the size of the page container object and all its associated embedded objects. In the case of the normal proxy settings and our `PC-Bundle` mechanism, the page traffic could be smaller in the case of page/object hits (where only validation and not the entire object/page is needed). Thus, for our study, we can regard the page traffic for the normal proxy settings and that of our `PC-Bundle` mechanism to be the same.

Table 7. Effect of Cache Size on Total Page Latency (GDS, Parallelism = 4)

Cache size	`N-to-1 Bundle` / Normal			`PC-Bundle` / `N-to-1 Bundle`			`PC-Bundle` / Normal		
	NLANR	Digital	Berkeley	NLANR	Digital	Berkeley	NLANR	Digital	Berkeley
5%	56.0%	24.9%	38.4%	21.2%	22.7%	27.4%	65.3%	41.9%	55.3%
10%	55.3%	22.4%	35.5%	22.3%	25.0%	31.3%	65.3%	41.9%	55.7%
15%	55.3%	21.5%	34.1%	22.9%	26.2%	33.6%	65.6%	42.0%	56.2%
20%	55.0%	20.9%	33.3%	23.5%	27.0%	34.8%	65.6%	42.2%	56.5%

In our study, the standard cache configuration and network environment for reference is:

- Cache size equal to 10% of the total size of unique objects in the trace under study
- Replacement policy being set to Greedy-Dual-Size [1] [2] [12]
- Parallelism width for simultaneous object fetching being set to 4 (typical in current browsers including Netscape and IE [5])

Each of these parameters will also be varied so that the sensitivity of our results and conclusion on each of these parameters can be investigated.

5.2 Results with respect to cache size

The size of a proxy cache effectively determines the number and size of objects that could be stored in the cache. In our study, we express the cache size as a percentage of the total size of unique objects in the dataset under examination. The values used in this study are 5%, 10%, 15% and 20%. To examine the effect of cache size on Total Page Latency and Total Page Traffic, we set the cache replacement policy to be the Greedy Dual-Size (GDS) algorithm and the degree of parallelism to 4.

Cache Size on Total Page Latency. Table 7 shows that the `N-to-1 Bundle` and `PC-Bundle` improve significantly on the Total Page Latency and that `PC-Bundle` outperforms `N-to-1 Bundle`. It shows that `N-to-1 Bundle` improves the Total Page Latency by about 55% for the NLANR dataset while for the Digital and Berkeley datasets, the it is about 20% to 38%. This difference could be explained by that the NLANR proxy is a public proxy while the Digital and Berkeley each served a specific user group. As such, the NLANR dataset would contain relatively a lesser number of partial page hits and hence the negative effects of `N-to-1 Bundle` on caching are less influential on the NLANR dataset. This is an indication of the influence that `N-to-1 Bundle` has over public proxies. We also observe that as the cache size is increased, `N-to-1 Bundle` degrades in performance. This should be due to that as the cache size is increased, there would be more partial page hits.

Compared to `N-to-1 Bundle`, `PC-Bundle` has a much superior performance. It outperforms the `N-to-1 Bundle` by about 25% for Digital, 30% for Berkeley, and 22% for NLANR. And these results are quite insensitive to the cache size. This is not surprising, as the performance gain comes from the partial hit pages.

Cache Size on Total Page Traffic. Table 8 shows that while `N-to-1 Bundle` gives a better Total Page Latency than the normal case does, it consumes excessive memory bandwidth. The increase in Total Page Traffic can range from about 18% to 36%, which is very significant. On the other hand, although `PC-Bundle` gives even better Total Page Latency performance than `N-to-1 Bundle`, it does not cause any additional bandwidth consumption even when compared to the normal case. This shows its superiority and practicability over `N-to-1 Bundle`.

Table 8. Effect of Cache Size on Total Page Traffic (GDS, parallelism = 4)

Cache Size	`N-to-1 Bundle` / (Normal or `PC-Bundle`)		
	NLANR	Digital	Berkeley
5%	17.2%	27.3%	24.6%
10%	18.4%	33.2%	30.2%
15%	19.3%	36.8%	33.8%
20%	20.5%	39.4%	36.4%

Table 9. Effect of Cache Replacement Policy on Total Page Latency (Cache Size = 10%, Parallelism = 4)

Cache Replacement Policy	`N-to-1 Bundle` / Normal			`PC-Bundle` / `N-to-1 Bundle`		
	NLANR	Digital	Berkeley	NLANR	Digital	Berkeley
GDS	55.3%	22.4%	35.5%	22.3%	25.1%	31.3%
LFU	54.2%	28.6%	44.1%	26.2%	22.4%	27.5%
LRU	55.0%	31.6%	18.0%	25.2%	20.9%	25.3%

As the cache size is increased, the Total Page Traffic from `N-to-1 Bundle` remains relatively constant while that of the normal proxy settings and `PC-Bundle` decreases. This decrease is most probably accounted for by the increase in the number of partial page hits in the corresponding increase in the cache size. This indicates an effective increase of Total Page Traffic for `N-to-1 Bundle` as the cache size is increased.

The additional page traffic from the Digital and Berkeley datasets using `N-to-1 Bundle` amount to about 24% to 40% respectively while that for the NLANR dataset is less than 20%. This reinforces our observation that there are relatively more partial page hits in the Digital and Berkeley datasets than in the NLANR dataset.

5.3 Results with respect to cache replacement policy

The cache replacement policies that are used in this study are the Greedy Dual-Size (GDS), Least-Frequently-Used (LFU) and Least-Recently-Used (LRU) algorithms. The cache parameters, cache size and degree of parallelism, are set to 10% and 4 respectively.

Cache Replacement Policy on Total Page Latency. Table 9 shows that the superior performance of `PC-Bundle` over `N-to-1 Bundle` is quite insensitive to the cache replacement policy. Similar performance gain of 22% to 27% is still observed for `PC-Bundle`. Further, it shows that LFU in the NLANR dataset achieves the lowest Total Page Latency while it is so for GDS in both the Digital and Berkeley datasets. Correspondingly, with reference to Table 9, `N-to-1 Bundle` performs the worst with LFU in the NLANR dataset and with GDS in both Digital and Berkeley datasets while `PC-Bundle` has the best performance over `N-to-1 Bundle` in the exact same order. Thus, the relative performance of `PC-Bundle` over `N-to-1 Bundle`, using different cache replacement policies, corresponds inversely to the performance of `N-to-1 Bundle`.

Cache Replacement Policy on Total Page Traffic. As illustrated in Table 10, the Total Page Traffic for `N-to-1 Bundle` remains relatively constant regardless of the cache replacement policy used. GDS in the NLANR dataset accounts for a larger Total Page Latency for that of normal proxy settings and `PC-`

Bundle while it is the case for LRU in both the Digital and Berkeley datasets although the differences here are very slight. In the case of the NLANR dataset, PC-Bundle is most effective when LRU is used and least effective when GDS is used. In the other two datasets, the choice of the cache replacement policy does not seem to have a significant effect on the effectiveness of PC-Bundle over N-to-1 Bundle. All these result are expected, just like the situation of changing cache size.

Table 10. Effect of Cache Replacement Policy on Total Page Traffic (Cache Size = 10%, Parallelism = 4)

Cache Replacement Policy	N-to-1 Bundle / (Normal or PC-Bundle)		
	NLANR	Digital	Berkeley
GDS	18.4%	33.2%	30.2%
LFU	27.0%	33.7%	29.6%
LRU	26.5%	31.8%	28.0%

Table 11. Effect of Parallelism on Total Page Latency (Cache Size = 10%, GDS)

Degree of Parallelism	N-to-1 Bundle / Normal			PC-Bundle / N-to-1 Bundle		
	NLANR	Digital	Berkeley	NLANR	Digital	Berkeley
4	55.3%	22.4%	35.5%	22.3%	25.1%	31.3%
8	55.0%	21.7%	34.9%	22.0%	24.5%	30.2%
16	54.9%	21.5%	34.9%	21.9%	24.4%	29.9%
32	54.9%	21.5%	34.8%	21.9%	24.4%	29.9%

5.4 Results with respect to degree of parallelism

We assume that there is a degree of parallelism in all web page retrievals. The degrees of parallelism examined in this study are 4, 8, 16 and 32. The cache parameters, cache size and cache replacement policy, are set to 10% and GDS respectively. As the degree of parallelism does not have any effect on the Total Page Traffic, we would not be discussing this aspect.

Parallelism on Total Page Latency. Table 11 shows that as the degree of parallelism is increased, the effectiveness of N-to-1 Bundle and PC-Bundle over N-to-1 Bundle decreases very slightly. This indicates that an increase in the degree of parallelism does not significantly affect the effectiveness of N-to-1 Bundle and PC-Bundle. A probable explanation for this is that the basic degree of parallelism (4) has already significantly reduced Total Page Latency so much so that further increases in the degree of parallelism would only return marginal improvements. This is verified by Table 12 which shows the relative improvements in Total Page Latency using various degrees of parallelism on the NLANR dataset.

Table 12. Effect of Various Degrees of Parallelism on Total Page Latency (NLANR Dataset, Cache Size = 10%, GDS)

Degree of Parallelism (DOP)	Relative Improvement in Total Page Latency over Previous DOP
1	N/A
2	16.5%
3	3.69%
4	1.36%
5	0.71%
6	0.38%
7	0.22%
8	0.18%
9	0.11%
10	0.10%

6. Conclusions

In this paper, we propose a proxy-cache aware bundling mechanism, the `PC-Bundle`, to accelerate the downloading process of embedded objects in a web page. `PC-Bundle` is a form of HTTP transfer encoding applied to a group of objects (in this case embedded objects) to effectively reduce the total number of `TCP` data packets and hence reduce their effective retrieval latency. The formal specifications of the `PC-Bundle` mechanism are presented. To show its effectiveness, we perform trace simulation study to compare our `PC-Bundle` with the latest bundling solution, the `N-to-1 Bundle` under varying cache environments using three sets of public proxy traces. Our result shows that:

- Using the same assumptions and environment that of `N-to-1 Bundle`, `PC-Bundle` outperforms `N-to-1 Bundle` in terms of Total Page Latency and Total Page Traffic by about 20% to 25%.
- An increase in the cache size will result in a decrease and an improvement in the effectiveness of `N-to-1 Bundle` and `PC-Bundle` in terms of Total Page Latency respectively. In addition, `N-to-1 Bundle` will have a corresponding increase in Total Page Traffic.
- LFU seems to favor public proxies (such as NLANR) using `PC-Bundle` in terms of both Total Page Latency and Total Page Traffic. In addition, applying GDS to public proxies would reduce slightly the effectiveness of `PC-Bundle` over `N-to-1 Bundle`.
- In proxies which are more centric to a group of users (Digital and Berkeley), the choice of the cache replacement policy does not significantly affect the effectiveness of `N-to-1 Bundle` and `PC-Bundle`.
- The degree of parallelism has only a slight influence on the Total Page Latency.

References

[1] Arlitt, M., Friedrich, R., Jin, T., Performance evaluation of Web proxy cache replacement policies, *Technical Report HPL-98-97, HP Laboratories,* June 1998.

[2] Arlitt, M., Williamson, C., Trace-driven simulation of document caching strategies for Internet Web servers, *Simulation Journal*, Vol. 68, No. 1, January 1997, pp. 23-33.

[3] UC Berkeley Web Proxy Traces, http://ita.ee.lbl.gov/

[4] Cardwell, N., Savage, S., Anderson, T., Model TCP latency, *Proceedings of the 2000 IEEE INFOCOM Conference*, Tel-Aviv, Israel, March 2000.

[5] Chi, C.H., Li, X., Lam, K.Y., Understanding the object retrieval dependence of Web page access, *Proceedings of the 10th IEEE International Symposium on Modeling, Analysis, and Simulation of Computer and Telecommunication Systems (MASCOTS),* October 2002.

[6] Fielding, R., Gettys, J., Mogul, J., Frystyk, H., Masinter, L., Leach, P., Berners-Lee, T., Hypertext Transfer Protocol -- HTTP/1.1, *IETF RFC2616*, June 1999.

[7] Franks, J., Proposal for an HTTP MGET Method, 1994. http://ftp.ics.uci.edu/pub/ietf/http/hypermail/1994q4/0260.html

[8] Gettys, J., Nielsen, H.F., The WebMUX protocol, *Expired Internet Draft*, August 1998. http://www.w3.org/Protocols/MUX/WD-mux-980722.html

[9] IRCACHE Proxy Traces, http://ircache.nlanr.net.

[10] Krishnamurthy, B., Rexford, J., *Web Protocols and Practice*, ISBN 0-201-71088-9, Addison-Wesley, 2001.

[11] Krishnamurthy, B., Willis, C.E., Analyzing factors that influence end-to-end Web performance, *Proceedings of the 9th World Wide Web Conference*, April 2000.

[12] Lorenzetti, P., Rizzo, L., Replacement policies for a proxy cache, *IEEE/ACM Transactions on Networking,* Volume 8, Issue 2, April 2000.

[13] Mah, B., An empirical model of HTTP network traffic, *Proceedings of IEEE INFOCOM Conference*, April 1997.

[14] Miu, A., Shih, E., Performance analysis of a dynamic parallel downloading scheme from mirror sites throughout the Internet, Term Paper, *LCS MIT*, December 1999.

[15] Mogul, J., Digital's Web Proxy Traces, 1996.

[16] Padmanabhan, V.N., Mogul, J.C., Improving HTTP latency, *Proceedings of the 2nd International World Wide Web Conference*, Oct. 1994. Also in *Computer Networks and ISDN Systems*, 28(1/2), December 1995, pp. 25-35.

[17] Rabinovich, M., Spatscheck, O., Web *Caching and Replication*, ISBN 0-201-61570-3, Addison-Wesley, 2002.

[18] Rodriguez, P., Biersack, E.W., Dynamic parallel-access to replicated content in the Internet, *IEEE/ACM Transactions on Networking*, August 2002.

[19] Rodriguez, P., Kirpal, A., Biersack, E.W., Parallel-access for mirror sites in the Internet, *Proceedings of IEEE INFOCOM 2000 Conference*, March 2000.

[20] IRCACHE proxy traces from http://ircache.nlanr.net: pb.sanitized-access.20020818-24

[21] Wessels, D., *Web Caching*, ISBN 1-56592-536-X, O' Reilly & Associates, 2001.

[22] Wills, C.E., Mikhailov, M., Shang, H., N for the price of 1: Bundling Web objects for more efficient content delivery, *Proceedings of the 10th International World Wide Web Conference*, May 2001.

A CASE FOR DYNAMIC SELECTION OF REPLICATION AND CACHING STRATEGIES

Swaminathan Sivasubramanian, Guillaume Pierre and Maarten van Steen
Dept. of Mathematics and Computer Science, Vrije Universiteit, Amsterdam, The Netherlands

Abstract Replication and caching strategies are being used to reduce user perceived delay and wide area network traffic. Numerous such strategies have been proposed to manage replication while maintaining consistency among the replicas. In earlier research, we demonstrated that no single strategy can perform optimal for all documents, and proposed a system where strategies are selected on a per-document basis using trace-driven simulation techniques. In this paper, we demonstrate the need for continuous dynamic adaptation of strategies using experiments conducted on our department Web traces. We also propose two heuristics, *Simple* and *Transition*, to perform this dynamic adaptation with reduced simulation cost. In our experiments, we find that *Transition* heuristic reduces simulation cost by an order of magnitude while maintaining high accuracy in optimal strategy selection.

1. Introduction

Web users often experience slow document transfers for Web documents. To reduce access time, many systems replicate or cache documents at servers close to the clients. This process allows load balancing among different servers and decreases the access latencies experienced by clients. However, if a document is updated, replicas must be updated to prevent clients accessing a stale copy.

Many strategies have been proposed to achieve replication while maintaining consistency among replicas. A replication strategy dictates the number and location of replicas and the choice of a protocol governing the creation of replicas and consistency enforcement. Different strategies may offer various levels of performance and consistency, so a system designer should be careful when selecting a replication strategy.

We showed in earlier research that no single strategy can universally perform optimal for all documents [8]. An important gain in performance can be obtained by associating each document with the strategy that suits it best. Moreover, the document-to-strategy associations must be re-evaluated from time to time, since changes in access and update patterns are likely to affect the performance characteristics of strategies.

In this paper, we make a case for re-evaluating this document-to-strategy association from time-to-time, as changes in documents' access and update patterns are

F. Douglis and B.D. Davison (eds.), Web Content Caching and Distribution, 275-282.

likely to affect system performance. We do not deal with adaptation of strategies during emergency situations, such as flash crowds, but rather focus on adaptation of relatively stable access patterns.

We employ an approach where the best strategy for each document is periodically selected among a set of candidate strategies. The choice of "best" strategy is made by simulating the performance that each strategy would have provided in the recent past. If necessary, the strategy for the concerned document is switched dynamically.

Each server adaptation requires $d \cdot s$ simulations, where d is the number of hosted documents and s is the number of candidate strategies. In our current system, the simulation of a single strategy takes several tens of milliseconds on a 1-GHz PIII machine. Considering that we apply traditional and well-known trace-driven simulation techniques, we expect comparable performance for systems similar to ours. Each adaptation may thus lead to significant computational load if the number of objects or the number of candidate strategies is high. In particular, the latter can happen if we want to integrate parameterized strategies. Such strategies have a tunable parameter affecting their behavior, such as a "time-to-live" value or a "number of replicas."

The contributions of this paper are as follows: (i) we demonstrate the need for continuous dynamic adaptation of replication strategies for Web documents; and (ii) we present techniques for the selection of an optimal replication strategy from a number of candidate strategies, while keeping the selection costs low. None of these contributions are reported in our earlier works [7],[8].

The rest of the paper is organized as follows: Section 2 describes our evaluation methodology. Section 3 demonstrates the need for continuous dynamic adaptation of replication strategies using our Web traces. Sections 4 discusses our two strategy selection heuristics and present their performance evaluation. Section 5 discusses the related work and Section 6 concludes the paper.

2. Evaluation methodology

We set up an experiment that simulates a system that is capable of switching its strategies dynamically for changes in the access and update patterns of its documents. Our experiment consists of collecting access and update traces from our department Web server, simulating different strategies and selecting the best one for a given period. With this setup, we observe the changes in the strategy adopted by the documents over the entire length of the traces. Here we present our simulation model in detail.

2.1 Simulation model

We assume that documents have a single source of update called the *primary server*, which is responsible for sending updates to replicas located at *intermediate servers*. For the sake of simplicity, we consider only static documents in our evaluations, that is documents that change only due to updates by the primary server.

In our experiments, we use a list of 30 strategies to choose from: (i) NR: No replication, (ii) CLV[p] (Cache with Limited Validation): Intermediate servers cache documents for a given time. Each document has a validity time (TTL) after which it is removed from the cache. Different strategies are derived with TTL value fixed at

p = 5%, 10%, 15%, and 20% of the age of document. (iii) SI (Server Invalidation): Intermediate servers cache the document for an unlimited time. The primary server invalidate the copies when a document is updated. (iv) SU[x] (Server Updates): The primary server for a document maintains copies at the x most popular intermediate servers (the top x servers sorted based on the total number of clients handled by them) for the document. When the document is updated, the primary server pushes update to the intermediate servers. Different strategies are derived for $x = 5, 10, 25, ..., 50$. (v) Hybrid[x] (SU[x] + CLV[10]) - The primary server maintains copies at the x most popular servers, where $x = 10, 15, 20, 25, 30, 40, 50$ and the other intermediate servers follow CLV strategy, with TTL fixed as $p = 10\%$ of the age of the document.

In our simulations, we group clients based on the Autonomous Systems (AS) to which they belong. We measure the available network bandwidth between the AS belonging to the primary server and other ASes as follows: We record the time t taken by our server to serve a document of b bytes to a client from an AS. We approximate the bandwidth between our AS and the AS to which the client belongs to as b/t.

We redirect the requests of a client to the replica located in the client's AS. If there is no such replica available, then the requests are redirected to the primary server. We did not implement more sophisticated policies due to the lack of inter-AS network measurements.

2.2 Adaptation mechanisms

As shown in [8], one can optimally assign a strategy to each object using a cost function. This function is designed to capture the inherent tradeoff between the performance gain by replication to performance loss due to consistency enforcement. In our experiments, we use a cost function that takes three parameters: (i) access latency, l, (ii) number of stale documents returned, c and (iii) network overhead, b, that is the bandwidth used by the primary server for maintaining consistency and serving clients from ASes without replicas. The cost function for a strategy s during a given period of time t is:$cost(t, s) = w_1 * l + w_2 * c + w_3 * b$, where w_1, w_2 and w_3 are constants determining the relative weight of each metric.

The performance of a strategy s, during a given period of time t, is represented by the value of the cost function $cost(t, s)$. This value is obtained by simulation of strategy s with past traces. The primary server periodically evaluates the performance of a set of candidate strategies for each document and selects the best as the one that had the smallest cost.

The Web trace used in the experiments covers the requests and updates made to the documents hosted in our Web server from June 2002 to March 2003. Numerical details about the trace are given in Table 1. We perform our evaluations only for objects that receive more than 100 requests a week, reducing the total number of objects to be evaluated in the order of thousands. We adopt the no replication (NR) strategy for the rest. We fix the adaptation period of the server to one day.

Table 1. Trace Characeteristics

Number of days	273
Number of GET requests	78,049,912
Number of objects	2,185,896
Number of updates	156,721
Number of unique clients	1,252,779
Number of different ASes	2853

3. The Need for Dynamic Adaptation

In this section, we demonstrate the need for continuous dynamic adaptation of replication strategies for Web documents. To do so, we measure the proportion of documents that change their strategy over the duration of the traces, determined by Adaptation Percentage (AP). AP is defined as the ratio of the total number of adaptations observed and the total number of possible adaptations. For example, in a 7 day period, if a document makes 2 adaptations out of the maximum possible 6 adaptations (as the adaptation period is fixed to be one day), then its AP is 33%. This metric shows the rate at which an object switches its strategy dynamically. A higher value of AP implies that the objects dynamically switch their strategy more often thereby showing a clear need for continuous adaptation.

We performed a complete evaluation of the popular documents (those receiving at least 100 requests a week) in our traces and plotted the average AP over all documents (aggregated every week). The results are given in Figure 1.

As seen in the figure, AP varies from 5% (objects switch their strategy only 5% of the times) to 50% (objects switch their strategy half of the time). This figure shows that the objects switch their strategy often, though the rate of adaptation varies over a period of time. This clearly demonstrates the need for continuous dynamic adaptation of strategies, if the adopted strategy for an object needs to remain optimal.

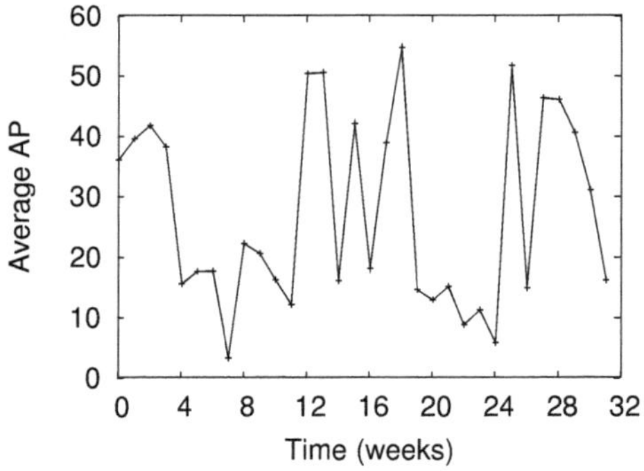

Figure 1. Time plot of AP of documents (aggregated weekly) for the entire length of traces

Table 2. Performance of *Non-adaptive* and *Adaptive* selection methods given in terms of (i) Total Client latency (TCL) for all requests, (ii) number of stale documents delivered to clients (NS) and (iii) network usage bandwidth (BW)

Strategy	TCL(hrs)	NS	BW (GB)
Non-adaptive	37.5	11	13.1
Adaptive	31.3	3	12.7

We further evaluated the need for continuous dynamic adaptation by comparing the performance of a system that would associate a strategy to each document once and never adapt (*Non-adaptive*), to one which would periodically adapt to the current-best strategy associations (*Adaptive*). Table 2 shows the performance of the two selection methods for the same time period in terms of the three cost function parameters: client latency, number of stale documents and bandwidth usage. It can be seen that the *Adaptive* selection method outperforms its *Non-adaptive* counterpart, according to all metrics, because of its ability to adapt to changes in client request and document update patterns.

From these simple experiments performed with our traces, it can be seen that documents need to continuously adapt strategies to maintain optimal performance, even for a simple Web server like ours. We believe that continuous dynamic adaptation will be more beneficial for replicated Web services handling more clients and hosting higher number of popular objects.

4. Strategy Selection Heuristics

In this section, we propose two selection heuristics, *Simple* and *Transition*, that aim to reduce the simulation cost by reducing the number of candidate strategies evaluated during the selection process.

4.1 Performance evaluation metrics

To evaluate the performance of strategy selection heuristics, we compared the selection accuracy and computational gain in comparison to the full evaluation method (evaluating all candidate strategies every period). The goal of a selection heuristic is to provide the same decisions as the full evaluation strategy, but with less simulation cost. We evaluate the performance of selection heuristics with the following metrics: (i) **Speedup**: This is defined as the ratio of total number of candidate strategies evaluated by the full evaluation method to that of the selection heuristic, (ii) **Accuracy**: This is defined as the percentage of times when the heuristic has selected the same strategy as the full evaluation strategy and (iii) Average Worst Case Ratio (**AWCR**): This metric indicates how bad a strategy is when the heuristic has made a non-optimal selection. The AWCR of a heuristic is computed as the average of WCRs for all non-optimal selections in an adaptation period, where WCR is defined as follows:

$$WCR(t) = \frac{|(cost(selected, t) - cost(best, t))|}{|(cost(worst, t) - cost(best, t))|}$$

AWCR can vary from 0 to 1. The greater its value, the worse is its choice of strategy.

4.2 Selection heuristics

The basic assumption behind the *Simple* heuristic is that a significant change in the request or update rate indicate the need for a strategy re-evaluation. In such cases, a full evaluation is performed. Otherwise, the current strategy is retained. Our evaluations showed that this heuristic yields poor accuracy with very little gain in speedup. We conclude that more sophisticated is necessary to give a better speedup without much loss in accuracy.

The *Transition* heuristic tries to predict the likely strategy transitions and evaluates only the most promising strategies. It aims to gain speedup by evaluating only this subset of likely strategies instead of the entire set of candidate strategies. Obviously, the size and constituents of the selected subset of strategies to evaluate determines the speedup and accuracy of the heuristic. The smaller the size of this subset, the higher will be the speedup. On the other hand, the heuristic will find the optimal strategy only if it belongs to the subset of evaluated strategies.

The *Transition* heuristic works in two phases. In the first phase, full evaluations are performed on the traces to build the *transition graph*. This graph captures the history of transitions between different strategies. It is a weighted directed graph, whose nodes represent the candidate strategies and weights are the number of observed transitions between strategies. No speedup is gained in this phase as full evaluations are performed.

The second phase of the heuristic uses the transition graph built during the first phase to determine the likely subset of strategies that need to be evaluated to perform strategy selection. This subset is determined as the set of target strategies whose estimated transition probability is greater than $y\%$.

The accuracy of this heuristic depends on two factors: (i) the value of y and (ii) the duration of phase 1.

Effect of y: We evaluated the accuracy and speedup of this heuristic for different values of y with a transition graph built from a week long trace. The results are presented in Table 3. As could be expected, the speedup of the heuristic increases when y increases, since less strategies are evaluated. At the same time, the accuracy decreases and AWCR increase. When $y = 10\%$, we obtain speedup of 12, with only a little loss in accuracy. Thus, *Transition* strategy drastically reduces the simulation cost making accurate strategy selections.

Effect of the length of Phase 1: We evaluate the accuracy of the heuristic for different durations of Phase 1, with y fixed at 10%. Results are given in Table 4. As seen from the table, a transition graph built out of just 5-day traces for each object already leads to an accuracy of 90.7%. Increasing the size of transition graph leads to better accuracy, however the gain stabilizes around 7 days. This corresponds to 6 full evaluations performed over thousands of documents.

An important issue in using *Transition* is to determine how often to rebuild the transition graph. In our experiments, we did not observe a degradation in accuracy for

Table 3. Performance of *Transition* for different values of y

y%	Accuracy	Speedup	AWCR
5%	97.8%	8	0.01
10%	95.5%	12	0.03
15%	93.4%	13	0.06
20%	92.2%	15	0.08
25%	88%	15.5	0.10

Table 4. Length of phase 1 vs. accuracy

No. of days	Accuracy	AWCR
5	90.7%	0.09
7	95.5%	0.03
9	96%	0.03
14	96%	0.03

9 month traces with a transition graph built from one week. Hence, we feel that this transition graph needs to be rebuilt only rarely.

One of the shortcomings of this heuristic is its inability to handle emergencies like flash crowds as the pre-built transition graph does not cover such drastic changes in access patterns. Such scenarios might call for fast detection of the onset of emergencies (also determining the access patterns of emergency) and triggering a strategy selection mechanism that evaluates only a small set of candidate strategies that can perform well in the given scenario.

5. Related work

A large number of proposals have been made in the past to improve the quality of Web services. Cache consistency protocols such as Alex [3] and TTL policies aim to improve the scalability of the Web. Invalidation strategies have been proposed to maintain strong consistency at relatively low cost in terms of delay and traffic [2]. Several replication strategies have been proposed in the past. Radar uses a dynamic replication protocol that allows dynamic creation/deletion of replicas based on the clients' access patterns [9]. In [6], the authors propose replication protocols that determine the number and location of replicas to reduce the access latency while taking the server's storage constraints into account. All these systems adopt a single strategy or single family of strategies for all documents, possibly with a tunable parameter. However, our earlier work advocated the simultaneous use of multiple strategies.

A number of systems select strategies on a per-document basis. In [1], the authors propose a protocol that dynamically adapts between variants of push and pull strategies on a per-document basis. Similarly, in [4] proposes an adaptive lease protocol that switches dynamically between push and pull strategies. Finally, in [5], the author propose a protocol that chooses between different consistency mechanisms, invalidation or (update) propagation, on a per-document basis, based on the document's past access and update patterns. These protocols perform a per-document strategy selection similar to ours but they are inherently limited to a small set of single family of strategies (e.g., push or pull, invalidation or propagation) and cannot incorporate different families of strategies as done in our system.

6. Conclusions and Future Work

The need for dynamically selecting a strategy on a per-document basis was shown in our earlier research. In this paper, we demonstrate the need for continuous dynamic adaptation of strategies with experiments performed on the traces of our department Web server. We find that continuous adaptation is beneficial even for seemingly relative stable access patterns. As a second contribution of this paper, we proposed two heuristics to perform this continuous adaptation with reduced simulation cost. In our experiments, we find that the *Transition* heuristic performs better than its simple counterpart, both in terms of accuracy and speedup. We conclude that, in our traces, evaluating strategies based on their past transition patterns is a good solution that yields high speedup with high accuracy. Another way of reducing simulation overhead would be to perform *object clustering*, i.e., to group objects with similar access and update patterns. We are also investigating schemes to handle emergency situations such as flash crowds in our system.

References

[1] M. Bhide, P. Deolasee, A. Katkar, A. Panchbudhe, K. Ramamritham, and P. Shenoy. Adaptive push-pull: Disseminating dynamic Web data. *IEEE Transactions on Computers*, 51(6):652–668, June 2002.

[2] P. Cao and C. Liu. Maintaining strong cache consistency in the world wide Web. *IEEE Transactions on Computers*, 47(4):445–457, Apr. 1998.

[3] V. Cate. Alex – a global file system. In *USENIX File Systems Workshop*, pages 1–11, May 1992.

[4] V. Duvvuri, P. Shenoy, and R. Tewari. Adaptive leases: A strong consistency mechanism for the world wide Web. In *19th INFOCOM Conference*, pages 834–843. IEEE Computer Society Press, Mar. 2000.

[5] Z. Fei. A novel approach to managing consistency in content distribution networks. In *6th Web Caching Workshop*, pages 71–86, June 2001.

[6] J. Kangasharju, J. Roberts, and K. W. Ross. Object replication strategies in content distribution networks. In *6th Web Caching Workshop*, June 2001.

[7] G. Pierre, I. Kuz, M. van Steen, and A. S. Tanenbaum. Differentiated strategies for replicating Web documents. *Computer Communications*, 24(2):232–240, Feb. 2001.

[8] G. Pierre, M. van Steen, and A. S. Tanenbaum. Dynamically selecting optimal distribution strategies for Web documents. *IEEE Transactions on Computers*, 51(6):637–651, June 2002.

[9] M. Rabinovich and A. Aggarwal. Radar: A scalable architecture for a global web hosting service. *Computer Networks*, 31(11–16):1545–1561, 1999.

LINK PREFETCHING IN MOZILLA: A SERVER-DRIVEN APPROACH

Darin Fisher[1] and Gagan Saksena[2]
[1]*IBM*, [2]*AOL*

Abstract This paper provides a synopsis of a server-driven link prefetching mechanism that we have designed and implemented for the Mozilla web browser, a popular Open Source web browser. The mechanism depends on the origin server or an intermediate proxy server determining the best set of documents for the browser to prefetch. The browser follows prefetch directives provided by the server, either embedded in an HTML document using the `<LINK>` tag or specified via `Link` HTTP response headers. The browser determines when best to prefetch the specified URLs based on its own heuristics. In this paper, we describe the mechanism and discuss some of the practical issues that impacted its design and implementation.

1. Introduction

Web caching is a common optimization used by web browsers to reduce latency for the user when they re-visit a web page. For narrow-band Internet connections, the performance gain can be dramatic. It is of interest therefore to consider mechanisms that would enable the web browser to pre-populate its cache of documents, such that when the user initially views a web page, the latency will be nearly reduced to that of loading the web page from the local cache. This technology is commonly referred to as prefetching and has been shown to be effective in reducing latency in the Web [9].

Despite the potential benefit of prefetching, it has not been widely deployed at this time. In addition to the family of Mozilla-based browsers, only a few production web browsers (e.g., WebTV and iCab) and personal web proxies (e.g., NetSonic Pro, Naviscope, and Wcol) support some form of link prefetching. The methods of prefetching vary considerably between products. For example, some of these products analyze the contents of pages or the user's browsing history to predict what the user might visit next. Indeed, a large body of research covering a variety of approaches to prefetching exist on this topic [2–4, 8, 10, 11]. Other applications rely on some cooperation between browser and server.

In addition, content authors have devised *ad-hoc* JavaScript-based techniques to prefetch content. These generally involve loading content into hidden `<IMG>` or `<IFRAME>` HTML elements so as to pre-populate the browser's document cache.

F. Douglis and B.D. Davison (eds.), Web Content Caching and Distribution, 283-291.

While these JavaScript-based approaches have the benefit of working with most browsers, they often result in complex JavaScript code that is difficult to maintain. With such techniques, the browser is not able to prioritize normal requests above prefetch requests. As a result prefetch requests may end up competing with normal requests for network bandwidth. Additionally, JavaScript-based approaches afford the server little opportunity to control traffic and limit server load resulting from prefetch requests.

In this paper we describe a link prefetching mechanism that involves modifications to both the browser and the server. It is designed to provide a solution to prefetching that is "friendlier" to both client and server, specifically addressing some of the shortcomings of existing *ad-hoc* methods. The browser follows special directives from the web server (or proxy server) that instruct it to prefetch specific documents. The server can inject these directives into an HTML document or into the HTTP response headers. The browser responds to the prefetch directives after it finishes rendering the document (or HTTP response) containing the directives. This mechanism allows servers to control precisely what is prefetched by the browser, and it allows the browser to determine when best to prefetch documents based on local network inactivity, browser idle time, etc.

We chose a server-driven approach to link prefetching because we believe that servers (including intermediate proxy servers) can better predict what users are likely to visit next. Since pages often contain numerous hyperlinks that the user is unlikely to follow, it is generally difficult for the browser to determine the best subset of hyperlinks to prefetch. For example, an origin server or content author may know that the user is browsing through a series of large images or documents (e.g., an image slideshow or pages in a book) in which case it would be advantageous for the browser to prefetch the next image or document in the series. Or, while the user is confronted with a login prompt, the site might like to have the browser prefetch the portions of the site that do not require user authentication. In such cases there is a static relationship between the previous and the next document, and static prefetch directives can be encoded into the documents to indicate what should be prefetched next. In addition to static prefetch directives, a server might dynamically inject prefetch directives into outbound HTTP responses for the benefit of a link prefetching enabled browser. For example, the server may want to instruct the browser to prefetch the most popular URLs on a given day. Because this technique works with HTTP headers, a proxy can very well generate such dynamically determined directives for popular links.

What follows is a description of the link prefetching mechanism along with a number of practical issues that shaped our design and implementation of link prefetching for the Mozilla browser, a popular Open Source web browser. We added our implementation to version 1.2 and 1.0.2 of the Mozilla browser. It is also included in Netscape 7.01 and other recent Mozilla-based browsers.

2. Prefetching Directives

The HTML 4.01 specification [13] briefly mentions an opportunity for browsers to prefetch URLs indicated by content authors via the `<LINK>` tag. The description of the "next" link relation type (from section 6.12 of [13]) says the following:

> [The "next" link relation type] refers to the next document in a linear sequence of documents. User agents may choose to pre-load the "next" document, to reduce the perceived load time.

We have implemented this mechanism in the Mozilla browser along with some variations on this same approach, which are described below. The WebTV browser also follows this recommendation [12]. (Our prior approach involved extending all linked tags with a prefetch marker attribute [16] and using *pathfiles* [15] to prefetch dynamic content. That technique had its limitations and has since evolved into our current implementation.)

Motivated by the fact that the "next" relation is intended to represent the "next" document in a linear sequence of documents, and given that web pages should be able to specify multiple documents to be prefetched, we chose to have Mozilla additionally recognize the link relation type "prefetch" to indicate a document that should be prefetched. This link relation type currently is not standardized; however, the XHTML 2.0 draft specification [1] declares the "prefetch" link relation type.

Examples of these two types of link prefetching HTML directives follow:

```
<LINK rel="next" href="next.html">
<LINK rel="prefetch" href="big.jpg">
```

Suppose `next.html` contains an `<IMG>` tag, which would cause `big.jpg` to be loaded. These tags might be found in a document in which `next.html` is likely to be the next document visited by the user. As a result of these tags, the browser would silently prefetch the two documents while the user is perhaps busy reading through the current document.

Link prefetching directives specified solely in the body of a document have limited utility in solving problems in which content to be prefetched might be dynamically determined by the origin server or an intermediate proxy server. To support such configurations, Mozilla additionally recognizes link prefetching directives specified via the `Link` HTTP header. The `Link` HTTP header appears in section 19.6.2.4 of RFC 2068 [5]; however, the updated version of the HTTP 1.1 standard, RFC 2616 [6] does not define it. Use of this header, for example to enable servers to apply style sheets to a group of documents, is described in section 14.6 of the HTML 4.01 specification [13]. The link prefetching example from above could be equivalently written as follows:

```
Link:  <next.html>; rel="next"
Link:  <big.jpg>; rel="prefetch"
```

Note that as with many HTTP response headers, the `Link` header may also appear in a `<META>` HTML tag via the `http-equiv` attribute. This is described in section 7.4.4 of the HTML 4.01 specification [13].

We recommend the HTTP-level `Link` header mechanism above the other mechanisms since it allows proxy servers to more easily intercept prefetch directives and, for example, process them on behalf of a browser that does not support link prefetching. Proxy servers could also remove prefetch directives from the headers before sending them to its clients as a way of limiting the prefetching done by its clients. This might be desirable if the clients connect to the proxy over a high bandwidth connection such that prefetching between client and proxy would have little benefit.

3. Determining When To Prefetch

Given these link prefetching directives, it is up to the browser to determine when best to load the specified documents. This can be based on any number of heuristics chosen by the browser. For documents already in the browser's cache, the browser need not prefetch the document; however, if those documents have expired, the browser may choose to validate the cached documents. For documents the browser chooses to prefetch, an obvious heuristic is to defer prefetching until the browser's network connection is otherwise underutilized. Determining such a state of the user's system is in general a complex problem and one that requires special knowledge of the underlying operating system. APIs to access such information generally vary from platform to platform and are non-existent on some platforms. A simplistic approximation to this, and indeed what Mozilla implements, is to defer prefetching until the browser is idle (i.e., when the browser has loaded all content on the page) and to stop prefetching when the user, or JavaScript on a page, directs the browser to load something else.

However, this approach ignores the problem of multiple applications on the user's system competing for network bandwidth. To minimize this problem the Mozilla browser prefetches documents sequentially, deferring the next prefetch until after the former has completed. While this is not an ideal solution, it minimizes the problem by reducing the prefetching application's demand on the operating system's networking subsystem.

Given that web pages are commonly composed of multiple HTML frames, the browser may see link prefetching directives from one or more of the frames. Therefore, the Mozilla browser does not begin prefetching until all frames, and any in-line content they contain, have finished loading. This approximately corresponds to the time at which the DOM's `onLoad` event handler for the HTML `<BODY>` tag would fire for the top-most frame in a frame hierarchy.

As stated above, the intent of link prefetching is to improve the user's perception of how quickly pages load. It is therefore paramount that link prefetching not interfere with normal user activity. Indeed, when the user clicks on a link, or when JavaScript executing on the page causes something new to be loaded, the browser must stop all link prefetching activity and respond to the new network request as soon as possible. For link prefetching this creates a situation in which the browser may be left with partially prefetched documents. Moreover, if the browser was utilizing a persistent connection for prefetching, the connection would have been closed.

Mozilla's support for partial cache entries and byte-range requests under HTTP/1.1 [6] to complete partial cache entries helps to minimize the impact of aborted prefetch attempts. Moreover, the serialization of prefetch requests minimizes the loss of persistent connections to at most one. This is important since browsers commonly keep at most two persistent connections open per origin server, as recommended in section 8.1.4 of RFC 2616 [6], and four per proxy server (which is how most common web browsers are configured).

As a result, if the link clicked by the user happens to be to the same host as the prefetch requests, chances are good that there will be a persistent connection in the browser's idle connection list available for immediate use. In such cases, the loss of a persistent connection due to an aborted prefetch request is minimized. We assert that the cost is justified by the possibility of correctly pre-populating the browser's cache, even if only with a partial document.

4. Browser Imposed Restrictions

Given that link prefetching is purely a mechanism to pre-populate the browser's cache, the expiration time calculation for prefetched documents must still adhere to the HTTP caching rules as described in section 13.2 of RFC 2616 [6]. This means that the server should only direct the browser to prefetch documents that have a non-zero freshness lifetime. Since the browser must honor the server specified freshness lifetime, it is crucial that the server specify a reasonable interval of time within which the user can view the prefetched document without having to wait for the document to be validated or reloaded by the browser. As a result of this requirement, the Mozilla browser will abort a prefetch if it determines from the HTTP response headers that the resulting document would not be reusable from the cache without a round trip to the server.

The `<LINK rel="next">` prefetching directive creates an interesting problem for the browser. Given that this header is already in use on the Internet for the purposes of site navigation, the authors of the original content may not have considered the possibility of these documents being prefetched. Therefore, it is likely that documents referenced by this tag may not be reusable from the browser's cache. Sites with dynamic content would be excessively burdened by a browser attempting to prefetch references to dynamically generated content.[1] In an attempt to minimize this problem, the browser will not prefetch documents referenced by an URL with a query string when the URL appears in a `<LINK rel="next">` tag. While this is clearly a sub-optimal solution, since certainly any URL could be served up with a zero freshness lifetime, this does catch a large number of instances where prefetching would be of no value. The same restriction is not applied to the "prefetch" relation.

The Mozilla browser further restricts prefetching to only `http://` URLs. It will not prefetch other protocols, with HTTPS being the significant protocol excluded. While HTTPS page load times are typically much higher than those for the same content delivered over raw HTTP, there are a number of compelling reasons not to prefetch HTTPS content. The main reason is that we do not want to load secure content without the user's explicit knowledge. The lock icon found in most browsers is an example of this philosophy. Moreover, for most browser implementations secure content is only

[1]Bugzilla (`http://bugzilla.mozilla.org/`), the Mozilla bug tracking system, serves dynamically generated, non-cacheable bug reports via CGI with `<LINK rel="next">` references to the next bug report. The URLs include a query string.

stored in the browser's memory cache (for privacy reasons), which is typically very limited in size.

An issue that has been widely debated is whether or not the browser should be permitted to prefetch documents from a different domain. The Mozilla browser allows for this, and the main reason for doing so is for parity with existing *ad-hoc* prefetching mechanisms. Moreover, the ability of one site to cause content from another site to be prefetched is no more or less of a user privacy concern than loading hidden images from another domain or JavaScript's ability to cause content to be loaded without the user's knowledge. However, in light of this issue, the browser offers the user the option to disable prefetching just as it offers the user the option to disable JavaScript. This is of course a heavy-handed solution to the problem of cross-domain prefetching, and other browsers might choose to offer the user more fine-grained control.

Finally, there is the issue of recursion and recursion depth. We took a simplified approach here by limiting the browser to only recognize prefetch directives in documents loaded explicitly by the user. This means that prefetch directives included in the HTTP response headers of a prefetched document will not be followed by the browser until the browser is directed by the user to load the prefetched document. While there is utility for recursive prefetching, we felt that it was prudent to limit the depth of prefetching to one level for simplicity. We hope to investigate recursive prefetching for the browser in the future. This may be a particularly useful approach for proxies prefetching from origin servers.

5. Identifying Prefetch Requests

It is often valuable for servers to track the origin of requests. The `Referer` HTTP request header (see section 14.36 of RFC 2616 [6]) is commonly used for this purpose. For the server's benefit, the browser may wish to distinguish prefetch requests from normal user initiated requests. The Mozilla browser sends the following additional request header with every prefetch request:

```
X-moz:  prefetch
```

Note that the name and format of this header is not finalized. In the future this header (or one serving the same purpose) may carry additional information about the context of the prefetch. For example, it may be valuable to distinguish whether or not the prefetch directive came from a document or from a HTTP header. This would allow an administrator of a website with public or user-provided HTML content to block prefetch directives originating from said content.

6. Practical Limitations

One of the most significant detractors of link prefetching is the potential cost in lost or misused bandwidth. This affects both users and servers. For servers the impact can be very high if enough browsers are hitting servers with prefetch requests for documents that the users never chooses to load. For users who pay for Internet access by the number of bytes transfered (or more commonly by the number of integral megabytes), link prefetching can potentially end up costing the user money. While the majority of Internet users in the United States pay a flat rate or are billed by the number of hours they spend online, this is not necessarily the case in other countries.

Because of these practical problems, our approach to link prefetching has been a decidedly conservative one. While other approaches involve the browser or personal web proxy blindly prefetching embedded anchor tags, we felt that such a technique would not be acceptable in practice. Moreover, by making the server responsible for choosing what the browser prefetches, and by giving servers context enough to recognize a prefetch request, we feel that servers can properly limit the use of prefetching to instances where it is most beneficial.

7. Applications and Results

As a proof of concept we investigated some simple prefetching scenarios. The first involved a simple slide-show of images with each new page containing two new images. Page size was between 17Kb and 18Kb, and image size was between 10Kb and 20Kb. Each page included prefetch directives for the next page and its images. With a modest connection speed of 128Kb/sec (typical upstream transfer rate of an ADSL connection), each full page took on average 3 seconds to load over the network, and only 300 milliseconds to load out of the browser's cache. This seems to be a case where prefetching is likely to help significantly, given that the user might very well spend 3 seconds looking at each page.

A second usage case involved taking the results page of a typical WWW search engine query and modifying it manually to have the browser prefetch the top-ranked pages when loading the search results. This is something that a WWW search engine could easily implement.

One thing we noticed with this approach is that it is somewhat common for the top-level page of a site to be served up non-cacheable. This seems to be used as a mechanism of tracking page views or rotating advertisements. This problem combined with the fact that we were not recursively prefetching content made prefetching in this scenario much less beneficial. Sites could be better designed (e.g., by making use of HTTP/1.1 cache control mechanisms that would trigger a revalidation on every visit instead of a fresh download) to avoid this problem while still enabling the tracking of individual page hits.

8. Conclusion

The link prefetching mechanism described in this paper is admittedly simplistic in a number of ways. In spite of this, it has proven to be an effective solution for a number of interesting applications and generally works well. In some cases, we've seen dramatic performance gains as a result of prefetching.

When this feature was added to Mozilla (in version 1.2, released November 2002), we received a lot of feedback, both positive and negative, from the web community. In most cases, the negative feedback came from folks concerned about increased load on their networks. However, the general response from the community was that a server-driven approach was the right approach because it puts the burden on servers to not misuse the network, and server administrators are generally cognizant of their network bandwidth utilization because of the associated monetary impact.

One of our immediate goals is to try to better inform the community about this mechanism and the implementations that exist. To this end, we have published a Mozilla Prefetching FAQ [7] on `www.mozilla.org` and there is an article on `devedge.netscape.com` [14] explaining how to make use of this link prefetching mechanism. The article includes a live example.

As for gauging the effectiveness of our link prefetching mechanism, it is a challenging problem, but we feel that there are opportunities for it where the gain is very apparent (e.g., Internet slide shows, login systems, etc.). It is likewise very apparent that it is inappropriate for other types of applications (such as whenever dynamic content is involved). Prefetching top-ranking search results may be an interesting application area; however, utility in the case of general websites is limited since many top-level pages of popular sites serve non-cacheable content.

As a future research direction, we hope to explore recursive prefetching, better browser-side heuristics for determining local network idleness, as well as mechanisms to order and prioritize link prefetching directives. The latter feature would potentially simplify the design of pages with many prefetch directives. As is, the order in which links are prefetched is unspecified, and indeed the order in which Mozilla prefetches documents may vary depending on a number of factors.

Acknowledgments

We would like to thank Prof. Brian D. Davison and Waldemar Horwat for their valuable feedback on this paper.

References

[1] J. Axelsson, B. Epperson, M. Ishikawa, S. McCarron, A. Navarro, and S. Pemberton. *XHTML 2.0 Working Draft*, May 2003. http://www.w3.org/TR/xhtml2/.

[2] J. Borges and M. Levene. Data mining of user navigation patterns. In *WEBKDD*, pages 92–111, 1999.

[3] B. D. Davison. Predicting web actions from HTML content. In *Proceedings of the The Thirteenth ACM Conference on Hypertext and Hypermedia (HT'02)*, pages 159–168, College Park, MD, June 2002.

[4] D. Duchamp. Prefetching hyperlinks. In *Proceedings of the Second USENIX Symposium on Internet Technologies and Systems (USITS '99), Boulder, CO*, Oct 1999.

[5] R. T. Fielding, J. Gettys, J. C. Mogul, H. Frystyk, and T. Berners-Lee. *RFC 2068: Hypertext Transfer Protocol, HTTP 1.1*, Jan 1997. Obsoleted by RFC 2616. http://www.ietf.org/rfc/rfc2068.txt.

[6] R. T. Fielding, J. Gettys, J. C. Mogul, H. Frystyk, L. Masinter, P. Leach, and T. Berners-Lee. *RFC 2616: Hypertext Transfer Protocol, HTTP 1.1*, Jun 1999. http://www.ietf.org/rfc/rfc2616.txt.

[7] D. Fisher. http://mozilla.org/projects/netlib/Link_Prefetching_FAQ.html.

[8] Z. Jiang and L. Kleinrock. An adaptive network prefetch scheme. In *IEEE Journal on Selected Areas in Communications, 16(3)*, pages 358–368, Apr 1998.

[9] T. M. Kroeger, D. D. E. Long, and J. C. Mogul. Exploring the bounds of web latency reduction from caching and prefetching. In *USENIX Symposium on Internet Technologies and Systems*, 1997.

[10] V. N. Padmanabhan and J. C. Mogul. Using predictive prefetching to improve World-Wide Web latency. In *Proceedings of the ACM SIGCOMM '96 Conference*, Stanford University, CA, 1996.

[11] J. E. Pitkow and P. Pirolli. Mining longest repeating subsequences to predict World Wide Web surfing. In *USENIX Symposium on Internet Technologies and Systems*, 1999.

[12] Prodigy. Designing for WebTV. http://pages.prodigy.net/guide/optimize/spec30.htm.

[13] D. Raggett, A. L. Hors, and I. Jacobs. *HTML 4.01 Specification*, Dec 1999. http://www.w3.org/TR/html4/.

[14] D. Rosenberg. Prefetching content for increased performance, Feb. 2003. http://devedge.netscape.com/viewsource/2003/link-prefetching/.

[15] G. Saksena. System and method for creating pathfiles for use to predict patterns of web surfing. US Patent 6,055,572, Netscape Communications Corp., Jan. 1998.

[16] G. Saksena. User configurable prefetch control system for enabling clients to prefetch documents over a network server. US Patent 6,023,726, Netscape Communications Corp., Jan. 1998.

A GENERALIZED MODEL FOR CHARACTERIZING CONTENT MODIFICATION DYNAMICS OF WEB OBJECTS

Chi-Hung Chi and HongGuang Wang
National University of Singapore

Abstract In this paper, we would like to investigate one of the most fundamental questions behind dynamic web caching, the pattern and modeling for the content modification dynamics of web objects. Since content providers cannot guarantee 100% accuracy about the life-span of a web object, a single-valued expire time is found to be too simple to describe its content modification dynamics for risk analysis to reuse copies of the data in the local cache. To capture the statistical features of the content modification dynamics of a web object, a novel mathematical quantity matrix is proposed. The three statistical metrics of measurement are: dynamic degree, predictability index, and safety bound for prediction. All our arguments and models are verified through active monitoring of the content change of real web objects.

1. Introduction

Current research efforts on dynamic web caching mainly focus on the object deposition for partial object caching [20] and on aggressive server-assisted revalidation [6] [9] [12] [17] [19] [23] [24] [25] [26], we would like to investigate one of the most fundamental questions behind dynamic web caching, the pattern and modeling for the content modification dynamics of web objects in this paper. It is observed that although the reusability of a web object should be driven by the content modification dynamics, current cacheability rules in proxy and explicit cache control tags in the HTTP header generated by the content servers are mainly determined by the content generation dynamics instead. Here, we refer the *content generation dynamics* as the frequency pattern of triggering of application execution in a web server to produce the final data sent out to a client upon receiving his web request and the *content modification dynamics* to the actual change pattern of the object content with respect to the previous version for the same request. Through our active monitoring of web object content on Internet, we observe that there is a big discrepancy between the content generation dynamics and modification dynamics of a web object. Using the former one to determine an object's cacheability and time-to-

F. Douglis and B.D. Davison (eds.), Web Content Caching and Distribution, 293-301.

live (TTL) is also found to be too conservative for content reuse, thus resulting in unnecessary network traffic.

To describe the content modification dynamics of a web object, we also argue that since its content provider cannot guarantee 100% accuracy about the object's life-span, reusing a cached object in proxy even within its TTL period actually involves risk of expired content. As a result, a single-valued expire time is definitely not enough for the proxy cache to make good content reuse decision. In this paper, we would like to address this problem through our novel mathematical quantity matrix to capture the statistical features of the content modification dynamics of a web object. The three key statistical metrics of study are the dynamic degree, predictability index, and the safety bound for prediction. With all these information, a client (either proxy or browser) can estimate the risk of content reuse in its local cache/disk; he can also balance the cost of validating (with retrieval, if necessary) the data with the original content server and the risk level of content inaccuracy that he can accept. In our study, we verify all our arguments and models through active monitoring of the content modification dynamics of real web objects on Internet.

To help our discussion in the paper, we define the following terms precisely here:

- *Life-Span of a Web Object*

  It refers to the time period during which the content of an object is “fresh” and is valid to be used.

- *Time-to-Live (TTL) of a Web Object*

  It refers to the time period during which a system server (such as proxy or browser cache) can use the content of a copy of an object it stores without contacting the original content server.

Note that while the life-span depends solely on the content nature, the TTL depends on many other factors such as server storage policy, and requirements for client behavior monitoring.

2. Related Work

To improve the performance of web caching for dynamic objects, researchers also investigate various criteria to decompose a web object into static and dynamic fragments for partial object caching. These include the markup language support (such as ESI [1] [13] [18] [20] [22]), dynamic template techniques [2] [3] [4] [10], and “reverse proxy-like” systems in front of the database and web server (e.g. XCache) [8] [11] [21]. There are also studies on the possibility of caching web query pages in proxy [14] [15] [16].

In web information system, the interest on the dynamics of content modification of web objects is due to the huge amount of effort to maintain the most recent information content in the system. This is particularly important to search engines because crawling over the Internet might take weeks to finish. Brewington modeled the change of web content as a renewal process. He proposed an up-to-date measure for indexing a large set of web resource. While his model can help to reduce the bandwidth usage of web crawlers, the relative dynamic nature of contents in the monitored population is not available. Cho [5] proposed and compared several estimators for the modification frequencies of web pages. In his model, he assumed that the change of a web resource can be modeled by a Poisson process, which might or might not always be acceptable [7]. Another limitation of his study is the monitoring time interval, which is chosen to be daily. Such a large time interval is too long to capture the essential modification pattern of web content for proxy caching.

3. Modeling Content Modification Dynamics of Web Objects

In this section, we would like to propose a new quantity matrix with multiple metrics to capture the key features of the life-span distribution of web objects.

3.1 Quantity matrix for content modification dynamics

There are many features that are related to the content modification dynamics of web objects. A quantity matrix is a measurement methodology to describe its key features so that effective content delivery services and proxy caching can be based on. In this paper, we propose three measurement metrics for the quantity matrix for web content modification dynamics. They are the *dynamic degree* for the expectation value, *predictability index* for the spreading variance, and the *safety bound* value for prediction. These features will be defined based on the distribution of the probability distribution function of the life-span of web object content.

- *Dynamic Degree*

  This parameter is to measure the central tendency of the life-span of an object. It can be determined based on the mean, median, or mode value of the PDF of an object's life-span value. Most likely, this value will be used directly in the prediction of the life-span of an object. To simplify its usage, the domain of this value ranges from 0 to 100, with the maximum value referring to the most rapid modification dynamics.

- *Predictability Index*

  This index is to describe the likelihood for the life-span's PDF to be aggregated around the expectation value. It is used to measure the spreading variance of the life-span's PDF. If the life-span distribution aggregates near the centre or expectation value, it will be appropriate to use the centre value as the predicted value for the next life-span. Here, we define the domain of this predictability index to range from 0 to 1, with larger value implying better predictability.

- *Safety Bound Value*
  This parameter is related to the boundary feature of the life-span value. Since the life-span only distributes over a certain range of values, the lower bound of the range can be viewed as the safety bound for prediction. It is safe because all the previously monitored life-span values always pass beyond this boundary.

3.2 Calculation of quantity matrix

With the three key metrics defined in the last section, we would like to define their calculation in this section. Below are the formulas we propose. Note that while there are still room for fine-tuning, the basic concept and idea are captured in the formulas. Validation of the result of the formulas will be given in Section 4 to support our proposal here.

Assume that the PDF of the life-span value of a web object k , $PDF(Obj_k)$ is given by $f(\Delta t)$, where Δt is the life-span between two consecutive content changes:

$$Dynamic\ Degree\ (Obj_k) = \frac{1}{\Delta T_{Typical} + 1}$$

$$Predictability\ Index\ (Obj_k) = \frac{\Delta T_{Typical}}{\sigma(\Delta T_{Typical}) + \Delta T_{Typical}}$$

$$Safety\ Bound\ Value\ for\ Prediction\ (Obj_k) = \Delta T_{Min}$$

where ΔT_{Min} is the lower bound value for the non-zero $PDF(Obj_k)$, $\Delta T_{Typical}$ is the expectation value to represent the central position of $PDF(Obj_k)$, and $\sigma(\Delta T_{Typical})$ is the standard derivation of $f(\Delta t)$ relative to $\Delta T_{Typical}$. In our model, the content modification dynamics of a given web object is represented by the probability distribution function $f(\Delta t)$ of the life-span values of the object. The modification interval (or life-span), Δt, is defined by the time period between two consecutive content change of the object. Once $f(\Delta t)$ is known, the three key metrics for the content modification dynamics of a web object given above can be found.

The first metric, the dynamic degree, is a measure to quantify the central tendency of $f(\Delta t)$. Possible statistical values to describe the central tendency of $f(\Delta t)$ include the mean, mode, and median values. The choice of selection depends on the domain of web applications under study as well as the statistical features of interest. For example, the median value is usually chosen when $f(\Delta t)$ is skew; the mode value is usually used to represent the peak position of a typical "mountain-like" distribution. In our definition of dynamic degree, we purposely use the term "central tendency" to give flexibility to the model usage. In the above formula, the central tendency is denoted by the symbol $\Delta T_{Typical}$.

Being a quantitative measurement parameter, the dynamic degree is a non-negative value ranging from 0 to 100. A larger value of dynamic degree corresponds to a smaller value of the life-span of an object; a smaller value of the dynamic degree implies that the object of study is very static. In other words, this dynamic degree is inversely proportional to (or a reciprocal function of) $\Delta T_{Typical}$. Under the active monitoring process for the life-span history of a web object, there is an intrinsic limitation on its resolution. Within one cycle period of monitoring, only one life-span value can be reported, independent of the actual value of the modification frequency. With the typical life-span ranging from zero to infinite, the formula for the dynamic degree will give values from 0 to 100.

The second metric, the predictability index, is a measure for the confidence level for the expectation value of the distribution to be the next life-span value. It tries to reflect the spreading variance of the PDF. The value of the predictability index ranges from 0 to 1, with larger value corresponding to a smaller derivation from the expectation value of the distribution. This implies that the predictability index is inversely proportional to $\sigma(\Delta T_{Typical})$. If two distributions have the same derivation, the less dynamic one should also be more predictable. Hence, a better measure will be $\sigma(\Delta T_{Typical})/\Delta T_{Typical}$ instead of $\sigma(\Delta T_{Typical})$. Furthermore, it is possible for the derivation to be zero, and this might map to the predictability index of infinity. To map the predictability index of this extreme situation to 1, we propose to shift the denominator by 1. The other extreme situation is when the derivation is very large. This will cause the predictability index to approach zero, which implies that the expectation value of the distribution is not suitable to be used as the next predicted life-span value.

The last metric, the safety bound for prediction, is another measure for the spreading variance of the PDF of the life-span value of a web object. Based on the history of previous life-span values, it covers the range of the distribution within which all monitored life-span values actually occur. Thus, its lower bound gives the minimum predicted life-span value that an object is still fresh whereas its upper bound gives the maximum life-span value that the object is likely to be stale. In proxy caching, researchers are more interested in the lower bound value than the upper bound value because the former one is directly related to the accuracy of the TTL setting of an object. As a result, we take the lower bound value as the safety prediction value in our model and represent it by ΔT_{Min}. Since the life-span of an object is non-negative, ΔT_{Min} (or the safety bound value for prediction) is also non-negative.

About the probability distribution function of the life-span of web objects, statistical approach is recommended. In statistics, the histogram of a random variable is usually chosen to summarize the distribution graphically. Curve fitting of the histogram will then be used to get the proper distribution model. While there can be multiple distributions that can satisfy our need, we found that the gamma distribution is actually a fairly good choice [ChZ03] (and results in the later part of this section also support our argument here). Let us assume the PDF of the life-span of a web object be a gamma distribution and our monitoring cycle period is one minute, we have the following:

$$f(\Delta t) = Gamma(\Delta t; \alpha, \beta, \mu) = \frac{(\Delta t - \mu)^{\alpha-1} e^{-(\Delta t - \mu)/\beta}}{\beta^{\alpha}\Gamma(\alpha)}$$

where α s the shape parameter, μ is the location parameter, β is the scale parameter, and $\Gamma(\alpha)$ is the gamma function with the following formula:

$$\Gamma(\alpha) = \int_0^{\infty} \Delta t^{\alpha-1} e^{-\Delta t} \partial \Delta t$$

And the statistics to characterize the central tendency and the derivation of gamma distribution are:

$$\Delta T_{Typical} = Mode(Gamma(\Delta t; \alpha, \beta, \mu)) = (\alpha - 1)\beta + \mu$$

$$\sigma(\Delta T_{Typical}) = \sqrt{(\alpha + 1)\beta^2}$$

Substituting these values into the definitions of the quantity metrics gives:

$$\textit{Dynamic Degree} = \frac{100}{\alpha\beta + \mu - \beta + 1}$$

$$\textit{Predictability Index} = \frac{(\alpha - 1)\beta + \mu}{(\alpha + \sqrt{\alpha + 1} - 1)\beta + \mu}$$

$$\textit{Safety Bound Value for Prediction} = \mu$$

The above calculation for the quantity matrix uses the mode value as the central location for the life-span distribution. If we use the mean value instead of the mode value, the formulas will be transformed into the followings:

$$\Delta T_{Typical} = E(\Delta t) = \alpha\beta + \mu$$

$$\sigma(\Delta T_{Typical}) = \sqrt{\alpha\beta^2}$$

$$\textit{Dynamic Degree} = \frac{100}{\alpha\beta + \mu + 1}$$

$$\textit{Predictability Index} = \frac{\alpha\beta + \mu}{(\alpha + \sqrt{\alpha})\beta + \mu}$$

$$\textit{Safety Bound Value for Prediction} = \mu$$

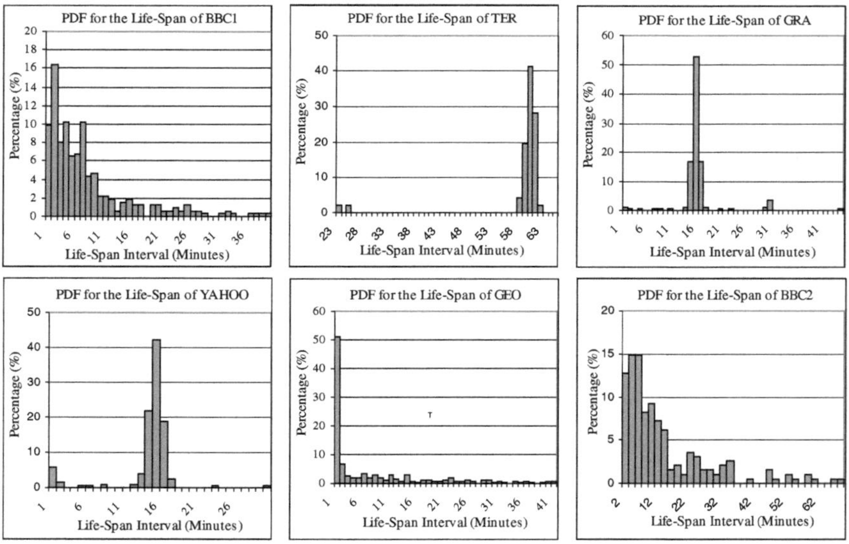

Figure 1: Probability Distribution Function for the Life-Span of Six Sample Objects

4. Verification of Quantity Matrix for Content Modification Dynamics of Objects

In this section, we would like to use real data to verify our proposed quantity matrix for content modification dynamics given in the last section. With consistent results obtained for hundreds of objects under active monitoring for content change, six typical objects are randomly selected from our dataset for presentation here and their data are used for curve fitting. To study the correctness of the model, we calculate the mean square error (MSE) between the observed and predicted values in each monitoring interval. If the MSE is less than or equal to 10^{-3} level, it is generally agreed that the fitted curve can approximate the actual distribution of the life-span's PDF of the observed object. In our curve fitting process, Gamma distribution is used as the basic curve family. And the parameters of the curve are determined by fine-tuning the value of the maximum likelihood estimation (MLE).

The PDF of the life-span values for each sample object is shown in Figure 1, with the parameters of its associated best-fitted gamma distribution given in Table 1. The table shows that the Gamma distribution is indeed a reasonable PDF to describe the content modification dynamics of web objects. With proper curve-fitting, all the mean square errors are about one order of magnitude less than the threshold 10^{-3}. The values for the three key metrics of the quantity matrix are also given in Table 2. The table shows that although BBC1 and GEO are more dynamic than the others, their predictability indexes are actually the worst. On the contrary, TER is the least dynamic but it is the most predictable one. It also has the largest safety bound for life-span prediction. Another interesting observation is that although YAHOO has similar dynamic degree as GRA does, its higher predictability index suggests its less variance nature. Finally, the typical life-span values of the six sample URLs are also given in the last column of Table 2. These values can be approximated to be the TTLs of the objects for proxy caching

Table 1. Parameters of the Best-Fitted Gamma PDFs for the Life-Span of Six Sample Objects

Sample URL	Curve Fitting Parameters			Mean Square Error of Curve Fitting
	Theta (μ)	Shape (α)	Scale (β)	
BBC1	0	1.75	2.7	$2.50*10^{-4}$
TER	57	12	0.273	$1.52*10^{-4}$
GRA	12	17	0.18	$2.19*10^{-4}$
YAHOO	11	17	0.24	$3.37*10^{-4}$
GEO	0	1.78	0.77	$9.21*10^{-4}$
BBC2	0	1.5	6.4	$1.35*10^{-4}$

Table 2. Values of the Three Key Metrics for the Content Modification Dynamics of the Six Sample Objects

Sample URL	Quantity Matrix			$\Delta T_{Typical} = (\alpha-1)\beta+\mu$
	Dynamic Degree	Predictability Index	Safety Bound for Prediction (μ)	
BBC1	33.00	0.3120	0	2.03
TER	1.64	0.9839	57	60.00
GRA	6.30	0.9512	12	14.88
YAHOO	6.31	0.9358	11	14.84
GEO	62.50	0.3185	0	0.60
BBC2	23.81	0.2402	0	3.2

5. Conclusions

In this paper, we address one of the most fundamental questions in dynamic web caching, the modeling for the content modification dynamics of web objects. With detail monitoring data of the content change of web objects, we show that the cost of object validation without object body fetching is actually not cheap. Current simple heuristics to approximate content generation dynamics to content modification dynamics are also found to be quite inefficient – the penalty is either the unnecessary consumption of network bandwidth or the potential of retrieving outdated object content. Instead of the single-valued expire time, we propose a quantity matrix to quantify the content modification dynamics of web objects. All the claims and insights are supported and verified with detail experimental data. These results are very important because it opens new ways to improve proxy caching and to reduce efforts to monitor web objects for content updating.

References

[1] Anton, J., Jacobs, L., Liu, X., Parker, J., Zeng, Z., Zhong, T., Web caching for database applications with Oracle Web cache, *Proceedings of the ACM SIGMOD International Conference on Management of Data*, June 03-06, 2002.

[2] Brabrand, C., Moller, A., Olesen, S., Schwartzbach, M.I., Language-based caching of dynamically generated HTML, *Journal of World Wide Web*, 5(4):305-324, 2002.

[3] Challenger, J., Iyengar, A., Dantzig, P., A scalable and highly available system for serving dynamic data at frequently accessed Web sites, *Proceedings of ACM/IEEE Supercomputing*, 1998.

[4] Challenger, J., Iyengar, A., Witting, K., Ferstat, C., Reed, P., A publishing system for efficiently creating dynamic Web content, *Proceedings of the INFOCOM*, pp. 844-853, 2000.

[5] Cho, J., Garcia-Molina, H., Estimating frequency of change, Technical Report, Stanford University, 2000, http://dbpubs.stanford.edu:8090/pub/1999-22.

[6] Cohen, E., Kaplan, H., Refreshment policies for Web content caches, Proceedings *of the IEEE INFOCOM Conference*, 2001.

[7] Crovella, M., Bestavros, A., Self-similarity in World Wide Web traffic: Evidence and possible causes, *IEEE/ACM Transactions on Networking*, 5(6):835-846, 1997.

[8] Daniela, R., Iyengar, A., Dias, D., Web proxy acceleration, *Journal of Cluster Computing*, 4(4):307-317, October 2001.

[9] Dilley, J., Arlitt, M., Perret, S., Jin, T., The distributed object consistency protocol: Version 1.0, Technical Report HPL-1999-109, Hewlett-Packard Laboratories, Palo Alto, CA, 1999.

[10] Douglis, F., Haro, A., Rabinovich, M., HPP: HTML Macro-Preprocessing to support dynamic document caching, *Proceedings of the USENIX Symposium on Internet Technologies and Systems*, December 1997, pp. 83-94.

[11] Dutta, K., Datta, A., Meer, D.V., Suresha, Thomas, H., Ramamritham, K., Proxy-based acceleration of dynamically generated content on the World Wide Web: An approach and implementation, *Proceedings of ACM SIGMOD*, June 2002.

[12] Duvvuri, V., Shenoy, P., Tewari, R., Adaptive leases: A strong consistency mechanism for the World Wide Web, *Proceedings of IEEE INFOCOM*, 2000.

[13] Liu, X., Developing high performance applications with Oracle 9i as Web cache and ESI, Oracle Corp., 2002.

[14] Luo, Q., Krishnamurthy, S., Mohan, C., Pirahesh, H., Woo, H., Lindsay, B., Naughton, J.F., Middle-tier database caching for e-business, *Proceedings of SIGMOD,* 2002.

[15] Luo, Q., Naughton, J.F., Krishnamurthy, R., Cao, P., Li, Y., Active query caching for database Web servers, *Proceedings of WebDB,* 2000.

[16] Luo, Q., Naughton, J.F., Form-based proxy caching for database-backed Web sites, *Proceedings of VLDB,* 2001.

[17] Ninan, A., Kulkarni, P., Shenoy, P., Ramamritham, K., Tewari, R., Cooperative leases: Mechanisms for scalable consistency maintenance in content distribution networks, *Proceedings of World Wide Web Conference*, May 2002.

[18] Rabinovich, M., Xiao, Z., Douglis, F., Kalmanek, C., Moving edge side includes to the real edge: the clients, *Proceedings of the 4th USENIX Symposium on Internet Technologies and Systems*, March 2003.

[19] Tewari, R., Niranjan, T., Ramamurthy, S., WCDP: A protocol for Web cache consistency, *Proceedings of the 7th WCW*, Boulder, Colorado, August 2002.

[20] Tsimelzon, M., Weihl, B., Jacobs, L., ESI language specification 1.0, 2000, http://www.esi.org.

[21] Xcache.http://www.xcache.com/home/default.asp?c= 32&p=165

[22] Xlinclude. http://www.w3.org/TR/xinclude

[23] Yin, J., Alvisi, L., Dahlin, M., Lin, C., Volume leases for consistency in large-scale systems, *IEEE Transactions on Knowledge and Data Engineering*, 11(4):563-576, 1999.

[24] Yin, J., Alvisi, L., Dahlin, M., Iyengar, A., Engineering server driven consistency for large scale dynamic Web services, *Proceedings of the Tenth International World Wide Web Conference*, May 2001.

[25] Yin, J., Alvisi, L., Dahlin, M., Iyengar, A., Engineering Web cache consistency, *ACM Transactions on Internet Technologies*, 2(3), August 2002.

[26] Yu, H., Breslau, L., Shenker, S., A scalable Web cache consistency architecture, *Proceedings of ACM SIGCOMM*, 1999.

SERVER-FRIENDLY DELTA COMPRESSION FOR EFFICIENT WEB ACCESS

Anubhav Savant and Torsten Suel
CIS Department, Polytechnic University

Abstract A number of researchers have studied delta compression techniques for improving the efficiency of web page accesses over slow communication links. Most of these schemes exploit the fact that updated web pages often change only very slightly, thus resulting in very small sizes for the transmitted deltas. However, these schemes are only applicable to a minority of page accesses, and require web or proxy servers to retain potentially many different outdated versions of pages for use as reference files in the encoding. Another approach, studied by Chan and Woo [4], encodes a page with respect to similar files located on the same web server that are already in the client's browser cache.

Based on the latter approach, we study different delta compression policies for web access. Our emphasis is on web and proxy server-friendly policies that do not require the maintenance of multiple older versions of a page, but only use reference files accessed by the client within the last few minutes. We compare several policies for identifying appropriate reference files and evaluate their performance on a set of traces. We show that there are very simple policies that achieve significant benefits over *gzip* compression on most web accesses, and that can be efficiently implemented at web or proxy servers. We also study the potential of file synchronization techniques such as *rsync* [28] for web access.

1. Introduction

Delta compression (delta encoding) is the process of encoding a *target file* with respect to one or several, usually similar, *reference files*. This encoding, called a *delta*, describes the target file in terms of the reference files, and a recipient that receives the encoding and already knows the reference files can thus efficiently reconstruct the target. Delta compression has numerous applications in scenarios where there are several versions of a file or many similar files, such as software revision control systems, distribution of software updates, content distribution networks, or efficient storage of related files. Several tools for delta compression, such as *bdiff*, *vcdiff* [10, 13], *Xdelta* [14], and *zdelta* [25], are freely available. We refer to [23] for an overview of delta compression techniques and applications.

F. Douglis and B.D. Davison (eds.), Web Content Caching and Distribution, 303-322.

1.1 Delta compression for Web access

A number of authors have proposed the use of delta compression techniques to improve the efficiency of web access [1, 4, 7, 9, 16, 17, 21, 27, 29]. In particular, when web pages are updated, they typically do not change by much, and thus delta compression can be used to very succinctly encode the difference between a new version of a web page and an outdated version already in the client's browser cache. Most proposals focus on encodings between different versions located at the same URL, which results in small sizes for the deltas but is restricted to pages that have been previously visited by the client. One exception is the work by Chan and Woo [4], which proposes to use as reference files other pages on the same site recently visited by the client, which tend to have a significant degree of similarity due to common layout features and HTML structure. In general, delta compression schemes for web access can be distinguished along the following axes:

- **End-to-end vs. proxy-based:** A recent proposal [16] discusses how to integrate delta compression into the HTTP/1.1 standard. This enables end-to-end use of delta compression techniques between web servers and end clients, potentially leading to significant savings in bandwidth over the internet backbone. On the other hand, proxy-based schemes allow savings over a bottleneck link, say a dialup connection of an end user, without requiring changes in the HTTP/1.1 protocol to be adopted by millions of servers. A common architecture is the *dual proxy* architecture, where a pair of proxies, one located on each side of the bottleneck link, use a usually proprietary protocol incorporating compression, image transcoding, and various other optimizations to increase performance over this link. A number of such systems have for example been deployed by the major cellular network providers, supplied by companies such as Bytemobile, Venturi Wireless, Slipstream Data, and others. In this paper, we focus on such dual-proxy architectures, and we do not discuss in detail how to integrate our approach with existing HTTP standards.
- **Standard vs. optimistic delta:** In the standard approach, both client and proxy have the same old version of a page. A proxy first waits for the updated version of the page to arrive from the server, and then transmits the delta between this and the old version to the client. In the optimistic delta approach in [1] only the proxy needs to hold the old version. Upon receiving a request, the proxy immediately starts sending the old version to the client, while waiting for the new version to arrive from the server; later, a delta between the versions is sent to the client. In a proxy-based environment, neither approach decreases the amount of traffic between server and proxy. The standard approach reduces the amount of data sent from proxy to client, thus reducing delay for low-bandwidth links. The optimistic delta approach does not reduce the amount of data sent to the client, but may decrease total delay in cases where server response delays are also significant compared with the transmission time over the bottleneck link. We focus on the standard approach.
- **Same URL vs. different URLs:** As described, we can limit delta compression to different versions of the same URL, or allow compression between pages

corresponding to different URLs. In the latter case, clients and proxies need to choose appropriate reference pages, and this is our main focus.

- **Delta compression vs. file synchronization:** In the standard delta compression scenario, the proxy needs a copy of the old and the new version to compute a delta. File synchronization techniques such as *rsync* [28], on the other hand, allow the proxy to send a delta without knowing the old version in the client cache, based on only a set of hash values sent by the client as part of the request. File synchronization can be seen as a special, more restricted, case of delta compression, and usually achieves compression ratios that are significantly worse than those of the best delta compression tools [23], particularly on files that share only short common substrings. (Formally, any file synchronization protocol can be used to produce a delta by storing all messages sent from target file to reference file.) We consider both techniques.

1.2 Discussion of known approaches

Delta compression between different versions of the same page typically achieves a high compression ratio, but suffers from two major shortcomings. First, it only gives benefits for pages that have been previously visited and that have since been updated. A 1996 study [17] found that this only applies to about 30% of page accesses. Second, it imposes significant costs on the proxy, or the server in an end-to-end approach, which needs to retain older versions of each web page for potential use as reference files in future accesses, possibly for a significant amount of time. Also, additional disk accesses may be required to fetch the reference files.

Delta compression between different pages typically achieves a more moderate compression ratio. One problem is how to identify appropriate reference files for a requested page. A simple scheme for doing this is proposed in [4], based on the directory paths of the URLs. However, the scheme requires several reference files for best compression, which could seriously slow down the throughput of the proxy due to the costs of fetching and processing the reference files for each requested file.

A preliminary evaluation of *rsync* for web access was performed in [27] as part of the *rproxy* project. However, results are limited to different versions of the same URL, and only a few numbers are provided. As mentioned, file synchronization achieves a more limited compression ratio than delta compression. For the case of related files, the result might not be better at all than using *gzip*, due to the more limited similarity between the files. File synchronization techniques typically require the client to send a set of hash values or other information about the reference file to the server, and there is a trade-off between the amount of this data and the achieved compression in the other direction. The primary advantage of file synchronization is that no old versions of pages have to be stored and then fetched from disk by the proxy.

An interesting alternative called *value-based web caching* was very recently proposed in [21]. As in file synchronization techniques, bandwidth savings are obtained by using hash values to refer to blocks of data already known to the client. However, the technique does not require the client or proxy to choose a particular set of reference files, but exploits similarity between the requested page and previously transmitted content independent of file boundaries. This is achieved by using Karp-Rabin fingerprints [11] to identify block boundaries in a consistent manner independent of position as proposed in [15], and keeping a limited amount of state for each client. Similar techniques have also recently been used in the *Low Bandwidth File System* [18] and the *Pastiche* distributed backup system [6].

Comparing *value-based web caching* to *rsync*, we would expect similar performance in cases where we can reliably select the best reference file, but better performance in cases where it is not clear which previously accessed file contains the similar content. In particular, *value-based web caching* resolves problems due to aliasing, i.e., different URLs returning the same content. Another advantage comes from caching hashes at the proxy, although one could also use an *rsync*-based approach to do the same (see Subsection 3.3).

1.3 Our approach and contributions

Our main focus is on studying web and proxy server-friendly schemes for delta compression between different pages that achieve good compression without adversely affecting throughput. In fact, we believe that delta compression between different pages has more practical potential than the more commonly studied case of delta compression between different versions of the same page. On the other hand, we conjecture that file synchronization techniques may be the most appropriate approach for delta compression between different versions of the same page, at least in a proxy environment. In this paper, we discuss techniques and provide experimental results for web access based on delta compression and file synchronization. In particular:

- We compare several policies for selecting appropriate reference files for delta compression, and evaluate the achieved compression ratio. Using a large set of *plausible* site visits by clients distilled from NLANR proxy traces, we show that significant average improvements over *gzip* are achieved by the best policy. Moreover, we show that very simple policies achieve close to optimal compression using only one or two reference files visited within the last few minutes, thus allowing for an extremely efficient main-memory based implementation in a web or proxy server.
- We discuss and evaluate the use of file synchronization techniques for efficient web access. We show that they are of limited use for delta compression between different pages, and currently studied improvements in file synchronization techniques are unlikely to change this. On the other hand, we show that file synchronization techniques have potential for delta compression between different versions of the same page, and we discuss how file synchronization tools could be reengineered for additional improvements.

While the chance for a broad adoption of the proposed delta compression schemes at clients and servers is very small, we believe that there is a significant potential for using such techniques in the context of proprietary proxy systems such as those deployed by wireless or dialup service providers. We note that the benefits of the techniques are limited to `text/html` files and do not apply to images and other multimedia objects, though there are other specialized techniques for such objects. According to the traces, `text/html` files made up 40 to 50% of the data going through the proxy, and about the same amount was due to image files. (Common statistics indicate that embedded images make up almost 70% of all data in a displayed page; however, images have better caching behavior and thus make up a relatively smaller part of the traffic over the internet.) The benefits are in addition to any benefits due to client-side caching, which do not appear in our traces. The schemes we propose are applicable to about 68% of all accesses to `text/html` files going through the proxy, and for those eligible pages we get up to a factor of 2.9 average improvement over *gzip*. For all pages, the average improvement over *gzip* is up to 1.7. Benefits are lower if duplicate URLs are removed from site visits; these duplicates in the proxy traces may be due to a changed page or a server that does not support the IF-MODIFIED-SINCE header.

In the next section, we study reference file selection policies for delta compression of different pages. In Section 3 synchronization techniques for web access. Finally, Section 4 provides some concluding remarks.

2. Delta Compression Schemes for Site Visits

This section contains the main results of this paper. We are interested in delta compression schemes that encode a requested page in terms of other pages that the client already has in its cache. We restrict ourselves to reference files from the same site that were very recently accessed by the client as part of the current *site visit*. We define a *site visit* of a client as a set of consecutive accesses to pages on the same site. We show that even with this limitation on the choice of reference files, we can obtain significant average compression benefits over *gzip*, by a factor of 1.7 for all pages and 2.9 for "eligible" pages. One challenge in the evaluation is that it is not easy to obtain client traces that can be used to evaluate our schemes, due to privacy concerns. In our case, we need a large number of site visits that are very recent, so that we can still obtain the pages from the origin servers. Since we were unable to get a representative set of end client traces, we decided to try to "distill" a plausible set of client visits from the publicly available NLANR proxy traces, as described later. We then fetched and stored all pages in those visits, and ran simulations based on the stored pages.

We used version 2.0 of the *zdelta* tool [25] to perform delta compression between pages (available at `http://cis.poly.edu/zdelta/`). The *zdelta* tool is carefully optimized for both compression and speed, and supports the use of up to four reference files. We note that in principle a similar compression ratio could probably also be obtained with the *vcdiff* compressor [13] provided that separate Huffman coders are applied to the different fields of the generated instruction stream; however, *vcdiff* currently only supports a single reference file. Both tools achieve speeds of several MB per second for encoding and tens of MB for decoding, similar to the speed of *gzip*.

2.1 Reference file selection policies

We now define a few simple policies for choosing reference files that we investigate. The policies are very simple and we make no claims of algorithmic originality.

Our primary goal is to evaluate their behavior on a large set of traces. Suppose that the client is performing the ith page access in the current site visit. The policies for choosing reference files from among the pages previously visited during the same site visit are as follows:

- **last-k:** choose the k pages most recently visited.
- **longest match-k:** choose the k pages whose URL has the longest directory path match with the requested URL; ties are broken by taking the more recently visited page.
- **best-k:** choose the k pages that each individually achieve the best compression when used as the only reference file for the requested page.
- **best set-of-k:** choose the set of k pages that provides the best compression.

We also consider two simple improvements over *last-k* and *longest match-k*, called *last-k^+* and *longest match-k^+*, that add the following two rules: (1) if the requested page was previously accessed, then that page is always selected as a reference file, (2) we make sure that all selected reference files are distinct, since there is no benefit in using two identical reference files. Note that if we eliminate duplicates in the traces, then these policies will be identical to the basic ones.

Of course, we can choose from at most $i-1$ reference files for the ith page; thus for $i \leq k$ all policies are the same. The *last-k* and *last-k^+* policies were chosen as the simplest heuristics we could come up with, and they also tend to minimizes the time a server or proxy has to retain pages for later use as reference files. The *longest match-k* and *longest match-k^+* policies are almost identical to the policy proposed by Chan and Woo [4], and should achieve very similar performance. The *best set-of-k* policy is clearly optimal, but inefficient since there are $\binom{i-1}{k}$ possible choices of sets. (Finding the best set is NP Complete due to a reduction from Set Cover, under reasonable assumptions about the delta compressor.) The third policy, *best-k*, is not guaranteed to be optimal, but more efficient to implement.

2.2 Experimental setup

To evaluate the policies, we downloaded traces from the NLANR proxy servers and partitioned the traces into "plausible" site visits as follows. In the traces, clients are identified by IDs that are unique during a given day and for a particular proxy. However, these clients are typically not end clients (browsers) but populations of users in a particular organization, and for our purposes we need access patterns for end clients. We defined a site visit with timeout t as a sequence of accesses to the same site by users with a common ID, with at most t minutes between any two consecutive accesses. If t is large, then a site visit to a popular web site is likely to consist of accesses by different end clients with a common ID. However, if we assume some degree of independence between the access times of the different users throughout the day, then by decreasing t we expect to eventually split up most of these visits into shorter visits by different users. We found that after decreasing t to less than an hour, subsequent decreases of t to a few minutes do not result in many additional splits, indicating that most of the site visits at this point are probably due to a single end user. Of course, we cannot claim that the site visits thus obtained are all due to single individuals, but we believe that our heuristic is close enough to allow an evaluation. We note that such *sessionizing* problems have been extensively studied, see, e.g., [22].

We downloaded all NLANR proxy traces for April 22, 2003, and extracted site visits with timeout 10 minutes. We then took a sample of site visits with a total of 74434 pages, which we downloaded and stored on April 24. We excluded any dynamic pages (with parameters in the URL), as those parameters are removed by NLANR for privacy reasons. We see no reason to believe that the approach would not work on dynamic pages; in fact, such pages may be particularly suitable for delta compression as they frequently change in minor ways and share a lot of content with other pages on the site. However, we have not yet obtained good traces to evaluate this. We made sure that our sample has the same distribution of site visit lengths (number of page accesses) as the entire set. Note that the proxy traces did not contain entries for accesses that were already satisfied by client-side caches; this is fine for our purpose since we are interested in benefits beyond those already obtained by client side caching.

The generated site visits did contain a significant number of duplicates, i.e., repeat accesses to the same URL in a single visit with return code 200. These accesses may be due to changed pages or due to servers that are not set up to support the IF-MODIFIED-SINCE feature in HTTP. While some of the duplicates may belong to different end clients, we believe most are due to the same client. In the following, we report results with duplicates included, and later discuss how the results are impacted if they are excluded.

We compare our results to *gzip* as a baseline compressor. Note that *gzip* is not optimized for HTML, and recent grammar-based approaches based on [12] can achieve significantly better results. (In fact, proprietary versions of these algorithms are the basis of the web acceleration system designed by Slipstream Data and deployed by NetZero.) These approaches can also be adapted to exploit similarity between pages.

2.3 Experimental evaluation of policies

In Figures 1 and 2 we see the size reductions obtained by the *last-k* and *longest match-k* policies, respectively, for different values of k. For the first page access in a site visit, all policies and also *gzip* achieve the same reduction, to about 22% of the original size on average. (If no reference file is given, *zdelta* becomes identical to *zlib*, which itself has a similar performance as the closely related *gzip*.) For the second access, pages are on average reduced to 10% of their uncompressed size. As there is only one available reference file, all policies except *gzip* achieve the same performance. On the third access, policies with $k \geq 2$ start outperforming those with $k = 1$, and on the fourth access policies with $k = 4$ start outperforming those with $k = 2$. Compression improves to about 5% of the uncompressed size after a few pages. (However, since most site visits are fairly short, the impact on the average benefit over all pages decreases as we move to the right, as shown later.) We also see that the policies *last-1*$^+$ and *longest match-1*$^+$ perform significantly better than the base policies. (Note that the base policy *longest match-k* only tries to match the directory part of the URL and does not distinguish between different pages in the same directory; otherwise, *longest match-1* and *longest match-1*$^+$ would be identical.)

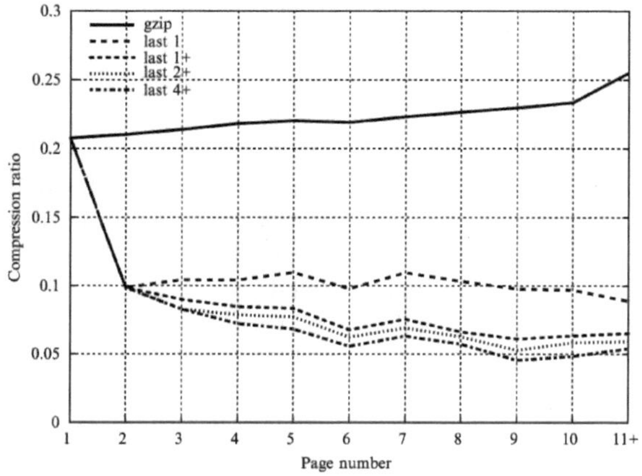

Figure 1. Average compressed size for *last-k* policies.

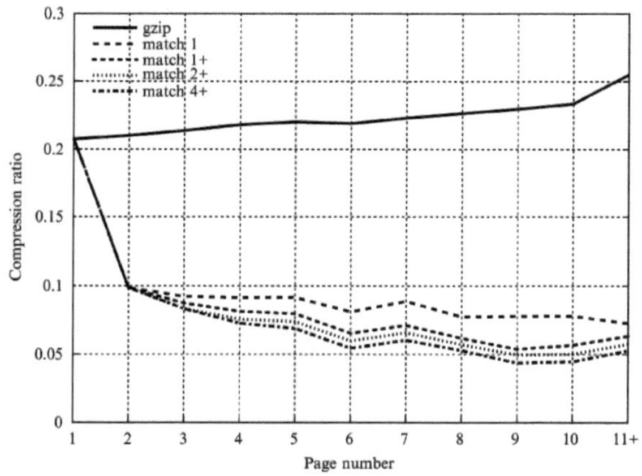

Figure 2. Average compressed size for *longest match-k* policies.

In summary, there is significant benefit in using delta compression versus *gzip*, and the benefit increases with the number of reference files and the length of the site visit. One curious detail is that *gzip* compression appears to get slightly worse as more pages on a site are visited; this is primarily due to a decrease in average page size as discussed further below.

In Figure 3 we show the performance of the *best-k* and *best set-of-k* policies, and in Figure 4 we show the performance of a selection of all policies. All of the *best-k* and *best set-of-k* policies have almost the same performance, independent of k. Note that while results are only plotted for $k = 1$ and $k = 2$, the benefit from adding a third file would clearly be less than that for adding the second file (which itself is close to

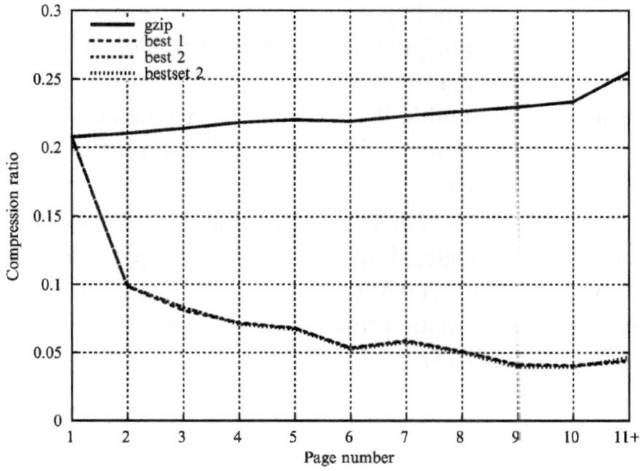

Figure 3. Average compressed size for *best-1*, *best-2*, and *best set-of-2* policies. The three policies result in almost identical graphs.

zero), due to the way current delta compressors works. The interpretation of these results is important. In general, using more than one reference file can improve delta compression for two reasons: it could be that each reference file contributes to the compression of the target file (e.g., a target file could be similar to one reference file in the first half, and to another reference file in the second half), or it could be that by using several files we simply have a better chance of including the one reference file that contributes most of the benefit.

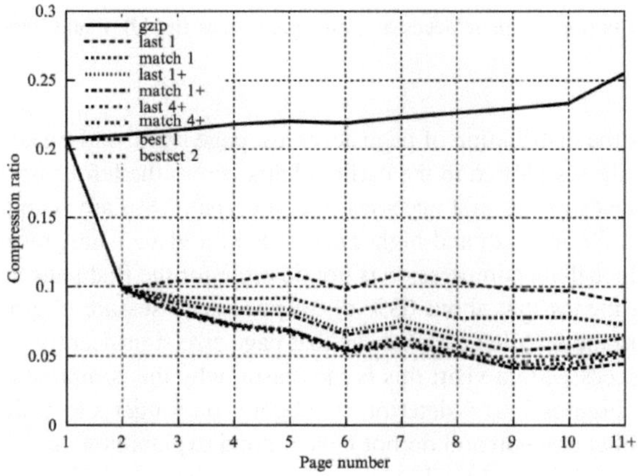

Figure 4. Average compressed size for various policies. The order of the graphs is (from the top) *gzip*, *last-1*, *longest match-1*, *last-1*$^+$, *longest match-1*$^+$, *last-4*$^+$ and *longest match-4*$^+$, and *best-1* and *best set-of-2*.

The above results implies that for our application it is primarily the second reason, since by choosing the best reference file we are getting essentially all of the possible benefit. A similar observation was also recently made in [8] in the context of file system compression. The *last-k* policy, and also the *longest match-k* policy proposed in [4], are not really good at identifying the best reference file, but by increasing k we improve our chances of including the best reference file in the list of selected pages. For efficiency it would be preferable if we could directly identify the best reference file, rather than use up to 4 reference files or try all different files as in the *best-1* policy; we address this issue later.

Comparing *last-k* and *longest match-k*, we see that *longest match-1* is better than *last-1* at identifying good candidates; however, this advantage largely disappears if we use the improved policies *last-k*$^+$ and *longest match-k*$^+$ instead. Thus, it seems that in practice the directory-based heuristic proposed in [4] really does not do much better than a trivial heuristic such as *last-k*$^+$.

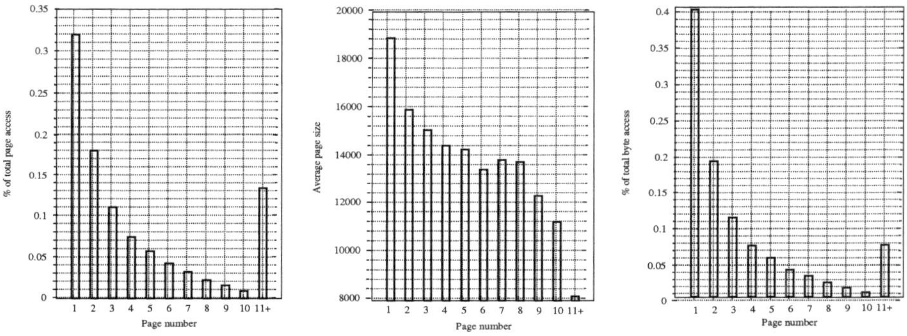

Figure 5. Distributions of page accesses (left), page sizes (middle), and total bytes (right) in our traces.

Figure 5 shows the distribution of page accesses, page sizes, and transmitted bytes over the different classes plotted in the earlier charts. From the left chart, we see that 32% of all page accesses are first accesses in a site visit, 18% are second accesses, and finally about 13% are 11th and higher accesses in a visit. Thus, most site visits are fairly short. Since delta compression is not possible for the first page access under our schemes, this means that about 68% of all page accesses are eligible for delta compression. From the middle chart, we see that page size significantly decreases for subsequent page accesses in a visit; this is the reason why the compression achieved by *gzip* shown in Figures 1 to 4 deteriorates slightly on longer site visits. (We are surprised by this phenomenon and do not have a good explanation yet.) In the right chart, we see the distribution of total bytes over the page accesses; due to the skew in page size the first pages in the site visits together account for about 40% of all bytes transmitted. Thus, about 60% of transmitted bytes are eligible for delta compression.

In Figures 6 and 7, we show the average compression factor, in terms of total transmitted bytes, over all eligible page accesses and all page accesses achieved by the various policies. Compression ranges from a factor of 4.5 for *gzip* to more than 13 for the best policy for eligible pages. We note that the various policies are actually much closer in terms of average benefit than suggested by Figures 1 to 4, since most

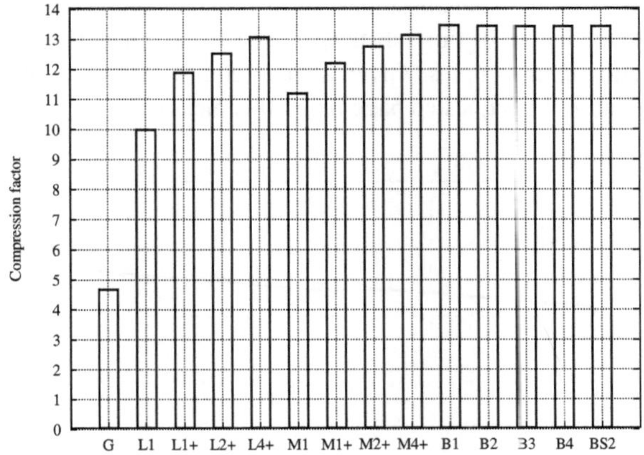

Figure 6. Average compression for policies on eligible pages. The policies are *gzip*, *last-1*, *last-k*$^+$, *longest match-1*, *longest match-k*$^+$, *best-k*, and *best set-of-2*.

transmitted bytes fall into the left half of those figures. In particular, even *last-1*$^+$ and *longest match-1*$^+$ perform quite well.

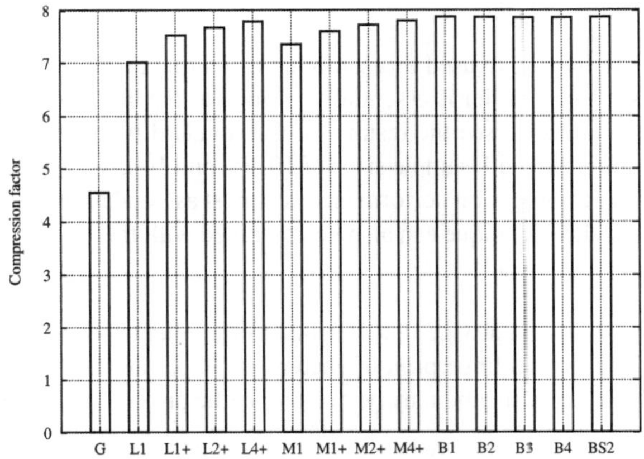

Figure 7. Average compression for policies on all pages.

2.4 An efficient and nearly optimal protocol

We now address the problem of how to efficiently identify the reference file that provides the best compression ratio. Our proposed solution is very simple, and uses random hash functions to create fingerprints for files according to a technique proposed by Broder in [2]. This technique has been previously used to cluster documents in [3], and was shown in [8, 20] to provide reasonable estimates for the size of a delta between two files. In particular, we hash all substrings of length 4 in a file to integer values, and then retain the s smallest hash values. (The method does not seem to be very sensitive to the length of this substring.) We then estimate the similarity of two files by intersecting their samples. Given a page request, we select the reference file that is most similar to the requested page under this measure.

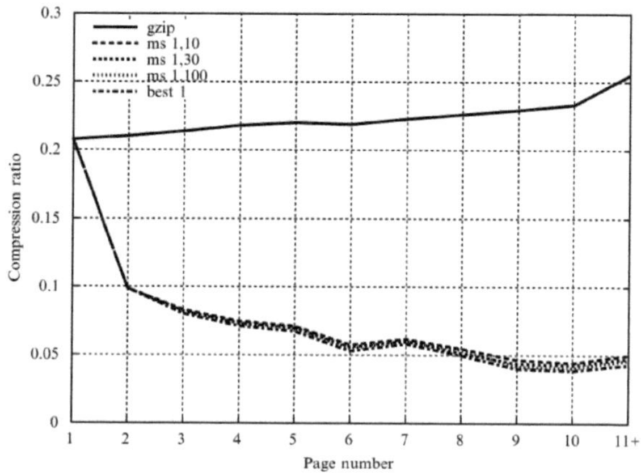

Figure 8. Performance of a sampling based approach for determining the most similar file, for sample sizes 10, 30, and 100.

In the first experiment, we used this sampling technique to approximate the *best-1* policy, which as we saw earlier is essentially as good as the best policy, *best set-of-k*, in practice. In Figure 8 we see the results for the standard *best-1* policy and for the sampling based approach with sample sizes of 10, 30, and 100. We find that even with sample size 10, the compression performance is almost as good as that of the standard *best-1* policy. This this technique can also be combined with the other policies to give the following simple protocol:

- The proxy stores each page that it sends to the client for a limited period of time, together with a fingerprint of each page. Pages are kept in main memory indexed by MD5 hash, and may be deleted by the proxy at any point in time.
- A client sends each page request to the proxy accompanied by several MD5 values of previously visited pages, called *candidate pages*. These candidate pages can be selected by the client based on *last-*k^+, *longest match-*k^+, or any other policy.
- The proxy checks which of the candidate pages it still holds in main memory, and uses the most similar of those as reference file.

Thus, if the client uses the *last-4* policy to identify candidates, then the resulting compression performance will closely track that of the standard *last-4* policy, using only a single reference file. Note from Figures 6 and 7 that both *last-4* and *longest match-4* achieve average compression quite close to the optimal, making them viable approaches. For the proxy (or server in an end-to-end approach), this scheme is highly

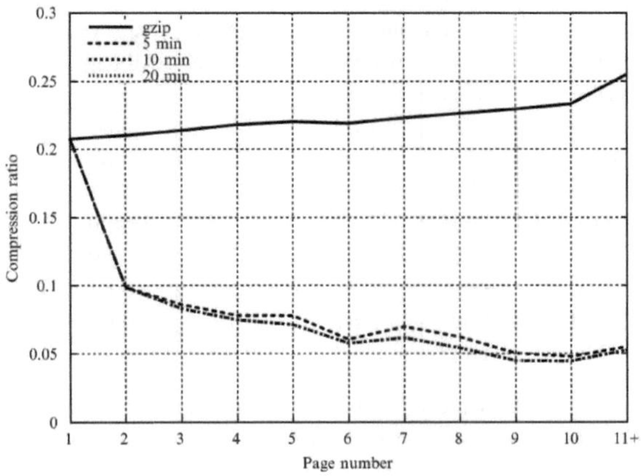

Figure 9. Performance of a sampling based approach for determining the most similar file, for histories of length 5, 10, and 20 minutes. The latter two cases are almost identical.

efficient as only a single reference file is used, and as pages only need to be retained for a few minutes.

In Figure 9 we show the performance for the case where only pages accessed in the preceding few minutes are used as candidate files; the achieved compression ratio is very close to optimal even for windows of 5 and 10 minutes. (While site visits were defined by us as having at most 10 minutes between consecutive accesses, most are actually much closer in time. This also supports our contention that our distilled site visits are mostly due to a single end client.)

We note two drawbacks of this scheme. First, the computation of the samples could become a bottleneck, though we believe this can be handled through careful optimization. Second, the most similar reference file can only be selected once the entire requested page has arrived at the proxy. In contrast, in policies such as *last-k* and *longest match-k* the proxy can select the reference files as soon as it receives the request. These can then be immediately inserted into the hash table of the delta compressor, and the requested page can be streamed through the compressor and forwarded as it arrives from the server. This second drawback might make schemes such as *last-*k^+ or *longest match-*k^+ for $k = 1$ or $k = 2$ more attractive in many scenarios.

2.5 Impact of duplicates

We now look at how results change if we remove any page accesses with return code 200 that go to a page previously accessed in the same site visit. Recall that such accesses may be due to changed pages or due to servers not supporting IMS. Table 1 shows the compression ratios achieved in this case. As we see, benefits of delta compression are reduced, but still significant. In particular, the average benefit over *gzip* is now a factor of about 1.4 instead of 1.7 for all pages, and 1.9 instead of 2.9 for eligible pages under the best policy.

	With duplicates		Duplicates removed	
Policy	Eligible	All pages	Eligible	All pages
GZIP	4.54	4.63	4.64	4.72
L1	9.95	7.00	7.91	6.31
L1+	11.85	7.51	7.91	6.31
L4+	13.03	7.77	8.78	6.62
M1	11.16	7.33	8.12	6.39
M1+	12.17	7.58	8.12	6.39
M4+	13.10	7.79	8.84	6.64
MS 1,10	13.02	7.82	8.73	6.61
B1	13.43	7.86	9.02	6.71

Table 1. Compression factors achieved with and without duplicates, for policies *last-1*, *last-k^+*, *longest match-1*, *longest match-k^+*, *most similar-1* with sample size 10, and *best-1*.

2.6 Summary of observations

The main observations from the experiments in this section are as follows: (1) even very simple schemes for selecting reference files achieve significant compression over *gzip* and come close to the optimum, (2) there seems to be only very limited benefit for the directory matching technique in [4] over other simple heuristics, (3) essentially all of the potential benefit can be achieved with a single reference file and there are simple sampling based methods for identifying this file, (4) however policies such as *last-1^+* and *longest match-1^+* perform quite well and might be preferable in practice for other technical reasons, and (5) benefits are somewhat lower if page size distribution is taken into account and duplicates are removed.

3. Utility of *rsync* for Web Access

In this section, we study the potential for using *rsync* and other file synchronization techniques for web access. By file synchronization, we refer to techniques where the server has the requested pages, but not the reference file held by the client. The most widely used tool for file synchronization is *rsync* [28], which uses a single round-trip between client and server as follows: the client partitions the reference file into blocks of a few hundred bytes, and sends a hash value for each block to the server. (The hash value has a strength of 6 bytes for common file sizes.) The server then sends the requested file to the client, but replaces any block that hashes to one of the received hash values by a reference to the hash. This method is then combined with standard *gzip* compression. The cost is given by the size of the hashes and the size of the encoded page, with a trade-off between the two. The size of the encoded page sent to the client is clearly lower-bounded by the size of a delta.

In fact, measurements in [23] show that the size of the data sent from server to client in *rsync* is usually significantly larger than a delta. This raises the question of whether file synchronization techniques are efficient enough for web access. In the following, we present results for related pages and versions of the same page.

3.1 Experiments for related pages

Figure 10 shows results for related pages, using the *last-1* and *longest match-1* policies and several block sizes in *rsync*. We also show results for the *most similar-1* policy, with and without duplicates, to get an upper bound on the possible benefit, even though this policy is not realistic as the selection of reference files for *rsync* is performed at the client. (We used *most similar-1* instead of *best-1* for efficiency but the result should be very similar. One caveat is that we did not adapt the substring size in the sampling step to the block size of *rsync*.)

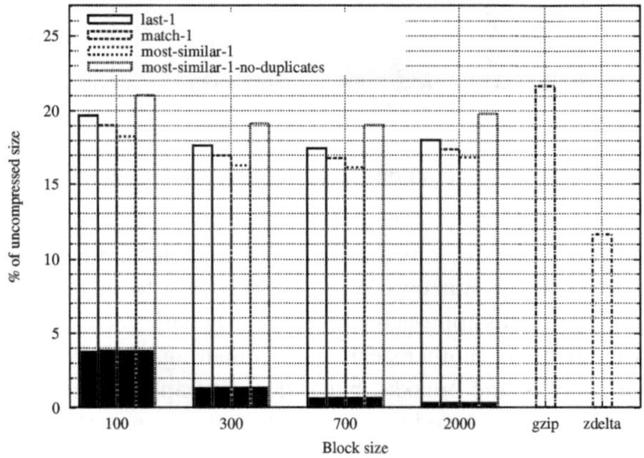

Figure 10. Performance of *rsync* under the *last-1*, *longest match-1*, and *most similar-1* policies for different block sizes.

We see from Figure 10 that *rsync* performs only slightly better than *gzip* if we consider the total data sent in both directions. Even for the *most similar-1* policy with block size 700, files only reduce on average to about 16% of their original size, compared with 22% for *gzip*. If we exclude duplicate URLs, then there is hardly any benefit over *gzip* at all even for the *most-similar-1* policy (about 19% vs. 22% for *gzip*). Performance is best for block sizes between 300 and 700 bytes.

In Figure 11 we investigate the use of more than one reference file in connection with *rsync*. To use *rsync* on two reference files, we concatenated the two files into one longer file. We do not show results for more than two reference files as there clearly would be no benefit in this. We note that the amount of data sent from client to proxy is proportional to the number of blocks, and thus increases with the inverse of the block size. We see that for block sizes of 700 and 2000 bytes, adding a second reference file gives negligible benefits under the *last-k* and *longest match-k* policies. Benefits are small because the increase in *read size* cancels out most of the decrease in *write size*.

In general, the reason for the disappointing performance of *rsync* on related pages on the same site is that these pages, while overall similar, differ in a number of places. While there is some benefit for aliased and very similar pages, in general the fine granularity of differences does not allow *rsync* to find large matches at the level of blocks of several hundred bytes, while block sizes of 100 bytes or less would greatly

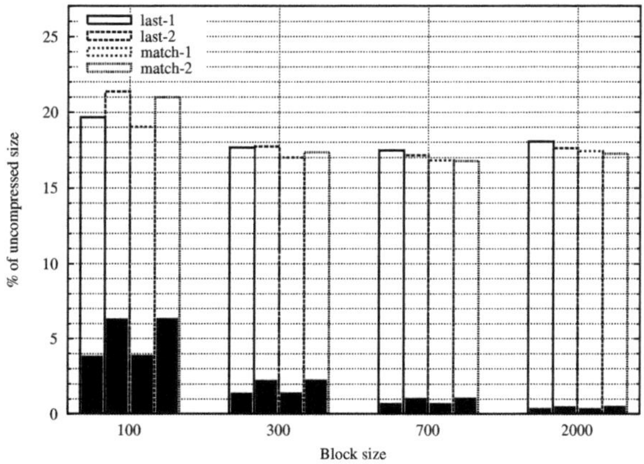

Figure 11. Performance of *rsync* for one and two reference files with different block sizes.

increase the number of hashes sent by the client. We note that Tridgell, in Section 4.4.3 of [26], proposes to use the Burrows-Wheeler transform to increase the locality of changes in updated files. This approach works well when files are updated by replacing all occurrences of a string by another string (e.g., renaming of a variable or a consistent change in format such as line breaks), but it does not appear to work well at all in our scenario. In fact, we observed a significant decrease in performance under this approach, as it also has the reverse effect of spreading out blocks of changed bytes all over the transformed files. We conclude that simple *rsync* is not a good approach for compression between related files.

3.2 Experiments for versions of a page

We now consider the case of several versions of the same page. Our NLANR traces are not very useful for this case since the client IP hashes change from day to day. Since we did not have alternative traces, we instead used data obtained from repeated web crawls of a certain subset of pages. In particular, we chose pages at random from two large page collections of more than 100 million pages each, one from the Internet Archive and one from our own crawls. The selected pages were than crawled every night for several weeks in Fall 2001. In the following, we use four versions of a subset of 10000 of these pages: one base set and three updated versions crawled 2, 20, and 72 days later. In the experiments, we measure the benefit of using *rsync* to fetch an updated page given that the client already has an outdated version from the base set. We give average results for all pages and for only those that have changed.

Figure 12 and 13 show the result for the different block sizes. As we see, even for pages that are revisited 72 days later, we still get significant benefit over *gzip*, by about a factor of 2 even if unchanged pages are excluded. On the other hand, delta compression does even better then *rsync*, by another factor of about 4. We note that user surfing behavior is likely to be biased towards sites that are frequently updated, and thus our results for a random sample of pages may be overly optimistic. Nonethe-

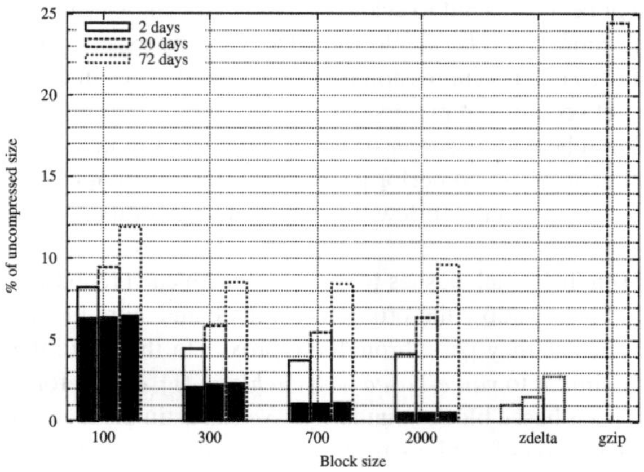

Figure 12. Performance of *rsync* on all pages for 3 different time intervals.

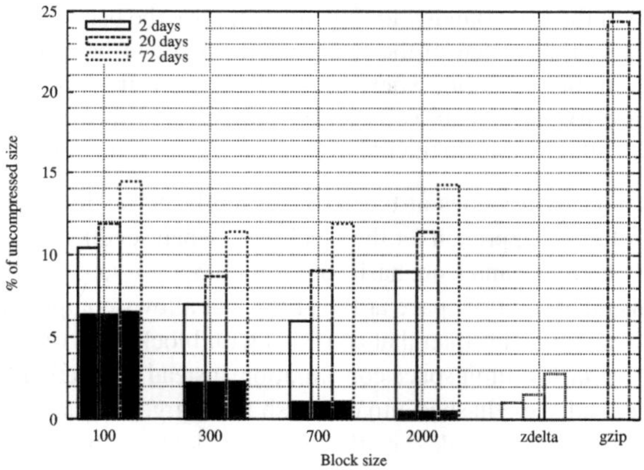

Figure 13. Performance of *rsync* on changed pages for 3 different time intervals.

less, there seems to be significant potential for using *rsync* between different versions of a page, as previously suggested by the *rproxy* project [27]. Note that examination of the pages shows that page updates tend to be highly clustered within a page; this explains the good performance of *rsync* in such cases.

3.3 Discussion and a generalized approach

The results indicate that file synchronization using *rsync* is efficient for different versions of the same page, but performs only slightly better than *gzip* on related pages from the same site. Of course, this could be due to shortcomings in the *rsync* algorithm, and there are several proposals for improved file synchronization techniques [5, 19, 24]. However, these proposals rely on additional roundtrips and would not be appropriate for a web access scenario since modem latencies are high and objects are not very large. It appears difficult to achieve significant bandwidth savings over *rsync* without incurring more than the single roundtrip used by *rsync*. This suggests that an approach that uses delta compression between related pages in a site visit, and file synchronization between different versions of the same page, might be a good combination. In such a scheme, the server or proxy only needs to keep old files for a few minutes, and still achieves good performance for different versions of the same page using *rsync*. This leads us to the following ideas and observations:

- Instead of sending block hashes to the proxy with each request under the *rsync* protocol, we could allow the proxy to compute and store block hashes for the references files as they pass through the proxy. In this case, the approach becomes very similar to *value-based web caching* [21], except for the use of fixed rather than hash-based block boundaries. We can estimate the benefit by deducting the dark portion of the bars from the charts in this section. Some additional benefits might arise since there is no need anymore to choose a small set of reference files for each access.
- In the case of the Low Bandwidth File System [18], the use of hash-based block boundaries is crucial due to the fact that files have to be transmitted in both directions. In our scenario, where files are only sent from proxy to client, we can also use fixed boundaries based on position and then try to match these blocks with all positions in the file to be encoded, as done in *rsync*.
- When caching hash values at the proxy, it might be interesting to consider a "multi-resolution" approach for the block size: when sending a file to the client, we initially compute and store hash values for blocks of fairly small size, say 32 bytes. This takes a lot of space at the proxy, but allows compression of similar files that is much better than that achieved with block sizes of several hundred bytes as in *rsync*. To limit space consumption, instead of evicting hashes we can combine two adjacent hashes into one hash for a larger block, provided that the hash function is *composable* (i.e., the hash of a block can be computed from the hashes of the left and right half). Alternatively, we could for each file generate hashes at different levels of granularity, and first evict fairly old hashes of small size. (This takes at most a factor of 2 more space than a combining approach and removes some other complications.)

Based on these ideas, we plan to investigate an approach that combines delta compression, value-based caching, and synchronization, as follows: Transmitted files are cached for a short time to allow delta compression, while block hashes for the files are cached for longer periods based on the above multi-resolution approach. Thus, hashes for small blocks are kept for a shorter period of time, and hashes for large blocks for longer periods. In addition, we could also allow the client to send hashes in some cases, such as when a page is revisited after several days and the old hashes are likely to have been evicted at the proxy. There are a number of questions to explore in this context, such as the advantages and disadvantages of fixed and hash-based block boundaries and the details of the multi-resolution approach, and we plan to address these in our future work.

4. Concluding Remarks

In this paper, we have studied the performance of simple delta compression and file synchronization schemes for efficient web access. Our focus was on web and proxy server-friendly schemes that do not require the storage and retrieval of old versions of web pages at the server. Our main conclusion is that there is significant benefit to using delta compression between pages on the same site, and we gave several low-overhead schemes that improve significantly over *gzip*. On the other hand, we found that single-roundtrip file synchronization techniques such as *rsync* obtain good compression between different versions of a page at low overhead, but do not significantly improve over *gzip* for related pages.

We are working on several extensions of this work. We plan to study the approach combining delta compression, value-based web caching, and file synchronization described in Subsection 3.3, and to integrate it into a dual proxy architecture called *SPAWN* (see `http://cis.poly.edu/spawn/` for more details) that we have implemented at Polytechnic University. We are also working on improved software tools for file synchronization and plan to evaluate these in the current context and for content distribution networks and large replicated document collections.

Acknowledgements

We thank Dimitre Trendafilov for help with experiments, and the anonymous referees for helpful comments. Research was supported by NSF CAREER Award CCR-0093400 and New York State through the Wireless Internet Center for Advanced Technology (WICAT) at Polytechnic University. Traces were provided by the IRCache Project, which is supported by the National Science Foundation (grants NCR-9616602 and NCR-9521745) and the National Laboratory for Applied Network Research.

References

[1] G. Banga, F. Douglis, and M. Rabinovich. Optimistic deltas for WWW latency reduction. In *USENIX Annual Technical Conference*, pages 289–303, 1997.

[2] A. Broder. On the resemblance and containment of documents. In *Compression and Complexity of Sequences*, pages 21–29. IEEE Computer Society, 1997.

[3] A. Broder, S. Glassman, M. Manasse, and G. Zweig. Syntactic clustering of the web. In *Sixth Int. World Wide Web Conference*, 1997.

[4] M. Chan and T. Woo. Cache-based compaction: A new technique for optimizing web transfer. In *Proc. of INFOCOM'99*, 1999.

[5] G. Cormode, M. Paterson, S. Sahinalp, and U. Vishkin. Communication complexity of document exchange. In *Proc. of the ACM–SIAM Symp. on Discrete Algorithms*, 2000.

[6] L. Cox, C. Murray, and B. Noble. Pastiche: Making backup cheap and easy. In *Proc. of the 5th Symp. on Operating System Design and Implementation*, 2002.

[7] M. Delco and M. Ionescu. xProxy: A transparent caching and delta transfer system for web objects. May 2000. unpublished manuscript.

[8] F. Douglis and A. Iyengar. Application-specific delta-encoding via resemblance detection. In *Proc. of the USENIX Annual Technical Conference*, June 2003.

[9] B. Housel and D. Lindquist. WebExpress: A system for optimizing web browsing in a wireless environment. In *Proc. of the 2nd ACM Conf. on Mobile Computing and Networking*, pages 108–116, November 1996.

[10] J. Hunt, K.-P. Vo, and W. Tichy. Delta algorithms: An empirical analysis. *ACM Transactions on Software Engineering and Methodology*, 7, 1998.

[11] R. Karp and M. Rabin. Efficient randomized patternmatching algorithms. *IBM Journal of Research and Development*, 31(2):249–260, 1987.

[12] J. Kieffer and E. Yang. Grammar based codes: A new class of universal lossless source codes. *IEEE Trans. on Information Theory*, 46(3):737–754, 2000.

[13] D. Korn and K.-P. Vo. Engineering a differencing and compression data format. In *Proc. of the Usenix Annual Technical Conference*, pages 219–228, 2002.

[14] J. MacDonald. File system support for delta compression. MS Thesis, UC Berkeley, May 2000.

[15] U. Manber and S. Wu. GLIMPSE: A tool to search through entire file systems. In *Proc. of the 1994 Winter USENIX Conference*, pages 23–32, January 1994.

[16] J. Mogul, B. Krishnamurthy, F. Douglis, A. Feldmann, Y. Goland, A. van Hoff, and D. Hellerstein. Delta Encoding in HTTP. 2002. IETF RFC 3229.

[17] J. C. Mogul, F. Douglis, A. Feldmann, and B. Krishnamurthy. Potential benefits of delta-encoding and data compression for HTTP. In *Proc. of ACM SIGCOMM*, 1997.

[18] A. Muthitacharoen, B. Chen, and D. Mazières. A low-bandwidth network file system. In *Proc. of the 18th ACM Symp. on Operating Systems Principles*, pages 174–187, 2001.

[19] A. Orlitsky and K. Viswanathan. Practical algorithms for interactive communication. In *IEEE Int. Symp. on Information Theory*, June 2001.

[20] Z. Ouyang, N. Memon, T. Suel, and D. Trendafilov. Cluster-based delta compression of a collection of files. In *Third Int. Conf. on Web Information Systems Engineering*, 2002.

[21] S. Rhea, K. Liang, and E. Brewer. Value-based web caching. In *Proc. of the 12th Int. World Wide Web Conference*, May 2003.

[22] M. Spiliopoulou, B. Mobasher, B. Berendt, and M. Nakagawa. A framework for the evaluation of session reconstruction heuristics in web usage analysis. *INFORMS Journal on Computing*, 15, 2003.

[23] T. Suel and N. Memon. Algorithms for delta compression and remote file synchronization. In *Lossless Compression Handbook*. Academic Press, 2002.

[24] T. Suel, P. Noel, and D. Trendafilov. Improved file synchronization techniques for maintaining large replicated collections over slow networks. In *Proc. of the Int. Conf. on Data Engineering*, March 2004.

[25] D. Trendafilov, N. Memon, and T. Suel. zdelta: a simple delta compression tool. Technical Report TR-CIS-2002-02, Polytechnic University, June 2002.

[26] A. Tridgell. *Efficient Algorithms for Sorting and Synchronization*. PhD thesis, Australian National University, April 2000.

[27] A. Tridgell, P. Barker, and P. MacKerras. rsync in http. In *Conference of Australian Linux Users*, 1999.

[28] A. Tridgell and P. MacKerras. The rsync algorithm. Technical Report TR-CS-96-05, Australian National University, June 1996.

[29] S. Williams, M. Abrams, C. Standridge, G. Abdulla, and E. Fox. Removal policies in network caches for World-Wide Web documents. In *Proc. of ACM SIGCOMM*, 1996.

EVALUATION OF ESI AND CLASS-BASED DELTA ENCODING

Mor Naaman, Hector Garcia-Molina, and Andreas Paepcke
Stanford University

Abstract The portion of web traffic attributed to dynamic web content is substantial and continues to grow as users expect more personalization and tailored information. Unfortunately, dynamic content is costly to generate. Moreover, traditional web caching schemes are not very effective for dynamically-created pages. In this paper we study two acceleration techniques for dynamic content. The first technique is Edge-Side Includes (ESI), and the second is Class-Based Delta Encoding. To evaluate these schemes, we present a model for the construction of dynamic web pages. We use simulation to explore how system, page and algorithm parameters affect the performance of dynamic-content delivery techniques, and we present a detailed comparison of ESI and delta encoding in two representative scenarios.

1. Introduction

A large number of web pages served today are dynamically generated based on the "profile" of the particular requestor, or on the characteristics of a particular request. For example, a user's page at Yahoo can contain stock prices from the user's portfolio, weather summaries for cities of interest to the user, and scores from selected sports events. Users love to get "personalized" content, and dynamic pages will clearly be a growing fraction of web traffic. However, dynamic content is expensive to generate and deliver, as page construction is resource intensive, and the pages are too dynamic or too personalized to be cached.

Thus, with a "naive" dynamic web delivery scheme, most requests for pages propagate to the server. In addition to all the assembly work and resulting higher latencies, network bandwidth consumption is high, causing this strategy not to scale well. A number of techniques have been proposed to accelerate the delivery of dynamic web pages. Some of them (e.g., [7, 8]) are only concerned with reducing (or handling) the computational load on the server, without any influence on the network traffic load. However, our interest is in techniques that incorporate network savings as well, enabling caching of significant parts of the dynamic content. There are two classes of (sometimes orthogonal) techniques suggested in the literature. Work in [1, 6, 9, 18] focused on deferred assembly of the dynamic page where the final page is assembled

F. Douglis and B.D. Davison (eds.), Web Content Caching and Distribution, 323-343.

from cacheable page fragments. On the other hand, delta encoding focused on serving deltas between the page and other (cacheable) pages.

In this paper we focus on two concrete proposed systems, as representatives for each of the techniques. One system is based on ESI (Edge Side Includes), a scheme proposed by Akamai and Oracle for describing page assembly. ESI provides a way to fragment the page such that it can be assembled on edge servers, which in addition are able to cache the page fragments.

The second system we evaluate in this paper is class-based delta encoding. Delta encoding (DE) was investigated in a web context in [4, 11, 14] and has recently been described and implemented in [2, 17]. In class-based DE [17], the server may encode the dynamic page as a "delta" from some "base file". The delta is sent to the client together with a reference to the base file. If the client does not have the appropriate base file, it asks for the base file from the server. Fortunately, the base files are similar to a static web page, and can be cached on the client side or on any network cache.

For a description of additional dynamic-page delivery systems and techniques, see our extended technical report [15].

In particular, the contributions of this paper are:

- A page-content model that captures the essential features of dynamic web page creation and assembly.
- First-published performance evaluation of ESI and DE delivery in some representative scenarios, highlighting the impact of key data and system parameters.
- A proposed variation for class-based DE. The variation, same-base reply, can significantly reduce client traffic as compared to traditional class-based DE, with only a small increase in server-side traffic.
- A detailed comparison of ESI and class-based DE. It should be noted that ESI and class-based DE are not functionally equivalent. For instance, our modeled ESI does not reduce client traffic, while DE often does. Yet, our comparison helps us understand the tradeoffs between solutions, and the impact of load, number of caches, link per-byte cost, and other factors on the system's performance.

To this end, we start in Section 2 by describing the ESI system. In Section 3 we describe the page-content model, and the physical system model is described in Section 4. Section 5 briefly describes the setup for the simulation, while Section 6 presents the simulation results and an analytical discussion of the results for ESI. In Section 7 we describe class-based DE. Simulation results for DE follow in Section 8, together with a description of our same-base reply enhancement. Section 9 compares ESI and DE systems as informed by our simulation results, and we conclude in Section 10.

2. ESI

The first acceleration scheme we consider is Edge Side Includes as described in [1]. ESI is a specification, suggested by an industry panel headed by Akamai and Oracle, for an XML-based language that is used to enable assembly of a dynamic web page from smaller fragments. The fragments can be independently delivered and cached closer to the client. Instead of generating a full HTML page, the server generates ESI code fragments, each containing the original HTML code for this fragment with addi-

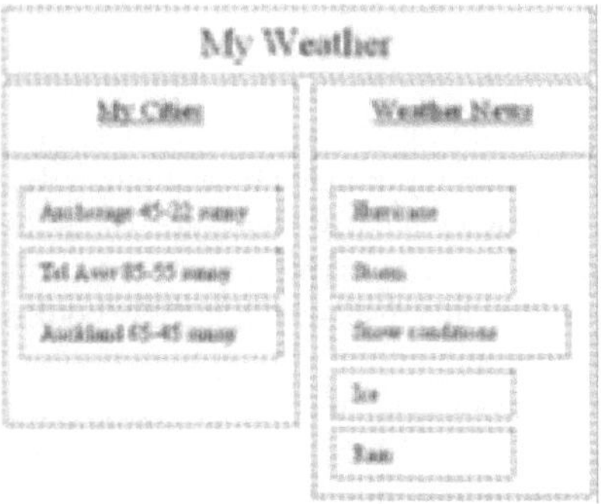

Figure 1. My Weather page

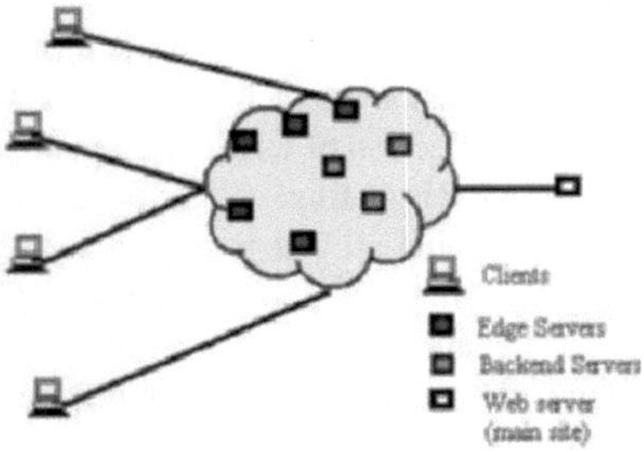

Figure 2. ESI architecture

tional ESI directives. Incidentally, ESI requires a complete revision of the website's code, as the code has to be re-written to use ESI language directives. A sample website and fragmentation of it are shown in Figure 1.

The fragments are cached on specialized edge servers. The edge servers are also where each page is assembled, based on the ESI directions in the fragments. The edge servers are assumed to be much closer to the client than the main server, thus not only saving the assembly cost on the main server, but also reducing the overall network and main server's bandwidth requirement.

As in [18], we differentiate between the ESI language and ESI-based systems. For the rest of this paper, "ESI" stands for our modeled ESI system, unless apparent from context we are referring to the language.

There are a number of possible architectures for an ESI system, differentiated by the number and position of the edge servers. Our modeled standard case involves thousands of edge servers spread across many different geographic locations[1], as shown in Figure 2. The exact number and location of those servers affects the performance in ways we discuss later. However, usually there is a need for backend servers, located closer to the main server, to compensate for the expected low hit rate at the edge servers (as the smaller client population for the edge servers entails poorer locality). The backend servers are part of the system we simulate in this paper, and the system architecture is detailed in Section 4.

3. A Page Content Model

Evaluation of page caching schemes typically requires a page access model. In our case, we need a much richer model, in order to represent what fragments are accessed by clients and how they are to be assembled into full pages. Our model strikes a balance between simplicity and the richness necessary to capture the main tradeoffs.

[1]Two special cases that we do not discuss are placing the servers on the main site, and on the other hand, pushing the assembly all the way to the clients [18].

The model consists of three parts: a model for the underlying data of a resource, a model for the construction of a dynamic page, and a model for the physical system (presented in the next section).

Up to now we have used the term "resource" loosely; we now define it more precisely. A resource is a homogeneous pool of web pages represented by a single URL, or by a collection of URLs. For example, our MyWeather and My Yahoo are resources represented by a single URL; the group of Amazon book pages is represented by a collection of URLs[2]. The term page is used as an instance of a resource - a specific book's page on Amazon; a client's actual MyWeather page, etc.

A dynamic page is constructed by a selection from available items. These data items are usually chosen from different groups. For example, the data for MyWeather is divided into two groups: the Weather News group and the MyCities group. The weather news group contains a number of different news items. The MyCities group contains the data items for all the cities available for display. In another example, a simplified model of the resource "Amazon books" has one group (a books group). The data items in this group correspond to all the books in the Amazon catalog.

Each group also has a static part, like the header of the MyCities fragment for example. In our model this part can be personalized or generic; i.e., can be shared by other users (e.g., the static part of the weather news group) or applies to only one user (e.g., the MyCities static part). To model the creation of an instance of a dynamic page (e.g., some user's MyWeather page) we simulate a two-phase selection. The first phase is a selection of groups to appear on the page. In the second phase we simulate a selection of data items from each group. Below are two examples for the selection process.

To generate an instance of MyWeather page, first both available groups are selected (news and MyCities appear on every client's page). Subsequently, the data items from each of the two groups that appear on a specific user's MyWeather page are selected. Selecting the news items for a page is simple, since in this case all the users get the same news items. The selection of items from the MyCities group, however, is based on user preferences; for user A the selection includes the Anchorage, Tel Aviv and Auckland data items.

A construction of an Amazon book page is slightly different in practice, but is still accommodated by our model. Unlike the previous example where the page is constructed based on the user's pre-defined preferences (that correspond in part to the selection of groups and items), here the user "creates" his page implicitly, by choosing a specific book's URL. However, we can still model this process as a selection from the "books" group - there are many possible data items to be picked, and one of them is selected to appear on the simulated page.

We model the selection of groups and items using a Zipf-like distribution [21], where the likelihood that an item is picked is proportional to $1/i^{\alpha}$, and i is the item rank based on popularity (most popular item is first, $i = 1$). The popularity curve is steeper for larger values of α. Many studies ([3, 5, 10, 16, 19]) suggest that web page requests follow Zipf-like distribution, although those studies do not agree on a single

[2] http://my.yahoo.com, http://www.amazon.com

value for α. We extend this hypothesis to items and groups, and use a range of values for α to ascertain its effect on the model.

The parameters for the model are discussed in greater detail in [15], and are summarized here in Table 1.

Table 1. Parameters of page model for Resource RS

Parameters	*Short description*
GLOBAL	(Global parameters)
N_{RS}	Number of available groups
g_{RS}	Distribution function describing the number of groups per page
f_{RS}	Distribution function describing the relative popularity of the available groups
GROUP	(Parameters for a specific group j)
$k_{RS,j}$	Number of available items in the group
$g_{RS,j}$	Distribution function describing the number of items picked for a page from group j
$f_{RS,j}$	Distribution function describing the relative popularity of items from group j
$ItemSize_{RS,j}$	Size in bytes of HTML/ESI code for items in group j
$StaticSize_{RS,j}$	Size in bytes of HTML/ESI code for group j without any items
$TTL_{RS,j}$	Time-to-live in seconds for items in the group

4. The System Model

The physical system described in this paper consists of:

- Clients that make requests for the resource, and receive page instances in response.
- Thousands of edge servers at the "edge" of the network, i.e. a few hops away from the clients. The number is chosen to resemble current CDNs, and is varied in our simulation.
- A few "backend" servers or caches located "closer" to the main site. The backend servers can fulfill requests from the edge servers, supplying another layer of caching between the edge servers (that handle a smaller number of clients and thus show lesser locality) and the main server. The better locality of the backend servers helps to further reduce load from the main site's servers. For simplicity, we assume a tree structure of the physical system: a client always accesses the same edge cache; an edge cache always accesses the same backend cache.
- A web server at the main site.

For this paper, bandwidth usage is the principal metric. We evaluate the total traffic on the different links: the total traffic (in bytes) sent by the server per day, as well as the total aggregate traffic per day received by all clients, and total traffic per day between the edge and backend servers. Metrics such as user-perceived latency and server CPU loads are, of course, closely related to the number of bytes transferred on each link, but are not explicitly evaluated in this paper. Instead, we use a simple

Table 2. Architecture model parameters and default values

Parameter	Default Value
Number of distinct clients	1,000,000
Page requests per day	8,000,000
Number of edge servers	4000
Number of backend servers	4

weighted-cost model of the traffic on each link when we compare the two systems, and discuss the other measure qualitatively in Section 10.

Factors such as CPU constraints, disk size and memory constraints on the edge, backend and main site servers are not modeled. All the caches modeled in our system are presumed to be of infinite size. We will present cache usage information for some of the simulations, to confirm that usage is modest.

The basic parameter values in modeling the architecture are detailed in Table 2.

5. Simulation

We use our model to simulate difference resources (or web applications). We consider one resource at a time, and show that the performance for different resources will be considerably diverse, due to the dependency on page model parameters. We have chosen to perform a separate evaluation for a number of specific, yet representative, resources as opposed to evaluating a "mean" behavior for a large number of mixed resources. This strategy lets us better understand why each scheme performs as it does. The representative resources we consider are:

- An online bookstore's book-page resource, similar, for example, to a book page on Amazon. This resource is an example of a resource with comparatively large, high TTL fragments and relatively little personalized content.
- A stock portfolio page, representing a resource with a large static portion and a personalized group selecting many small data-item fragments (many stocks on each page).
- My Page, a My Yahoo-style resource that combines many highly personalized as well as generic groups, and also both low and high TTL groups and items.
- Our hypothetical resource, MyWeather, represents a simple dynamic page, with a relatively large static portion, and two statistically different groups - the news, which is the same for all clients, and a personalized MyCities fragment.

Our software simulates the client requests and the way requests are processed at the different caches levels and at the main server. For each request, a page skeleton is generated using the given resource's parameters, listing all groups and items that appear on the page. This skeleton is used to generate an item request pattern at the simulated caches. The client request arrival rate follows a Poisson distribution. The simulated time period in each case was 24 hours, with a prior 10-hour "dry run" before collection of statistics starts to allow the system to reach a steady state.

6. ESI Simulation Results

As mentioned above, the main metric we examine is the amount of traffic on the different links. This metric is obviously directly related to the hit rate for page fragments on the edge servers. In our model there are two types of fragments: fragments that represent data items and fragments that represent static parts of the page. The latter (unless personalized) appear on most pages, and thus are almost always cached. For the following discussion, hit rate is therefore defined for data items only.

Before showing our simulation results, we derive two equations that yield useful insights, even though they are based on some simplifications. The equations are based on the work of Breslau et al in [5]. Assuming that item requests are independent, and that all items are homogeneous in size and time-to-live parameters, we show in [15] that when $1 \ll \lambda \cdot TTL \ll N$, the hit rate in each cache is expected to follow

$$HR_1 = \sum_{i=1}^{N} (1 - (1 - p(i))^{\lambda \cdot TTL}) \cdot p(i) \qquad (1)$$

where $p(i)$ is is the probability for item i to appear on any page (i.e., the item's popularity); λ is the item request arrival rate for a single cache, TTL is the time-to-live value for items and N is the number of items available to choose from. We also show that Equation 1 can be approximated in closed form using $HR_2 = \Omega ln(\lambda \cdot TTL \cdot \Omega)$ (for $\alpha = 1$) or

$$HR_2 = \frac{\Omega}{1-\alpha}(\lambda \cdot TTL \cdot \Omega)^{\frac{1}{\alpha}-1} \qquad (0 < \alpha < 1) \qquad (2)$$

where $\Omega = (\sum_{i=1}^{N} \frac{1}{i^\alpha})^{-1}$ is the Zipfian distribution constant.

This analysis suggests we can expect a logarithmic increase in a cache's hit-rate with an increase of the arrival rate λ or the TTL. This assumption is not always true for the edge servers because of the lower arrival rate λ ($1 \ll \lambda \cdot TTL$ does not hold). In fact, we witnessed in the simulation that the item hit-rate at the edge caches grows linearly with λ and with the TTL value when their product is small. However, we show below that the equations are good approximations for the *overall system hit-rate*: the percentage of all item requests satisfied by some cache from both levels (edge and backend), calculated as (number of hits at backend caches + number of hits at edge caches) / (number of total item requests). If we look at the entire cache system as a single cache, then Equations 1 and 2 are estimations for the overall system hit-rate.

6.1 Simulation Results - Online Bookstore's Book Pages

A "Bookstore book" resource can describe the collection of all individual product pages on a web site, for example, music-album pages on allmusic.com[3], book pages on Amazon, or similar resources. See Table 3 for a summary of the main parameter values. This resource has two groups that appear on each page instance. The first

[3]The model parameters of Table 3 are loosely based on this resource, found at http://www.allmusic.com.

Table 3. Default parameter values, bookstore resource

Num of Groups	2
Group 1	Appears on every page Static portion size: 10Kb Item size: 17Kb Items per page: 1 Item pool size: 400000 Item time-to-live: 1 hour $\alpha = 0.8, 1.1$
Group 2	Appears on every page Static size: 3Kb Personalized and un-cacheable

group's static part models the page header, static links etc. The items in this group are the items (e.g., books) available for this resource; the selection in the group is for one item per page out of the collection of available items. The second group on the page includes some personalized user information, such as a user name, specialized recommendation, shopping cart etc. Notice that for this resource, all data items belong to a single group; therefore, Equations 1 and 2 are directly applicable.

Figure 3 shows the hit-rate for the data-item fragments (i.e., the book fragments) vs. the items' TTL value. We show the system hit rate and the edge-cache hit rate for a single α value (the affect of α will be shown in Figures 5 and 6) as the TTL ranges from 60 seconds to 7200 seconds (two hours). In addition to the simulation results, Figure 3 shows the estimates for the system hit-rate using Equation 1 and Equation 2. We can see that both equations are relatively accurate, demonstrating the logarithmic nature of the TTL effect on the hit rate.

According to Figure 4, that shows the total traffic on each link as a function of the TTL, the edge servers are useful in caching the static fragments of the page despite

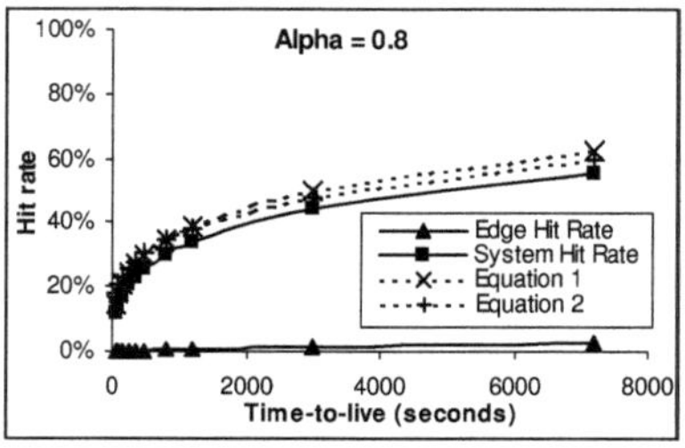

Figure 3. Edge and backend servers' item hit-rate vs. item TTL, bookstore-style resource (simulation results and estimations)

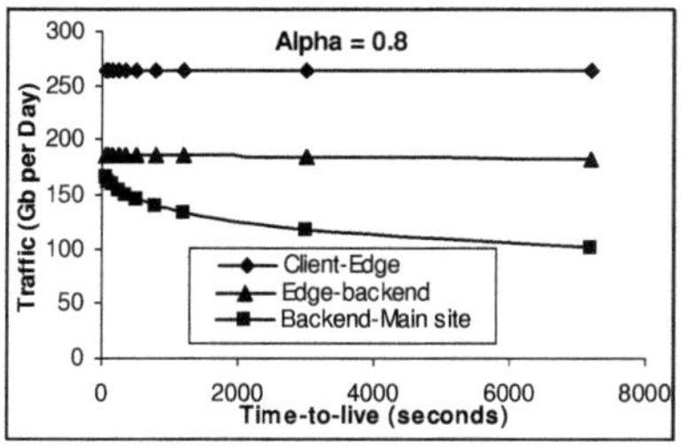

Figure 4. Traffic vs. item TTL, bookstore-style resource

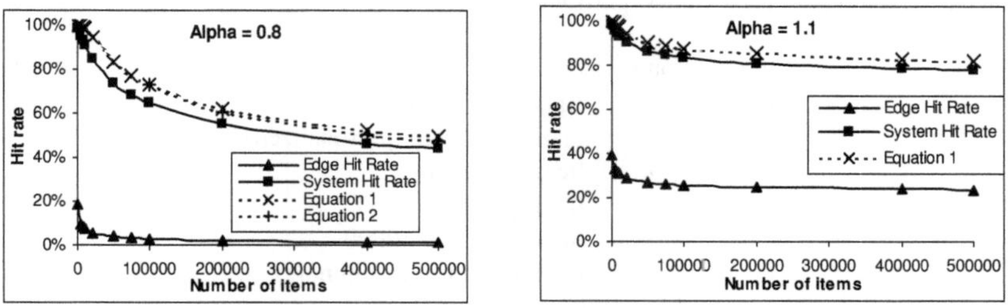

(a) (b)

Figure 5. Edge and backend servers' data-items hit-rate vs. number of available items, bookstore-style resource (simulation results and estimations)

the low item hit-rate, witnessed by the savings for the edge-backend link. On the client-edge link there are no savings since in our modeled system, clients receive fully-assembled pages.

Figure 5 illustrates the effect of the number of data items available for selection, N, on the hit-rates. The figure shows the curves for edge hit-rate, system hit-rate, and the approximations for system hit-rate using Equation 1 and Equation 2 (the latter is missing from Figure 5(b) since Equation 2 is not valid for $\alpha > 1$). Here we show results for two different α values. The low end of the item count, 1000 items, is representative for, e.g., a newspaper-articles resource. The higher end (500,000 items) is characteristic for product pages on a medium-scale online store. We can see from Figure 5 that ESI performance improves for a lower number of items; however, for larger values of α this effect is not as critical.

Figure 6 summarizes the effects of the slope of the Zipf-like curve (determined by the value of α) on the edge servers and system item hit-rates. The value of α ranges from 0.7 to 1.5 - values that were witnessed in numerous web-caching studies ([5, 10, 16] and others). Especially noteworthy is the magnitude of change in edge hit-rates, from close to 0% to about 60%. To give an idea, for 400,000 items, when $\alpha = 0.8$ the first three items account for 3.23% of all requests; when $\alpha = 1.1$, the first three items account for 23.12% of the requests.

For the basic settings of the bookstore resource, the simulation resulted in traffic savings of 30% for the edge-backend link, and 56% for the backend-main site link. The static parts of the page are the main contributor to the savings on the edge-backend link, as they are frequently served from the edge caches. The backend servers do a better job than edge servers in serving the data items, accounting for most of the additional 26% savings. Finally, the edge cache usage for this resource was around 1.2Mb per cache, consisting mostly of data items.

In Section 9 we present more variations on the model parameters, including experiments with different request arrival rates, which are analogous to varying the number of caches in the system.

Table 4. Default parameter values, portfolio resource

Num of Groups	2
Group 1 - stock entries	Appears on every page Static size: 2Kb (personalized) Item size: 1Kb Items per page: 1-50 Item pool size: 10000 Item time-to-live: 5 minutes $\alpha = 0.8$
Group 2 - header	Appears on every page Static size: 18Kb No items (static part only)

6.2 Simulation Results - Other Resources

We evaluated three resources in addition to the Bookstore resource. Due to space limitations, we only discuss here the key differences from the Bookstore resource (an extended discussion is found in our technical report [15]). For these new resources, a given user always sees a fixed set of groups and data items. For example, when users ask for their portfolio, they see the same stocks every time. Of course, the stock values may still change between requests. The first resource is indeed a brokerage portfolio resource (e.g., users' personal financial portfolio web page), described in Table 4. The second resource is the above-mentioned MyWeather. Table 5 lists the basic parameters for this resource. The third is the most complicated among our resources, the personalized My Page, a simplified model of a MyYahoo-style resource. My Page parameters are summarized in Table 6.

Table 7 lists the simulation results, as average savings and edge server cache size requirements (assuming evacuation of stale fragments only), for all the resources we simulated. The cumulative traffic on the edge-backend link and backend-main server link is measured in the simulation, and the percentages in Table 7 represent the reduction in traffic on these two links from the total traffic between server and client without ESI (the client link has no savings in ESI; it is not listed in the table).

Some comparative observations about Table 7 follow. The portfolio resource exhibits greater savings for both types of links compared to the bookstore resource, suggesting better caching at the edge servers. Since the relative size of the static portion of the two resources is similar, the improvement stems from better caching of the data items, in this case the stock fragments. The reason for that is twofold; first, there are fewer data items to choose from in the group; second, every page request translates into many data-item requests since a page has 25 stocks on it on average. The reduction is despite the lower TTL for the data items in the portfolio resource (5 minutes compared to 1 hour). The low TTL does affect the hit-rate at the edge; but the backend servers, having a much larger request arrival rate, are more effective in caching data items. The cache size usage for the portfolio resource is very low, and is in fact dominated by the cost of caching the personalized static part of the page.

Table 5. Default parameter values, MyWeather resource

Num of Groups	2
Group 1 - MyCities	Appears on every page Item size: 2Kb Items per page: 1-10 Item pool size: 5000 Item time-to-live: 1 hour $\alpha = 0.8$
Group 2 - Weather News & page header	Appears on every page Static portion size: 30Kb Item size: 3Kb Items per page: 5 Item pool size: 5 Item time-to-live: 5 minutes

Table 6. Default parameter values, My Page resource

Groups	14 groups Groups per page: 4-14 Group selection zipfian parameter $\alpha = 0.7$ Average page size: 42Kb
Group Names	News modules, movies, email notification, stock portfolio, yellowpages, personalized weather . . .

Table 7. Result summary

Resource	Savings edge-backend	Savings backend-main site	Edge cache usage
Bookstore	30%	56%	1.2
Portfolio	48%	95%	0.5Mb
MyWeather	84%	98%	1.3Mb
My Page	62%	75%	1.5Mb

For MyWeather we see even more savings than the portfolio resource. Compared to the portfolio resource, MyWeather has, in both groups, a smaller number of data-items with higher TTL. Especially, the data items in the 'news' group are few, and the selection from them is much more restricted. The result is the resource being very well cached in both edge and backend caches, as can be seen in Table 7. As a result of a larger TTL and a larger size of the personalized static part, the cache usage is higher than for the portfolio page. The cache contents in this case are split more evenly between data item fragments and personalized fragments.

My Page bandwidth savings from Table 7, although not as significant as the savings for the portfolio or MyWeather resources, are still better than the bookstore resource's savings, despite the large amount of personalized information. Notice that the savings on the edge-backend link are greater than for the portfolio resource, but it is the opposite case for the backend-main site savings. This difference is caused by the larger relative portion of personalized information for the My Page resource. While data items can be cached better using cooperative caching (i.e., backend servers)[20], cooperative caching does not add any benefits for the personalized parts of the page. The cache usage for My Page is the largest, mainly due to the many personalized static fragments, but also due to a large number of data items cached.

7. Class-Based Delta Encoding

Class-based delta encoding (DE) exploits the fact that two different versions of the same dynamic page will often bear great similarity. A "Condenser" is a machine (or a number of machines) located between the organization's web server and the Internet. The Condenser serves as a proxy for client access to the web server, forwarding the requests from clients to the web server, and applying DE to responses from the web server to the clients. Class-based DE allows the Condenser to exploit spatial locality, in this case similarity between different pages from the same resource viewed by different clients.

When the web server receives a request, it produces the complete page, and delivers it to the Condenser. The Condenser holds a number of candidate base files that can simply be described as previously generated pages. The Condenser chooses one of these base files, encodes the current page as a delta from the chosen file, and sends the delta to the client along with the base file's identifier. The details concerning how base files are selected for use in servicing a particular request appear in [17].

The main benefit of class-based DE is the reduction in traffic, since the base files are static resources and can be cached on traditional network caches (and on end-users' machines). In many cases, caching will reduce the traffic out of the Condenser to deltas, which are usually very small compared to the original page size. Notice that DE does not reduce the computational load on the web server, since every request propagates all the way to the server, and the server generates the entire page for each request.

7.1 System Model

The physical system model for DE simulation is very similar to the model for ESI described in Section 4. The only difference is the omission of the backend servers from the system model. Class-based DE can utilize regular proxy caches, installed by independent bodies such as ISPs and are usually found at the "edge" of the Internet, and not at the backend. In summary, the DE system includes clients, proxy caches (instead of "edge caches" as they appear in the ESI model), and a main site with a web server and a Condenser.

Some new parameters are needed for the simulation of class based DE, like the number of possible base-files generated by the Condenser. We do not have an estimate for this number, which is dependent on the clustering algorithm and the resource. Therefore, instead of predicting the number of base files created, we used a range of values, studying the effect on performance. In addition, we used a time-to-live parameter for base files, because the algorithm can replace them over time.

Another parameter is p_{same}, the probability of a client getting a delta from a base file it already has cached ("same" base file). The value of p_{same} is determined by the resource type. For example, for a resource like My Page, the client essentially accesses the same page every time. It is likely for the Condenser to use the same base file for the reply, unless this base file has timed out. Therefore, in this case, p_{same} is close to 1. On the other hand, consider a resource like the online bookstore book pages. Each request a client makes for a page will be subject to a process of finding the new best base file by the Condenser. The Condenser is not likely to pick a base file the client has already cached, so p_{same} is close to 0.

8. Delta Encoding Simulation

To simulate DE, we first generated pages with content, as opposed to the page skeletons we used so far. Recall that the page content model (Section 3) describes the components of the page, but not their textual content. Two methods were used for creating this content:

1. Download actual pages from the web that fit the model. This is the option taken for the bookstore-style resource.
2. If the model is not based on an existing resource, like My Page, we generate page instances. First, we manually generate templates for all the groups and items of the resource. Each template provides the HTML text that is constant for each group, and placeholders for fields that vary between instances. For example, the template for a Weather item contains placeholders for the city name and temperatures. Each template is generated to match the model parameters such as size. We then use these templates to generate item instances, replacing the placeholders with text, words or numbers generated by a random text generator. To produce full pages, we run our simulation to output page skeletons (containing identifiers for the constituent groups and items), which are then populated with group and item instances.

We used the following procedure to estimate the expected size of deltas d_{base}. We generated 2000 instances of each resource to serve as content pages, and we selected a subset S ($|S| = 20$) of pages to represent the Condenser's base files. Then, for each page P in the initial 2000 pages, we applied the $vcdiff$ algorithm to compute the delta between P and each S page ($vcdiff$ [13] is based on the $vcdelta$ algorithm [12] studied comparatively in web-context in [14], and used in [2, 17]). For each page P we picked the minimum delta size between P and any page from S. The average over all minimum values for pages P is the value d_{base} we used for the simulation.

In the rest of this section we present simulation results for the on-line bookstore resource, and touch on the performance difference between it and My Page resource. The significant difference shows the effect of a high p_{same} value versus low p_{same} value, especially on the client traffic. We later propose a variation on the algorithm for resources with a low p_{same}, and evaluate its potential benefits.

8.1 Simulation Results - Online Bookstore

The online bookstore resource represents a collection of URLs, in this case, all the books in the store's database. A user can access a different page (i.e., a URL of a different book) every time. For this class of resources there is no guarantee for the user to get a delta from the same base file on subsequent requests ($p_{same} \approx 0$). As a result, the traffic to clients is usually greater than the traffic without class-based DE, since the base file and delta are both retrieved for a total size greater than the original page size. On the other hand, traffic from the main site is still reduced since the base

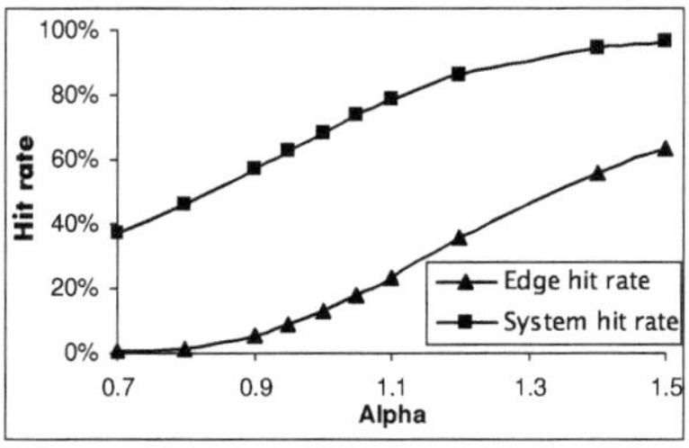

Figure 6. Hit-rate vs. value of α, bookstore-style resource

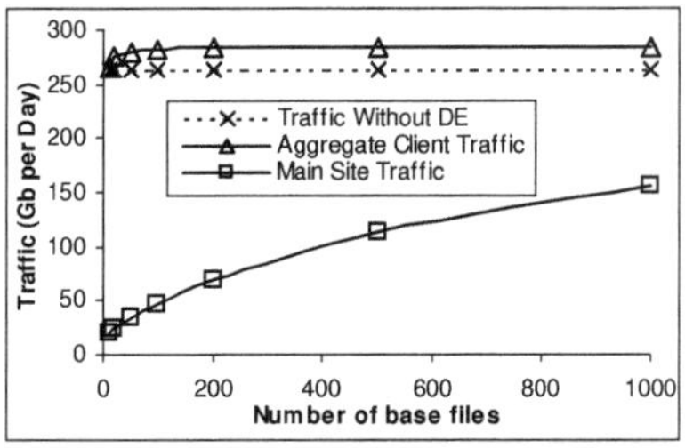

Figure 7. Online bookstore, class-based DE traffic vs. number of base files

files are requested more often and as a result are better cached on proxy caches. The size of delta between pages used in the simulation was $d_{base} = 1922$ bytes, generated as described above using real pages downloaded from allmusic.com.

Figure 7 shows the aggregate client traffic and total main site traffic as a function of the number of base files. The figure also shows the original volume of traffic without class-based DE. We can see that the client-side traffic almost always exceeds the original traffic but there are considerable savings on the server side. The hit-rate for base files goes down from almost 100% (10 base files) to around 52% (1000 base files), and the effect on traffic is seen in the figure.

In summary, with resources of this type, where a client is likely to download a different base file on every access, class-based DE usually reduces only main-site traffic and not client traffic. In that, the reduction is highly dependent upon the hit rates for base files on the proxy caches.

8.2 Simulation Results - My Page

For this type of resource we assume a client is likely with some high probability p_{same} to get the same base file on subsequent accesses, since the page viewed is not fundamentally different from one access to the other. We used d_{base}=1050 bytes, generated as described above, as the delta size for the simulation. The high p_{same} generates much larger client-side traffic savings. DE saves 66% of aggregate client traffic under the basic settings for this resource, while the Bookstore resource offers no savings for clients at all. On the server side, the reduction in traffic amounts to 87%. The extended results of this simulation are found in our technical report [15]. Some variations on the basic settings are done in Section 9.1.

8.3 Same-Base reply for Online Bookstore

In this section we introduce a simple modification that may improve class-based DE performance for the online bookstore resource and other resources where the clients with high probability receive a different base file on every access. We propose simply to set a threshold for a "same-base reply": The client request includes information,

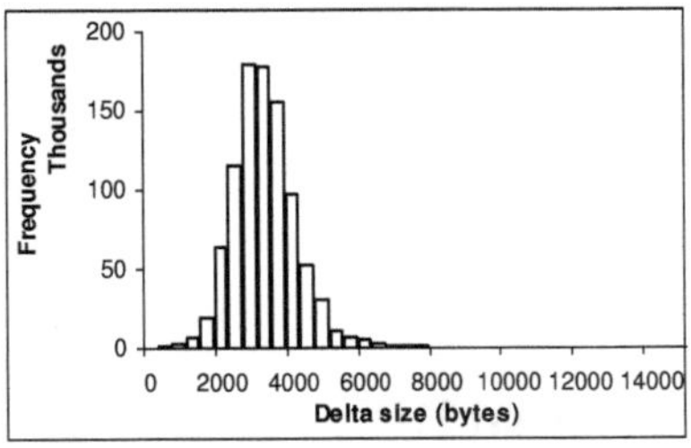

Figure 8. ds_{all} distribution for online bookstore resource

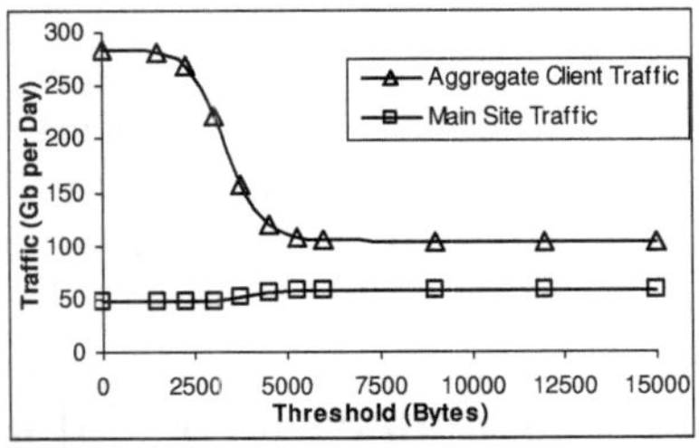

Figure 9. Traffic vs. threshold for Same-base Reply

possibly in a cookie, about the most recent base file in its cache. The DE Condenser first computes the delta of the new client page from the base file in the client's cache. If the delta is lower than a certain pre-defined threshold, the Condenser sends the computed delta to the client, which in turn applies it to the base file in its cache. In this case, the process of classifying the new page and finding a base file candidate on the Condenser is avoided. Note that we have the client send only the identifier of the most recent base file because identifying the entire set of its cached base files would require multiple delta computations on the Condenser.

This scheme is of course sensitive to the distribution of the delta sizes between pages. We used the set of 2000 downloaded pages to compute ds_{all}, the distribution of delta sizes for any pair of pages. The results are displayed in Figure 8. We used the ds_{all} distribution to compute the probability that a delta will be lower than the threshold, and the average size of those deltas. For the case where the threshold was exceeded, we plugged in the average delta size of the regular scheme, d_{base}.

Figure 9 shows the aggregate traffic to clients, and total traffic at the main site, as a function of the threshold set for the algorithm (the number of base files for this simulation is 100). For comparison note that the scheme is equivalent to the regular DE scheme when the threshold is set to 0. We can see this scheme reduces the aggregate client traffic since many times we replace downloading a new base file and a small delta, with downloading a slightly bigger delta, but without a base file. At the same time, traffic from the main site increases, mainly because the main site is supplying larger deltas in a lot of cases. The increase is less due to the reduction in the number of base-file requests (since caching was already keeping this traffic to minimum).

The breakdown of client traffic is a little different with same-base reply. The clients now receive greater deltas (transferred from the main site) instead of smaller deltas and base files (the latter are transferred, hopefully, from a closer proxy cache). Thus, even though the aggregate client traffic is lower, the reduction in latency may not be as low – fewer bytes are transferred, but they travel a longer way on average. A more careful study of the system's latency factors would be needed to show that same-base is indeed superior to the regular scheme.

Table 8. My Page comparison

	Savings - client link	Savings - server link	Edge cache Usage
ESI	0%	62%	1.5Mb
DE	66%	87%	3.2Mb

9. Comparison of ESI and Class-Based DE

This section presents a quantitative and qualitative discussion of ESI and class-based DE. The quantitative discussion is not straightforward, since the performance of the two techniques is dependent on parameters that may not be equivalent in each system. Furthermore, physical models for the two systems are somewhat different, with a cascaded architecture for ESI, and only one level of caching for DE. Thus, in some respects an ESI-DE comparison is like comparing apples to oranges. Nevertheless, we believe a comparison is useful because both solutions have the same ultimate goal: the reduction of traffic, and caching of content useful for dynamic pages. We overcome the physical system difference by eliminating the backend caches for ESI from the model; we will mention the additional savings using backend servers when relevant.

The main metric for evaluation continues to be network traffic. We present traffic measures for the different links (clients/edge servers, edge servers/main site, and also clients/main site). In addition, we offer a simple, weighted cost formula that combines traffic measures on the different links. The weights account for varying costs associated with the links. Finally, we also show the cache space usage for both schemes.

9.1 My Page Resource

Table 8 sums the reduction in traffic and cache usage for the My Page resource with ESI and class-based DE. As we mentioned in Section 8, in DE with this resource a client is likely to get the same base file on every access. Two points are worth repeating: First, our modeled ESI system does not introduce savings on the client link since the page is assembled as a whole on the edge servers. Second, there are further savings to ESI, when using backend caches, which are not part of the system model for this part (see Table 7 for the complete ESI results).

Figure 10 shows the variation in traffic on the client and main site links, for ESI and DE, as a function of the request arrival rate. The measure we use is bytes per request (total traffic in a day divided by the total number of requests) to better show the impact of a higher load (where overall traffic grows but traffic-per-request may drop). In both cases the traffic per request on the main site drops with the arrival rate due to better caching, but for DE the *client* traffic (per request) is reduced too. Since clients cache base files, and are likely to use the same base file on every access, the benefits of caching grow with the request rate. In conclusion, higher request rate entails better caching for both systems on the edge caches, but also means more benefits for client base-file caching in DE.

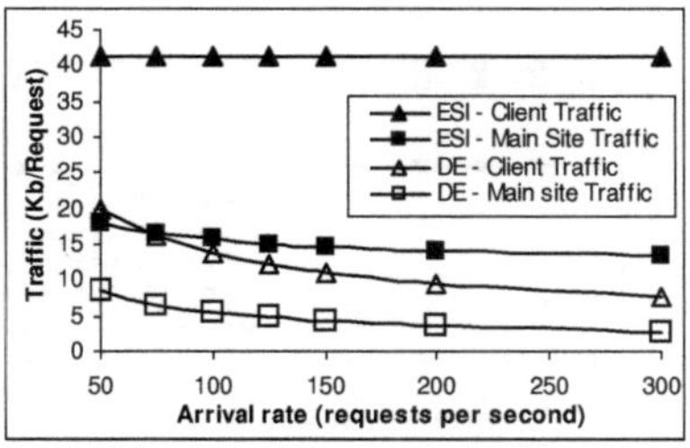

Figure 10. My Page resource, traffic for ESI and class-based DE vs. request arrival rate

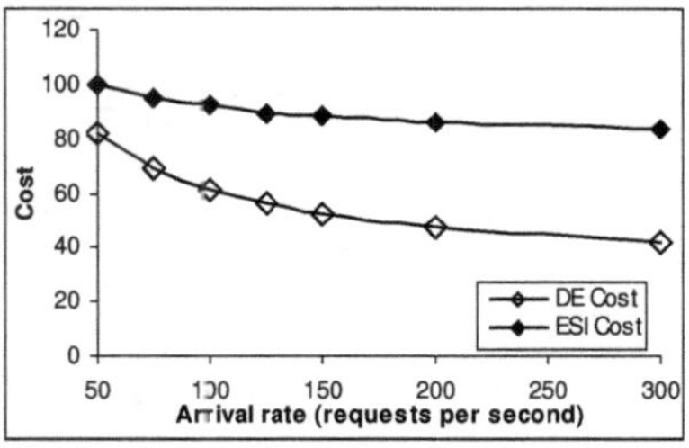

Figure 11. My Page resource, cost of ESI and class-based DE vs. request arrival rate

Notice that with DE, clients receive data from the main site "directly" (deltas), as well as base files from the proxy cache. The DE client curve of Figure 10 shows this aggregate traffic. On the other hand, with ESI, clients only receive data from the edge server, i.e., ESI client traffic in Figure 10 is exclusively edge-server-to-client traffic (of course, the edge server may get required data from main server when necessary). The client to main server traffic occurring with DE may be less desirable because it has to travel a longer distance (incurring more latency) and because it generates load (delta computation) at the server.

To take into account this and other differences, we propose a traffic cost model where each type of traffic is given a different weight. In particular, we use the cost function Cost = $C_1 \times$ client-server traffic + $C_2 \times$ cache-server traffic + client-cache traffic. Factor C_2 represents the cost ratio between cache-server and client-cache traffic. A high value for C_2 means that server traffic is more costly or less desirable than client traffic. Factor C_1 is only valid for DE since in ESI there is no "server-client" traffic. A high value of C_1 indicates that server-client traffic is more costly than other server traffic, because it incurs computational costs or because of larger latencies.

To illustrate, Figure 11 compares the ESI and DE costs for a scenario where $C_1 = 25$ and $C_2 = 5$. We chose a relatively large C_1 value to reflect what we think are the significantly higher costs of computation and latency of server-to-client traffic. With a C_2 value of 5, one byte of cache-server traffic is as costly as 5 bytes of client-cache traffic, perhaps because the cache is much closer to the client (this may not always be the case; these numbers are just illustrative). For this scenario, Figure 11 allows us to compare the desirability of each scheme. In this case we see that DE is preferable (lower overall cost) and the gap grows (DE even more desirable) as the arrival rate increases.

Figure 12 shows when DE is superior to ESI, as a function of the C_1 and C_2 cost factors, for the different arrival rates. The area under each line is where DE cost is lower than ESI cost for this arrival rate. For example, when $C_2 = 3$, DE is the better scheme as long as C_1 is less than or equal to 44 (for arrival rate = 1). Notice the high value of C_1 needed to equalize the cost for a given C_2. In other words, when the

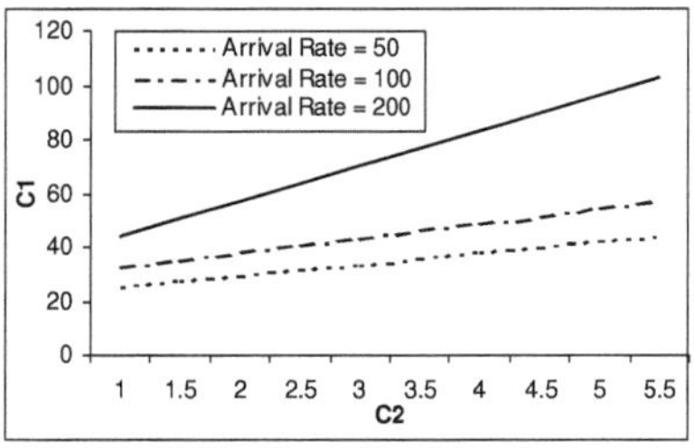

Figure 12. My Page, values for C_1, C_2 where DE cost is equal to ESI cost

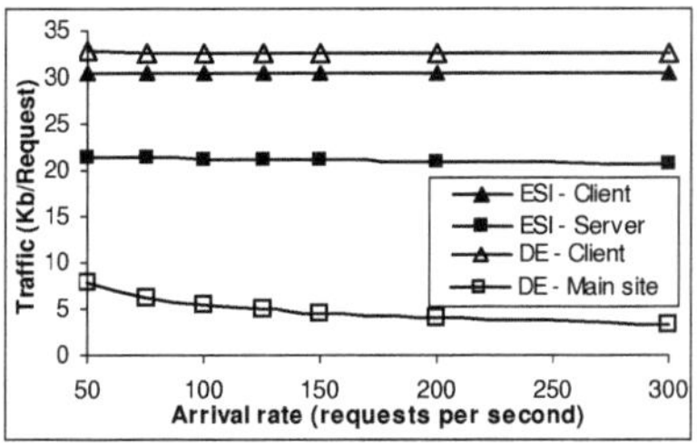

Figure 13. Online Bookstore resource, ESI and class-based DE traffic vs. request arrival rate

Table 9. Online bookstore comparison

	Savings - client link	Savings - server link	Edge cache Usage
ESI	0%	30%	1.2Mb
DE	-8%	82%	3.3Mb

cost of DE (generating a full page, computing a delta, transmitting to the client and applying the delta) are dominant, the schemes are comparable.

A variation in the number of users-per-cache does not have a considerable influence on performance for this resource, neither for ESI nor for class-based DE, and due to space limitation is omitted from this subsection. We will just mention that a 5-fold increase in the number of users-per-cache brings only a slight decrease in server traffic for both systems.

9.2 Online Bookstore Resource

Table 9 summarizes the savings and cache usage for an online bookstore type resource. Remember, for this resource, in DE, a client is likely to get a *different* base file on every access. Thus, the negative results for the client link savings with DE reflect more generated traffic than the regular scheme or ESI. On the server side, however, DE reduces traffic more than ESI. If we add backend servers, ESI server traffic reduction goes from 30% to 56%, still less beneficial than DE.

Figure 13 shows the traffic per request on the client and server links as a function of request arrival rate. Since in this case, for class-based DE, almost every client request forces the client to get a new base file. Therefore, the clients' traffic does not benefit from increase in arrival rate like we have seen for My Page. However, both ESI and DE main site traffic per request is again reduced as arrival rate increases, due to better caching.

Figure 14 illustrates the overall cost for the same example values of C_1 and C_2 (25 and 5, respectively) as a function of the request arrival rate. The more distinct

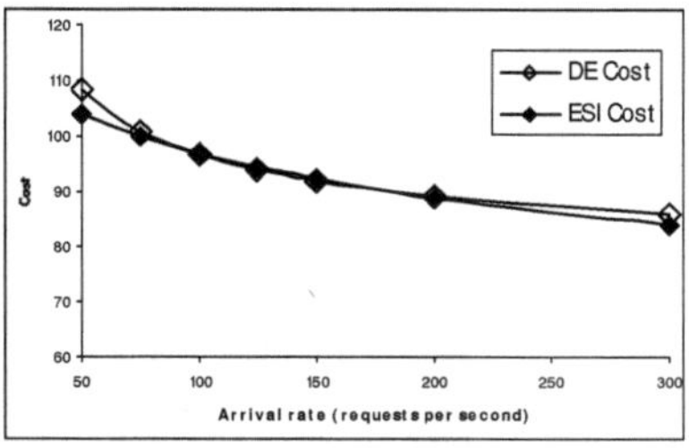

Figure 14. Online Bookstore, cost of ESI and class-based DE vs. request arrival rate

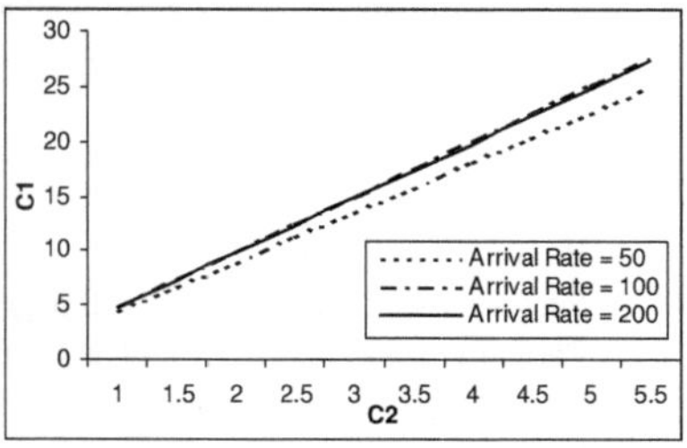

Figure 15. Online Bookstore, values for C1, C2 where DE cost is equal to ESI cost

improvement in DE (compare to the improvement in ESI) is due to better caching on the proxy caches as the request rate grows, reducing the expensive traffic from the servers. For ESI, however, the hit-rate does not improve as dramatically with the arrival rate, mainly due to the much larger number of items in the "pool" (corresponding to 400,000 books). For higher request rates, or for other resources, maybe with fewer data-items, the savings can possibly grow more considerably. For this resource, in both systems, the effect of the number of users-per-cache (or number of caches) is analogous to the effect of the arrival rate on performance. Finally, Figure 15 shows when DE outperforms ESI, as a function of the C_1 and C_2 factors, for three different arrival rate values. Again, the area underneath each line is where DE cost is lower. Note the ratio between C_1 and C_2 needed for equality is much lower in this case than in My Page case. For instance, when the arrival rate is 1 and $C_2 = 3$, we only need $C_1 = 15$ for both schemes to perform equally.

10. Additional Considerations

Table 10 summarizes some of the conclusions and additional considerations for comparing class-based DE to ESI. A major concern is the computational costs and latencies introduced by both schemes. Class-based DE requires generation of the entire page for each request. Moreover, the page then needs to be delta-encoded, which includes the process of finding a good base file and computing the delta. On the client side, the delta should be applied to the base file accounting for more delay. ESI, on the other hand, requires only assembly of the fragments of the page on edge servers into a full page that is served to the client. The assembly does not introduce new computational costs since it had to be done by the web server even without ESI. The web server benefits twice under ESI: not only does it not have to assemble the page; in many cases it is required to deliver only small parts of the page.

Another consideration is the transparency of both systems. While class-based DE as offered in [17] requires installation of hardware or software, usually near the web server, it does not require any change to web-pages code, and works transparently with existing network infrastructure of proxy caches and clients. ESI, on the other hand,

Table 10. Summary: Excellent *, Good +, Bad –, Sometimes ∼

	ESI	*DE*
Reduces server traffic	+	*
Reduces client traffic	–	∼
Reduces load on web server	*	–
Performance dependent on web page structure	Yes	Yes
Performance dependent on characteristics of data	Yes	No
Benefits greater when popularity rises	Yes	Less
Requires main site hardware/software installation	No	Yes
Requires web-page code changes	Yes	No
Requires network infrastructure (CDN services)	Yes	No
Can exploit information available from CDN for page construction	Yes	No

requires changes to web-pages code, as ESI code must be added over the original HTML. In addition, ESI requires specialized edge servers (i.e., the services of a CDN provider) since the assembly directives are not implemented on proxy caches.

The distribution of work in ESI between the main site and the CDN enables the web site to use information that is available to the CDN provider only, or not maintained by the main site, such as physical location of the clients. This can be exploited in the page construction (e.g., automatically inserting relevant weather) using the ESI language.

In conclusion, we showed that both ESI and DE can reduce traffic and improve caching for dynamic page delivery. However, the benefits of the techniques are highly dependent on the resource. For example, ESI is beneficial when there are a small number of items on the page, the item popularity is skewed, and their time-to-live is high. DE, while always good for reduction of main site traffic, may not always reduce client traffic.

Acknowledgments

We would like to thank Peter Danzig for helpful discussions, Phong Vo for support with the vcdiff software, and the reviewers for their helpful comments.

References

[1] ESI 1.0. http://www.w3.org/TR/esi-lang.

[2] FineGround Networks. Breaking new ground in content acceleration. http://www.fineground.com/pdf/FGCWhitepaper.pdf.

[3] V. Almeida, A. Bestavros, M. Crovella, and A. de Oliveira. Characterizing reference locality in the WWW. In *Proceedings of PDIS'96: The IEEE Conference on Parallel and Distributed Information Systems*, 1996.

[4] G. Banga, F. Douglis, and M. Rabinovich. Optimistic deltas for WWW latency reduction. In *Proceedings of USENIX Technical Conference*, pages 289–303, 1997.

[5] L. Breslau, P. Cao, L. Fan, G. Phillips, and S. Shenker. Web caching and zipf-like distributions: Evidence and implications. In *Proceedings of Infocom*, 1999.

[6] P. Cao, J. Zhang, and K. Beach. Active cache: Caching dynamic contents on the Web. In *Proceedings of IFIP International Conference on Distributed Systems Platforms and Open Distributed Processing (Middleware '98)*, pages 373—388, 1998.

[7] J. Challenger, P. Dantzig, and A. Iyengar. A scalable system for consistently caching dynamic Web data. In *Proceedings of the 18th Annual Joint Conference of the IEEE Computer and Communications Societies*, New York, New York, 1999.

[8] J. Challenger, A. Iyengar, K. Witting, C. Ferstat, and P. Reed. A publishing system for efficiently creating dynamic Web content. In *Proceedings of IEEE INFOCOM 2000, Tel Aviv, Israel*, 2000.

[9] F. Douglis, A. Haro, and M. Rabinovich. HPP: HTML macro-preprocessing to support dynamic document caching. In *Proceedings of the USENIX Symposium on Internet Technologies and Systems, Monterey, California*, 1997.

[10] S. Glassman. A caching relay for the World Wide Web. In *Proceedings of the First International World-Wide Web Conference*, 1994.

[11] B. Housel and D. Lindquist. WebExpress: A system for optimizing Web browsing in a wireless environment. In *In Proceedings of the Second Annual International Conference on Mobile Computing and Networking, pages 108–116, November 1996*, 1996.

[12] J. J. Hunt, K.-P. Vo, and W. F. Tichy. Delta algorithms: An empirical analysis. *ACM Transactions on Software Engineering and Methodology*, 7:192–214, 1998.

[13] D. G. Korn and K.-P. Vo. Engineering a differencing and compression data format. In *Proceedings of Usenix 2002*. USENIX, 2002.

[14] J. C. Mogul, F. Douglis, A. Feldmann, and B. Krishnamurthy. Potential benefits of delta encoding and data compression for http. In *Proceedings of ACM SIGCOMM*, pages 181–194, 1997. An extended version appears as Research Report 97/4, Digital Equipment Corporation Western Research Laboratory July, 1997.

[15] M. Naaman, H. Garcia-Molina, and A. Paepcke. Evaluation of delivery techniques for dynamic Web content (extended version). Technical report, Stanford University, June 2002. Available at http://dbpubs.stanford.edu/pub/2002-31.

[16] V. N. Padmanabhan and L. Qiu. The content and access dynamics of a busy Web site: Findings and implications. In *Proceedings of ACM SIGCOMM, Stockholm, Sweden*, 2000.

[17] K. Psounis. Class-based delta-encoding: a scalable scheme for caching dynamic Web content. In *22nd International Conference on Distributed Computing Systems Workshops*, 2002.

[18] M. Rabinovich, Z. Xiao, F. Douglis, and C. Kalmanek. Moving edge-side includes to the real edge – the clients. In *USENIX Symposium on Internet Technologies and Systems*, 2003.

[19] C. Roadknight, I. Marshall, and D. Vearer. File popularity characterisation. *SIGMETRICS Perform. Eval. Rev.*, 27(4):45–50, 2000.

[20] A. Wolman, G. M. Voelker, N. Sharma, N. Cardwell, A. Karlin, , and H. M. Levy. On the scale and performance of cooperative Web proxy caching. In *Proceedings of the 17th ACM Symposium on Operating Systems Principles (SOSP)*, 1999. Published as Operating Systems Review 34(5):16-31, December 1999.

[21] G. K. Zipf. Relative frequency as a determinant of phonetic change. *Reprinted from the Harvard Studies in Classical Philology*, XL, 1929.

Author Index

Aggarwal, Amit, 57
Akhmarov, Rustam, 119
Al Hamra, Anwar, 129
Amiri, Khalil, 79

Biersack, Ernst W., 129
Bustamante, Fabian E., 233

Canali, Claudia, 205
Cao, Pei, 159
Cardellini, Valeria, 205
Chen, Mao, 187
Chen, Songqing, 171
Chi, Chi Hung, 257, 293
Colajanni, Michele, 205
Cranor, Chuck, 119

Davison, Brian D., ix
Desai, Arun, 109
Douglis, Fred, ix

Fei, Zongming, 247
Fisher, Darin, 283
Fridman, Vitali, 19

Garcia-Molina, Hector, 323
Gausman, Paul, 119
Ge, Zihui, 139
Guo, Yang, 139
Gupta, Rajeev, 39

Hua, Zhigang, 91

Jain, Ravi, 1

Kawahara, Toshiro, 1
Ku, William, 257
Kumar, Apurva, 39

Lancellotti, Riccardo, 205
LaPaugh, Andrea 187
Lemon, Jonathan, 159
Li, Dan, 109

Morris, Stephen, 109
Mueller, Kenneth, 109
Mukherjee, Sarit, 223

Naaman, Mor, 323

Padmanabhan, Sriram, 79
Paepcke, Andreas, 323
Pierre, Guillaume, 275

Qiao, Yi, 233

Rabinovich, Michael, 57
Rangarajan, Sampath, 223
Rodriguez, Pablo, 19, 223

Saksena, Gagan, 283
Savant, Anubhav, 303
Shen, Bo Shen, 171
Shenoy, Prashant, 139
Singh, Jaswinder Pal, 187
Sivasubramanian, Swaminathan, 275
Sprenkle, Sara, 79
Stavisky, Dmitry, 109
Suel, Torsten, 303

Tariq, Muhammad Mukarram Bin, 1
Tewari, Renu, 79
Towsley, Don, 139

Urgaonkar, Bhuvan, 139
Urvoy-Keller, Guillaume, 129

van Steen, Maarten, 275
Van der Merwe, Jacobus, 119

Wang, HongGuang, 257, 293
Wang, Zhe, 159
Wee, Susie, 171

Xiao, Zhen, 57

Yang, Mengkun, 247
Yang, Zheng, 109, 159
Yu, Philip S., 205
Yuan, Chun, 91

Zhang, Xiaodong, 171
Zhang, Zheng, 91

MIX
Papier aus verantwortungsvollen Quellen
Paper from responsible sources
FSC® C105338

If you have any concerns about our products,
you can contact us on
ProductSafety@springernature.com

In case Publisher is established outside the EU,
the EU authorized representative is:
Springer Nature Customer Service Center GmbH
Europaplatz 3, 69115 Heidelberg, Germany

Printed by Libri Plureos GmbH
in Hamburg, Germany